Genetically Modified Foods

Contemporary Issues

Series Editors: Robert M. Baird
Stuart E. Rosenbaum

Volumes edited by Robert M. Baird and Stuart E. Rosenbaum
unless otherwise noted.

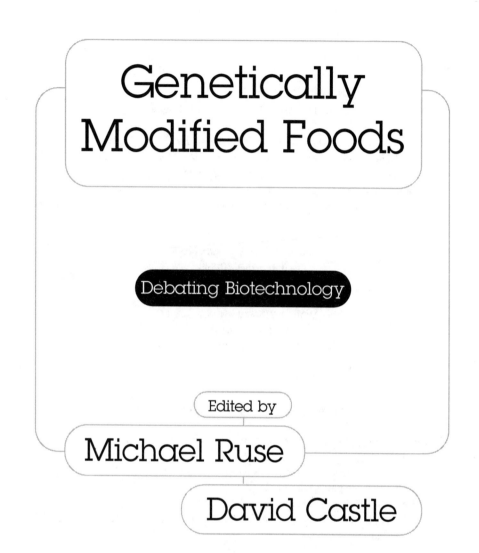

Genetically Modified Foods

Debating Biotechnology

Edited by

Michael Ruse

David Castle

Prometheus Books

59 John Glenn Drive
Amherst, New York 14228-2197

Published 2002 by Prometheus Books

Inquiries should be addressed to
Prometheus Books
59 John Glenn Drive
Amherst, New York 14228–2197
VOICE: 716–691–0133, ext. 207
FAX: 716–564–2711
WWW.PROMETHEUSBOOKS.COM

06 05 04 03 02   5 4 3 2 1

Library of Congress Cataloging-in-Publication Data

Genetically modified foods : debating technology / edited by Michael Ruse, David
 Castle.
 355 p. cm. — (Contemporary issues series)
 Includes bibliographical references.
 ISBN 1–57392–996–4 (alk. paper)
 1. Gentically modified foods. 2. Agricultural biotechnology. I. Ruse, Michael.
II. Castle, David, 1967– III. Series.

TP248.65.F66 G458 2002
363.19'29—dc21
                                                                    2002070510

Printed in the United States of America on acid-free paper

Contents

Prologue

Editors' General Introduction

Part 1. Biotechnology Case Study: Golden Rice

Part 2. Ethics in Agriculture

Part 3. Religion

Part 4. Labeling

Part 5. Law

Part 6. Food Safety and Substantial Equivalence

Part 7. Risk Assessment and Public Perception

Part 8. Precautionary Principle and Genetically Modified Foods

Part 9. Developing Countries

Part 10. Assessing Environmental Impacts

Glossary   343

Suggestions for Further Reading   351

Contributors   353

Prologue

Editors' Introduction

The Reith Lectures are a prestigious set of six lectures given annually over the British Broadcasting Corporation (BBC) network. In the year 2000, to celebrate the millennium, rather than the usual single person giving all six lectures, the topic of sustainable development was given to six different speakers, the last of whom was Charles, Prince of Wales, oldest son of Queen Elizabeth II and heir to the throne of England. Well known as an enthusiast for organic gardening, the prince here makes very clear his opposition to all kinds of artificial modification of the genetics of crops intended for our tables—or indeed, apparently, to all kinds of genetic modification. "I believe that if we are to achieve genuinely sustainable development, we will first have to rediscover, or re-acknowledge, a sense of the sacred in our dealings with the natural world, and with each other."

This talk brought forth a typically fiery response from Oxford biologist Richard Dawkins, well-known popular writer on matters evolutionary (*The Selfish Gene*, *The Blind Watchmaker*) and notorious atheist (*River Out of Eden*, *Climbing Mount Improbable*). For all the respect he claims to have for the prince's aims, Dawkins is scathing about the prince's argumentation: "Your embracing of an ill-sorted jumble of mutually contradictory alternatives will lose you the respect that I think you deserve." As far as Dawkins is concerned, to talk of "a sense of the sacred in our dealings with the natural world" is just nonsense. Worse, it leads to mystification and bad decisions, and this applies particularly to decisions about the genetic modification of organisms.

So who is right and who is wrong? We offer you this prologue simply to show you that there are very strong and very conflicting views on the question of genetically modified organisms. The rest of the collection will draw us closer to an answer to this difficult question.

Reith Lecture 2000

HRH The Prince of Wales

Like millions of other people around the world, I've been fascinated to hear five eminent speakers share with us their thoughts, hopes, and fears about sustainable development based on their own experience. All five of those contributions have been immensely thoughtful and challenging. There have been clear differences of opinion and of emphasis between the speakers, but there have also been some important common themes, both implicit and explicit. One of those themes has been the suggestion that sustainable development is a matter of enlightened self-interest. Two of the speakers used this phrase, and I don't believe that the other three would dissent from it, and nor would I.

Self-interest is a powerful motivating force for all of us, and if we can somehow convince ourselves that sustainable development is in all our interests, then we will have taken a valuable first step toward achieving it. But self-interest comes in many competing guises—not all of which I fear are likely to lead in the right direction for very long, nor to embrace the manifold needs of future

"A Royal View," by the Prince of Wales, BBC Reith Lectures 2000 (Home/Respect for the Earth). Reprinted with permission.

generations. I am convinced we will need to dig rather deeper to find the inspiration, sense of urgency, and moral purpose required to confront the hard choices which face us on the long road to sustainable development. So, although it seems to have become deeply unfashionable to talk about the spiritual dimension of our existence, that is what I propose to do.

The idea that there is a sacred trust between mankind and our Creator, under which we accept a duty of stewardship for the earth, has been an important feature of most religious and spiritual thought throughout the ages. Even those whose beliefs have not included the existence of a Creator have, nevertheless, adopted a similar position on moral and ethical grounds. It is only recently that this guiding principle has become smothered by almost impenetrable layers of scientific rationalism. I believe that if we are to achieve genuinely sustainable development, we will first have to rediscover, or reacknowledge, a sense of the sacred in our dealings with the natural world, and with each other. If literally nothing is held sacred anymore—because it is considered synonymous with superstition or in some other way "irrational"—what is there to prevent us treating our entire world as some "great laboratory of life" with potentially disastrous long-term consequences?

Fundamentally, an understanding of the sacred helps us to acknowledge that there are bounds of balance, order, and harmony in the natural world which set limits to our ambitions, and define the parameters of sustainable development. In some cases nature's limits are well understood at the rational, scientific level. As a simple example, we know that trying to graze too many sheep on a hillside will, sooner or later, be counterproductive for the sheep, the hillside, or both. More widely we understand that the overuse of insecticides or antibiotics leads to problems of resistance. And we are beginning to comprehend the full, awful consequences of pumping too much carbon dioxide into the earth's atmosphere. Yet the actions being taken to halt the damage known to be caused by exceeding nature's limits in these and other ways are insufficient to ensure a sustainable outcome. In other areas, such as the artificial and uncontained transfer of genes between species of plants and animals, the lack of hard, scientific evidence of harmful consequences is regarded in many quarters as sufficient reason to allow such developments to proceed.

The idea of taking a precautionary approach, in this and many other potentially damaging situations, receives overwhelming public support, but still faces a degree of official opposition, as if admitting the possibility of doubt was a sign of weakness or even of a wish to halt "progress." On the contrary, I believe it to be a sign of strength and of wisdom. It seems that when we do have scientific evidence that we are damaging our environment, we aren't doing enough to put things right, and when we don't have that evidence, we are prone to do nothing at all, regardless of the risks.

Part of the problem is the prevailing approach that seeks to reduce the natural world including ourselves to the level of nothing more than a mechanical process. For whilst the natural theologians of the eighteenth and nineteenth centuries like Thomas Morgan referred to the perfect unity, order, wisdom, and design of the natural world, scientists like Bertrand Russell rejected this idea as rubbish. "I think the universe," he wrote, "is all spots and jumps without unity and without continuity, without coherence or orderliness. Sir Julian Huxley wrote in "Creation a Modern Synthesis" that "modern science must rule out special creation or divine guidance." But why?

As Professor Alan Linton of Bristol University has written, "Evolution is a manmade theory to explain the origin and continuance of life on this planet without reference to a Creator." It is because of our inability or refusal to accept the existence of a guiding hand that nature has come to be regarded as a system that can be engineered for our own convenience or as a nuisance to be evaded and manipulated, and in which anything that happens can be fixed by technology and human ingenuity. Fritz Schumacher recognized the inherent dangers in this approach when he said that "there are two sciences—the science of manipulation and the science of understanding."

In this technology-driven age, it is all too easy for us to forget that mankind is a part of nature and not apart from it. And that this is why we should seek to work with the grain of nature in everything we do, for the natural world is, as the economist Herman Daly puts it, "the envelope that contains, sustains and provisions the economy, not the other way round." So which argument do you think will win—the living world as one or the world made up of random parts, the product of mere chance, thereby providing the justification for any kind of development? This, to my mind, lies at the heart of what we call sustainable development. We need, therefore, to rediscover a reference for the natural world, irrespective of its usefulness to ourselves—to become more aware in Philip Sherrard's words of "the relationship of interdependence, interpenetration, and reciprocity between God, Man, and Creation."

Above all, we should show greater respect for the genius of nature's designs, rigorously tested and refined over millions of years. This means being careful to use science to understand how nature works, not to change what nature is, as we do when genetic manipulation seeks to transform a process of biological evolution into something altogether different. The idea that the different parts of the natural world are connected through an intricate system of checks and balances which we disturb at our peril is all too easily dismissed as no longer relevant.

So, in an age when we're told that science has all the answers, what chance is there for working with the grain of nature? As an example of working with the grain of nature, I happen to believe that if a fraction of the money currently being invested in developing genetically manipulated crops were applied to understanding and improving traditional systems of agriculture, which have stood the all-important test of time, the results would be remarkable. There is already plenty of evidence of just what can be achieved through applying more knowledge and fewer chemicals to diverse cropping systems. These are genuinely sustainable methods, and they are far removed from the approaches based on monoculture, which lend themselves to large-scale commercial exploitation, and which Vandana

Shiva condemned so persuasively and so convincingly in her lecture. Our most eminent scientists accept that there is still a vast amount that we don't know about our world and the life-forms that inhabit it. As Sir Martin Rees, the Astronomer Royal, points out, it is complexity that makes things hard to understand, not size. In a comment which only an astronomer could make, he describes a butterfly as a more daunting intellectual challenge than the cosmos!

Others, like Rachel Carson, have eloquently reminded us that we don't know how to make a single blade of grass. And St. Matthew, in his wisdom, emphasized that not even Solomon in all his glory was arrayed as the lilies of the field. Faced with such unknowns, it is hard not to feel a sense of humility, wonder, and awe about our place in the natural order. And to feel this at all stems from that inner heartfelt reason which sometimes despite ourselves is telling us that we are intimately bound up in the mysteries of life and that we don't have all the answers. Perhaps even that we don't have to have all the answers before knowing what we should do in certain circumstances. As Blaise Pascal wrote in the seventeenth century, "It is the heart that experiences God, not the reason."

So do you not feel that, buried deep within each and every one of us, there is an instinctive, heartfelt awareness that provides—if we will allow it to—the most reliable guide as to whether or not our actions are really in the long-term interests of our planet and all the life it supports? This awareness, this wisdom of the heart, may be no more than a faint memory of a distant harmony, rustling like a breeze through the leaves, yet sufficient to remind us that the earth is unique and that we have a duty to care for it. Wisdom, empathy, and compassion have no place in the empirical world, yet traditional wisdoms would ask "Without them are we truly human?" And it would be a good question. It was Socrates who, when asked for his definition of wisdom, gave as his conclusion, "knowing that you don't know."

In suggesting that we will need to listen rather more to the common sense emanating from our hearts if we are to achieve sustainable development, I'm not suggesting that information gained through scientific investigation is anything other than essential. Far from it. But I believe that we need to restore the balance between the heartfelt reason of instinctive wisdom and the rational insights of scientific analysis. Neither, I believe, is much use on its own. So it is only by employing both the intuitive and the rational halves of our own nature—our hearts and our minds—that we will live up to the sacred trust that has been placed in us by our creator—or our "sustainer," as ancient wisdom referred to the creator. As Gro Harlem Brundtland has reminded us, sustainable development is not just about the natural world, but about people, too. This applies whether we are looking at the vast numbers who lack sufficient food or access to clean water, but also those living in poverty and without work. While there is no doubt that globalization has brought advantages, it brings dangers, too. Without the humility and humanity expressed by Sir John Browne in his notion of the "connected economy"—an economy which acknowledges the social and environmental context within which it operates—there is the risk that the poorest and the weakest will not only see very little benefit but, worse, they may find that their livelihoods and cultures have been lost.

So if we are serious about sustainable development, then we must also remember that the lessons of history are particularly relevant when we start to look further ahead. Of course, in an age when it often seems that nothing can properly be regarded as important unless it can be described as "modern," it is highly dangerous to talk about the lessons of the past. And are those lessons ever taught or understood adequately in an age when to pass on a body of acquired knowledge of this kind is often considered prejudicial to "progress"? Of course our descendants will have scientific and technological expertise beyond our imagining, but will they have the insight or the self-control to use this wisely, having learned both from our successes and our failures?

They won't, I believe, unless there are increased efforts to develop an approach to education which balances the rational with the intuitive. Without this, truly sustainable development is doomed. It will merely become a hollow-sounding mantra that is repeated ad nauseam in order to make us all feel better. Surely, therefore, we need to look toward the creation of greater balance in the way we educate people so that the practical and intuitive wisdom of the past can be blended with the appropriate technology and knowledge of the present to produce the type of practitioner who is acutely aware of both the visible and invisible worlds that inform the entire cosmos. The future will need people who understand that sustainable development is not merely about a series of technical fixes, about redesigning humanity or reengineering nature in an extension of globalized, industrialization—but about a reconnection with nature and a profound understanding of the concepts of care that underpin long-term stewardship.

Only by rediscovering the essential unity and order of the living and spiritual world—as in the case of organic agriculture or integrated medicine or in the way we build—and by bridging the destructive chasm between cynical secularism and the timelessness of traditional religion will we avoid the disintegration of our overall environment. Above all, I don't want to see the day when we are rounded upon by our grandchildren and asked accusingly why we didn't listen more carefully to the wisdom of our hearts as well as to the rational analysis of our heads; why we didn't pay more attention to the preservation of biodiversity and traditional communities or think more clearly about our role as stewards of creation? Taking a cautious approach or achieving balance in life is never as much fun as the alternatives, but that is what sustainable development is all about.

An Open Letter to Prince Charles

Richard Dawkins

Sunday May 21, 2000

Your Royal Highness,

Your Reith lecture saddened me. I have deep sympathy for your aims, and admiration for your sincerity. But your hostility to science will not serve those aims; and your embracing of an ill-assorted jumble of mutually contradictory alternatives will lose you the respect that I think you deserve. I forget who it was who remarked, "Of course we must be open-minded, but not so open-minded that our brains drop out."

Let's look at some of the alternative philosophies which you seem to prefer over scientific reason. First, intuition, the heart's wisdom "rustling like a breeze through the leaves." Unfortunately, it depends whose intuition you choose. Where aims (if not methods) are concerned, your own intuitions coincide with mine. I wholeheartedly share your aim of long-term stewardship of our planet, with its diverse and complex biosphere.

Richard Dawkins, Charles Simonyi Professor of the Public Understanding of Science

But what about the instinctive wisdom in Saddam Hussein's black heart? What price the Wagnerian wind that rustled Hitler's twisted leaves? The Yorkshire Ripper heard religious voices in his head urging him to kill. How do we decide *which* intuitive inner voices to heed?

This, it is important to say, is not a dilemma that science can solve. My own passionate concern for world stewardship is as emotional as yours. But where I allow feelings to influence my aims, when it comes to deciding the best method of achieving them, I'd rather think than feel. And thinking, here, means scientific thinking. No more effective method exists. If it did, science would incorporate it.

Next, Sir, I think you may have an exaggerated idea of the naturalness of "traditional" or "organic" agriculture. Agriculture has always been unnatural. Our species began to depart from our natural hunter-gatherer lifestyle as recently as 10,000 years ago—too short to measure on the evolutionary timescale.

Wheat, be it ever so wholemeal and stoneground, is not a natural food for *Homo sapiens*. Nor is milk, except for children. Almost every morsel of our food is genetically modified—admittedly by artificial selection not artificial mutation, but the end result is the same. A wheat grain is a genetically modified grass seed, just as a pekinese is a genetically modified wolf. Playing God? We've been playing God for centuries!

The large, anonymous crowds in which we now teem began with the agricultural revolution, and without agriculture we could survive in only a tiny fraction of our current numbers. Our high population is an agricultural (and technological and medical) artifact. It is far more unnatural than the population-limiting methods condemned as unnatural by the Pope. Like it or not, we are stuck with agriculture, and agriculture—*all* agriculture—is unnatural. We sold that pass 10,000 years ago.

Does that mean there's nothing to choose between different kinds of agriculture when it comes to sustainable planetary welfare? Certainly not. Some are much more damaging than others, but it's no use appealing to "nature," or to "instinct" in order to decide which ones. You have to study the evidence, soberly and reasonably—scientifically. Slashing and burning (incidentally, no agricultural system is closer to being "traditional") destroys our ancient forests. Overgrazing (again, widely practiced by "traditional" cultures) causes soil erosion and turns fertile pasture into desert. Moving to our own modern tribe, monoculture, fed by powdered fertilizers and poisons, is bad for the future; indiscriminate use of antibiotics to promote livestock growth is worse.

Incidentally, one worrying aspect of the hysterical opposition to the possible risks from GM [genetically modified] crops is that it diverts attention from definite dangers which are already well understood but largely ignored. The evolution of antibiotic-resistant strains of bacteria is something that a Darwinian might have foreseen from the day antibiotics were discovered. Unfortunately, the warning voices have been rather quiet, and now they are drowned by the baying cacophony: "GM GM GM GM GM GM!"

Moreover, if, as I expect, the dire prophecies of GM doom fail to materialize, the feeling of letdown may spill over into complacency about real risks. Has it occurred to you that our present GM brouhaha may be a terrible case of crying wolf?

Even if agriculture could be natural, and even if we could develop some sort of instinctive rapport with the ways of nature, would nature be a good role model? Here, we must think carefully. There really is a sense in which ecosystems are balanced and harmonious, with some of their constituent species becoming mutually dependent. This is one reason the corporate thuggery that is destroying the rainforests is so criminal.

On the other hand, we must beware of a very common misunderstanding of Darwinism. Tennyson was writing before Darwin, but he got it right. Nature really is red in tooth and claw. Much as we might like to believe otherwise, natural selection, working within each species, does not favor long-term stewardship. It favors short-term gain. Loggers, whalers, and other profiteers who squander the future for present greed are only doing what all wild creatures have done for three billion years.

No wonder T. H. Huxley, Darwin's bulldog, founded his ethics on a repudiation of Darwinism. Not a repudiation of Darwinism as science, of course, for you cannot repudiate truth. But the very fact that Darwinism is true makes it even more important for us to fight against the naturally selfish and exploitative tendencies of nature. We can do it. Probably no other species of animal or plant can. We can do it because our brains (admittedly given to us by natural selection for reasons of short-term Darwinian gain) are big enough to see into the future and plot long-term consequences. Natural selection is like a robot that can only climb uphill, even if this leaves it stuck on top of a measly hillock. There is no mechanism for going downhill, for crossing the valley to the lower slopes of the high mountain on the other side. There is no natural foresight, no mechanism for warning that present selfish gains are leading to species extinction—and indeed, 99 percent of all species that have ever lived are extinct.

The human brain, probably uniquely in the whole of evolutionary history, can see across the valley and can plot a course away from extinction and toward distant uplands. Long-term planning—and hence the very possibility of stewardship—is something utterly new on the planet, even alien. It exists only in human brains. The future is a new invention in evolution. It is precious. And fragile. We must use all our scientific artifice to protect it.

It may sound paradoxical, but if we want to sustain the planet into the future, the first thing we must do is stop taking advice from nature. Nature is a short-term Darwinian profiteer. Darwin himself said it: "What a book a devil's chaplain might write on the clumsy, wasteful, blundering, low, and horridly cruel works of nature."

Of course that's bleak, but there's no law saying the truth has to be cheerful; no point shooting the messenger—science—and no sense in preferring an alternative world view just because it feels more comfortable. In any case, science isn't all bleak. Nor, by the way, is science an arrogant know-all. Any scientist worthy of the name will warm to your quotation from Socrates: "Wisdom is knowing that you don't know." What else drives us to find out?

What saddens me most, Sir, is how much you will be missing if you turn your back on science. I have tried to write about the poetic wonder of science myself, but may I take the liberty of presenting you with a book by another author? It is *The Demon-Haunted World* by the lamented Carl Sagan. I'd call your attention especially to the subtitle: *Science as a Candle in the Dark*.

Editors' General Introduction

Michael Ruse and David Castle

And Jacob took him rods of green poplar, and of the hazel and chestnut tree; and pilled white strakes in them, and made the white appear which *was* in the rods. And he set the rods which he had pilled before the flocks in the gutters in the watering troughs when the flocks came to drink, that they should conceive when they came to drink. And the flocks conceived before the rods, and brought forth cattle ringstraked, speckled, and spotted.

Jacob loved Rachel, and to win her hand, he worked for seven years for Rachel's father, Laban. When Laban underhandedly put his older daughter, Leah, in Rachel's stead, Jacob worked another seven years to earn the wife that he wanted. Then, somewhat understandably, Jacob demanded wages as well as wives, and Laban agreed to let his son-in-law have all of the herd animals that were mottled or otherwise miscolored. But, again trying to cheat Jacob, Laban moved all of his mottled and miscolored animals far from the main flocks. At which point, Jacob took matters into his own hands, with, as we learn from Genesis, results entirely favorable to his own ends. "And it came to pass, whensoever the stronger cattle did conceive, that Jacob laid the rods before the eyes of the cattle in the gutters, that they might conceive among the rods. But when the cattle were feeble, he put *them* not in: so the feebler were Laban's, and the stronger Jacob's."

At least since biblical times, farmers and herdsmen have striven to mold their crops and their animals to their own ends, maximizing the quantity and quality of the end product and minimizing the labor to produce it. The merits of rotating crops was long known. In medieval times, serfs had moved in a three-year cycle from wheat to oats to fallow, and then back to wheat. But it was not until the time of the industrial revolution, beginning in the second half of the eighteenth century, that scientific principles of breeding and of crop management began to be uncovered and practiced. Some of the key techniques, like selective breeding, which today are regarded as so obvious, were learned only slowly and with difficulty. Amazingly to us, one finds that it was standard practice for a shepherd to castrate his best male lambs, for then the resulting wethers would be sure not to roam too far from the flock!

An industrial revolution demands an agricultural revolution, for now you have to feed many more mouths—many more urban mouths—from the labor of proportionately fewer rural workers. Those who first learned the skills kept them confidential, for they could profit from their esoteric knowledge. But eventually the techniques became widespread, leading to the improvements in highly domesticated animals and plants on which we rely today. Cows were made more milk-efficient, sheep gave better quality wool, and pigs were fatter and, as the taste for fat declined, leaner. Vegetables and fruits were improved, too—better quality cabbages, larger and rosier apples, and, especially, a bigger and better turnip. Traditionally, most of the pigs and cattle were killed in the fall, for there was no way of keeping them over winter. Now, with improved root crops, they could live without outside fodder, and there was no longer the annual bottleneck that prevented any significant long-term husbandry. By the time of Charles Darwin, indeed, agricultural technology, if we may so call it, had reached levels of considerable sophistication, and was a major factor in the continuing prosperity of the Old World and—increasingly—the New.

The twentieth century saw ongoing developments and triumphs in agriculture. Genetics—the science of heredity—began in earnest as the new century opened, thanks to the rediscovery of the principles of the Moravian monk Gregor Mendel. In the second decade of the century, at Columbia University in New York City, Thomas Hunt Morgan and his associates developed the so-called classical theory of the gene, locating the units of heredity on the chromosomes in the cells of all organisms. Now it was possible to learn precisely why it was that animals and plants develop as they do, to see truly how characteristics are transmitted from one generation to the next, and most important of all in agriculture, to control these processes. It is no surprise to find that, initially and for many subsequent years, it was in institutes of agriculture that genetics found its strongest advocates. Two of the greatest theoretical thinkers of the first half of the last century—Ronald Fisher in England and Sewall Wright in America—both started their careers affiliated with organizations connected directly with agriculture. Fisher worked for many years as the statistician at Rothhampstead Research Station, and Wright worked on longhorn cattle genetics for the U.S. Department of Agriculture.

The really major breakthrough, of course, came in 1953 with the discovery of the double helical structure deoxyribonucleic acid (DNA) by the young researchers James Watson and Francis Crick. This led to the development of molecular biology and to a profoundly new understanding of the nature of living things. At first, theory outstripped practice and technological application. But in the early 1970s techniques were developed whereby pieces of DNA could be broken apart and recombined to produce new properties in organisms. Recombinant DNA technology first started with the transfer of genes from one part of an organism's chromosomes to another, and then more drastically from one organism to another. Suddenly, an incredibly new and powerful way of modifying organisms genetically was at hand, one that could be applied equally to plants and to animals (even mixing genetic material between the two!) with the prospect of virtually unlimited design and change. One could breed bigger and better produce, maturing in less time, with less effort; one could produce crops with their own built-in herbicides and insect repellants; one could harness and harvest bacteria to yield an endless supply of cheap drugs; and much, much more. And all of this was not just for the already overprivileged First World, but for all humans. Developing countries would now have the opportunity to buy, develop, or receive through donation their own technology. With the ever-expanding arsenal of techniques, developing countries would be able to custom-tailor their agricultural output to their specific needs.

Apart from a brief moratorium in the early 1970s in response to concerns about the risks posed by recombinant bacteria, medical and agricultural biotechnology's growth has been exponential. Drugs like insulin are now produced by genetically engineered bacteria. This year's soybean crop in the United States is nearly 70 percent genetically modified. New technologies that merge agriculture and medicine are being pursued—such as plants that contain vaccines for use in disease-stricken developing countries. But with the new technology has come controversy. Or, let us more accurately say that, with the new developments in agriculture has come renewal of that controversy that has long plagued and troubled the whole field of genetics and biotechnology. Notoriously, when people have moved forward in developing and applying new technologies, there have been those who have resisted and objected, sometimes violently. At the time of the agricultural revolution at the beginning of the nineteenth century, farm laborers and others invented the notorious "Captain Swing," who burned ricks and smashed machinery, as the rural workers saw their traditional ways of living and working disrupted and destroyed. In the twentieth century, only too well known is the sad story of Soviet agriculture, both the enforced removal of the peasants from their lands combined with the introduction of massive collective farms, and the official endorsement of the charlatan ideas of Stalin's favorite agriculturalist, Trofim Lysenko. Some thirty years ago when the green revolution was introduced, bringing new crops for underdeveloped countries, there were complaints that the input costs—fertilizers and herbicides in particular—far outstripped the benefits. The only ones truly to gain, supposedly, were the large commercial interests of the West selling the inputs to developing countries.

All of these controversies look fair to be dwarfed by the disruptions being raised and caused by the new techniques—by the new techniques that produce "genetically modified" foods (GM foods). In North America, the United States and Canada specifically, GM foods have been introduced with little fanfare or fuss. However, in Europe and elsewhere on the globe—including Third World countries like India—there has been opposition and rejection. In Britain, supermarkets find it politic to refuse to carry modified foods; in Europe, there are violent demonstrations, and farmers growing such modified produce have been threatened and their crops destroyed; and, in Asia, articulate voices decry the technology and implore their governments to reject and to forbid the products. As we have already seen in the prologue, even that supporter of the wholesome and healthy, Prince Charles, has weighed in with his blessing—and financial support—to organic alternatives to GM foods. And, as tends to be the case on occasions such as these, the proponents of the new technology look now somewhat bruised and beleaguered, having lost public support for their science, and for having not received broad consumer acceptance of their products. At greater extremes, crops have been ploughed, laboratories have been burned, and scientists and corporate executives have been vilified.

What are the objections of the critics of genetically modified foods? As you will learn in the pages of this collection, they range over many factors and issues. They go from general distrust of anything supported and produced by big business or big science, to a deep conviction that any tampering with the natural is a violation of God's ordinance, something liable to incur—either here on earth or in the world thereafter—his righteous wrath. Some find GM foods to be morally distasteful. Others worry about their potential to poison us, or damage the environment. Those in the First World tend to fear that such products are part and parcel of a general contamination of pure food products. Those in the Third World tend to fear that such products will mean the end to traditional ways and lifestyles—perhaps, indeed, causing yet more oppression of the already-oppressed, like women, who now will be excluded even from those activities for which they were formally cherished. And then there are those who look down the road and see future problems, both culturally and biologically, as we increase yet more our dependence on the made and artificial, and as we change and reject yet again that which nature has made, slowly, over literally billions of years.

We cannot address every issue in a reasonably sized collection such as this, nor do we pretend to. And indeed, let us stress, it is not our aim to give you any prepackaged answer to anything. Rather, we want to introduce you to some of the main issues and then to let you make up your own mind. We have our own opinions, but it is our aim here to let you come to your opinions, from your own reasoning, in the light of the pertinent information and understanding. For this reason, having used the prologue as a teaser, as it were, to show you the strengths of the emotions involved, we begin the collection proper by turning to the technology. We give you material on one of the supposed major breakthroughs of the new technology, a rice with a natural supplement (provitamin A) needed by people as part of good nutrition but often not available—certainly not in sufficient quantities—in less industrialized parts of the globe. And then, at once, we show you how even something as apparently altruistic as this runs straight into controversy, with vehement critics and no less vehement defenders.

We move next to ethical and religious concerns. Note that nothing is simple or straightforward, and that here, as always, much depends on how you understand the state of the science and its possible implications for theoretical and practical matters. Much depends also on how one judges right and wrong, and to this end, philosophers have developed a number of theories to analyze ethical situations. Two secular theories are often at the fore in these discussions. First, there is some form of the system due to the great German philosopher Immanuel Kant, a system that urges you to judge actions according to the extent that they respect individuals as beings with intrinsic worth in their own right ("Treat people as ends rather than means"). Second, there is some form of the system due to the great English "utilitarian" philosophers that urges you to try to evaluate in terms of happiness or pleasure ("Act always so as to maximize happiness and minimize unhappiness"). Both of these approaches have virtues. Whether either is ultimately adequate to the task is for you to judge. And the same is true of religion. Some, like the Prince of Wales, seem to think that what God has made is in itself sacred and should not be altered. Others, including those included in our collection, see things as a little more complex. After all, God himself gave us our powers of reason. Does he expect us not to use them? Whether turning to the Bible is the answer is a moot point. On the one hand, there is the story of Jacob and Laban. On the other hand, Leviticus forbids the wearing of a garment made from a linen/wool mix. What then of a sheep with an inserted vegetable gene?

We move next to some more practical issues. Does the consumer have the right to know? Well, "yes" you say, but then comes the response, "the right to know what?" It may seem simple and obvious that people should have the right to know when their food is being modified by new technology, but, in the era of molecular technology, the simple and obvious soon become the very nonsimple and nonobvious. This is true even when you start to add qualifiers like "the right to know the relevant." At once, questions are raised about what is and what is not relevant. Suppose you are marketing vegetable oil made from crops that were genetically modified. Do you need to tell this fact to the consumer, even though the oil itself has been so refined that no DNA (that is, no genetic material) remains in it?

At one time, farmers would reclaim seed in the fall from their crop for next year's planting. Now they sign annual "terms of use agreements" with seed companies for modified seed each spring. Many opponents to GM foods see agri-food biotechnology companies monopolizing agriculture with such agreements, and point the finger at the ownership and control of agriculture through patents. Here we offer, first, a historical account of patents in agriculture showing that while agricultural patents are nothing new, they are more prevalent and this generates other concerns. Patents have also created agricultural biotechnology industries that, quite reasonably, model their business practices on patent law, but do so with serious social implications. Going beyond patents, we have included another piece that discusses the complex legal and ethical issues that arise in the environmental release of genetically modified organisms. As you can see, agriculture is good business for lawyers because they have to stretch the law, make new law, and rethink legal theory as biotechnology generates novel situations.

We then move on to a trio of sections that deal with the risks and benefits of genetically modified foods. One pressing question is whether or not the processes used to put novel traits into plants and animals change those organisms in ways that might make our food unsafe to eat. There are regulatory steps by which any novel food, GM included, are introduced into the food system. Who are the experts, and how sound are their theories? Can people really talk knowledgeably about new technology, or is it really all whistling in the dark and speaking with a confidence that is not really justified? We have included readings that consider whether the tests being performed inspire confidence in the regulation of genetically modified foods. Some, feeling uncertain, would prefer to take precautions against exposing themselves to potential risks. New technology generates new fears, and the precautionary approach is supposed to allay them. But regardless of whether GM food risks are under- or overstated, it only seems appropriate that people are kept informed. Advocates of GM foods and the public do not always speak the same language of risk, and our selections in this section captures the "two ships passing in the night" nature of this dialogue. There is public debate on genetically modified food, but whose voices are being heard, and is the truth always being told amid the rhetoric? Once we sift through these considerations, we must also think about what the outcome of public debate on GM foods will be: Do we expect perfect consensus at the end of the debate, or is it inevitable that some action will have to be taken, but without full endorsement?

Drawing to the end, we lift our gaze and look at some of the broader questions. How can, or how should, GM foods affect peoples and their ways in nondeveloped countries? Are the optimists right in thinking that, for all of the difficulties, the way is now open (as never before) to feed the multitudes and to eliminate poverty and hunger? Or are the pessimists right in thinking that this is one more false hope, and that truly the new agriculture will bring in its wake tragedy and dissension—that it was an ill day when people started to use molecular biology to transform crops and animals into strange and new forms? Let us simply say that we present both sides of the debate and that the discussion is bitter and intense. The same is true also of the final section of this collection, in which we turn to the environment and to such questions as the dangers posed to the rest of the organic world when we cultivate modified organisms. Should we fear the side effects, perhaps, from the possible escape of the new life-forms? Or are fears like these wildly exaggerated? There are people on both sides of the fence expressing deeply held opposing positions.

By now, we expect you yourself are primed and want to enter the debate, so at this point we prepare to leave you on your own. To light your way, we offer a short guide to the literature in the field and a brief glossary so that you will have some help with the technical terms used by some of the authors of the pieces of our collection. Do not make hasty judgments, and always remember Jacob. "And the man increased exceedingly, and had much cattle, and maidservants, and menservants, and camels, and asses."

Part 1

Biotechnology Case Study

Golden Rice

Introduction

If you have insufficient vitamin A in your diet, you suffer from a wide range of ailments, of which impaired vision and diarrhea are but two. The lack of this essential component of the human diet leads to widespread misery. In Southeast Asia alone over 70 percent of the children under five are afflicted, and if everyone got the required amount, then every year up to two million deaths of young people would be prevented. Rice, which is the staple of the diet of half the world's population, contains no vitamin A, unlike many common vegetables. What better idea then than to take genes from plants that produce provitamin A, a precursor to vitamin A, and to insert them into rice, so that then everyone will stand a good chance of getting enough? And this is precisely what a number of researchers have done, using daffodils as one of the sources of the required genes to produce "golden rice," given its name because the presence of provitamin A colors the rice golden.

Yet not everyone is happy. Indian activist Vandana Shiva writes of golden rice as a "hoax," and Greenpeace compares golden rice to "fool's gold." They complain that in order to ingest enough provitamin A, it would be necessary to consume vast—perhaps impossible—quantities of rice. All of agriculture and more would have to be given over to rice cultivation, and this would be devastating to already-existing native and traditional practices. Biodiversity would be wiped out, even though indigenous plants can themselves provide the needed vitamin. The only ones truly to gain would be the large Western corporations like Monsanto. Far from the technology producing golden rice being a boon to India, "It is a very effective strategy for corporate takeover of rice production, using the public sector as a Trojan horse."

Nonsense, reply the defenders of golden rice. Groups like Greenpeace, they charge, make their objections mainly because they will thereby attract funds, support, and sympathy. Golden rice was never expected to provide all of the necessary dietary provitamin A, and Shiva and Greenpeace's calculations are bogus and fraudulent. Surely some vitamin supplementation must be better than none. There is no reason to think that golden rice will supplant traditional foods—the aim is to supplement such foods. And, in any case, the defenders remind us, not everyone is able to grow and to consume traditional foods as advocated by Shiva—ignorance, climate, and other factors make alternatives, traditional or otherwise, simply out of reach for many. It would be most wrong, therefore, simply to let the critics have the last word. Golden rice and like foods offer a wonderful opportunity for humankind.

1

Biotechnology Food

From the Lab to a Debacle

Kurt Eichenwald et al.

In late 1986, four executives of the Monsanto Company, the leader in agricultural biotechnology, paid a visit to Vice President George Bush at the White House to make an unusual pitch.

Although the Reagan administration had been championing deregulation across multiple industries, Monsanto had a different idea: the company wanted its new technology, genetically modified food, to be governed by rules issued in Washington—and wanted the White House to champion the idea.

"There were no products at the time," Leonard Guarraia, a former Monsanto executive who attended the Bush meeting, recalled in a recent interview. "But we bugged him for regulation. We told him that we have to be regulated."

New York Times on the Web, 25 January 2001.

This article was reported by Kurt Eichenwald, Gina Kolata, and Melody Petersen and was written by Mr. Eichenwald.

Government guidelines, the executives reasoned, would reassure a public that was growing skittish about the safety of this radical new science. Without such controls, they feared, consumers might become so wary they could doom the multibillion-dollar gamble that the industry was taking in its efforts to redesign plants using genes from other organisms—including other species.

In the weeks and months that followed, the White House complied, working behind the scenes to help Monsanto—long a political power with deep connections in Washington—get the regulations that it wanted.

It was an outcome that would be repeated, again and again, through three administrations. What Monsanto wished for from Washington, Monsanto—and, by extension, the biotechnology industry—got. If the company's strategy demanded regulations, rules favored by the industry were adopted. And when the company abruptly decided that it needed to throw off the regulations and speed its foods to market, the White House quickly ushered through an unusually generous policy of self-policing.

Even longtime Washington hands said that the control this nascent industry exerted over its own regulatory destiny—through the Environmental Protection Agency, the Agriculture Department, and ultimately the Food and Drug Administration—was astonishing.

"In this area, the U.S. government agencies have done exactly what big agribusiness has asked them to do and told them to do," said Dr. Henry Miller, a senior research fellow at the Hoover Institution, who was responsible for biotechnology issues at the Food and Drug Administration from 1979 to 1994.

The outcome, at least according to some fans of the technology: "Food biotech is dead," Dr. Miller said. "The potential now is an infinitesimal fraction of what most observers had hoped it would be."

While the verdict is surely premature, the industry is in crisis. Genetically modified ingredients may be in more than half of America's grocery products. But worldwide protest has been galvanized. The European markets have banned the products, and some American food producers are backing away. A recent discovery that certain taco shells manufactured by Kraft contained Starlink, a modified corn classified as unfit for human consumption, prompted a sweeping recall and did grave harm to the idea that self-regulation was sufficient. The mighty Monsanto has merged with a pharmaceutical company.

How could an industry so successful in controlling its own regulations end up in such disarray?

The answer—pieced together from confidential industry records, court documents and government filings, as well as interviews with current and former officials of industry, government, and organizations opposing the use of bioengineering in food—provides a stunning example of how management, with a few miscalculations, can steer an industry headlong into disaster.

For many years, senior executives at Monsanto, the industry's undisputed leader, believed that they faced enormous obstacles from environmental and consumer groups opposed to the new technology. Rather than fight them, the original Monsanto strategy was to bring in opponents as consultants, hoping their participation would ease the foods' passage from the laboratory to the shopping cart.

"We thought it was at least a decade-long job, to take our efforts and present them to environmental groups and the general public, and gradually win support for this," said Earle Harbison Jr., the president and chief operating officer at Monsanto during the late 1980s.

But come the early 1990s, the strategy changed. A new management team took over at Monsanto, one confident that worries about the new technology had been thoroughly disproved by science. The go-slow approach was shelved in favor of a strategy to erase regulatory barriers and shove past the naysayers. The switch invigorated the opponents of biotechnology and ultimately dismayed the industry's allies—the farmers, agricultural universities, and food companies.

"Somewhere along the line, Monsanto specifically and the industry in general lost the recipe of how we presented our story," said Will Carpenter, the head of the company's biotechnology strategy group until 1991. "When you put together arrogance and incompetence, you've got an unbeatable combination. You can get blown up in any direction. And they were."

Biology Debate: New Microbes Bring New Fears

In the summer of 1970, Janet E. Mertz was working at Cold Spring Harbor Laboratory, picking up tips on animal viruses from Dr. Robert Pollack, a professor at the private research center on Long Island and a master in the field. One day she began to explain to Dr. Pollack the experiment she was planning when she returned to her graduate studies in the fall at Stanford University with her adviser, Dr. Paul Berg. They were preparing to take genes from a monkey virus and put them into a commonly used strain of bacteria, E. coli, as part of an effort to figure out the purposes of different parts of a gene.

Dr. Pollack was horrified. The virus she planned to use contained genes that could cause cancer in rodents, he reminded her. Strains of E. coli live in human intestines. What if the viral genes created a cancer-causing microbe that could be spread from person to person—the way unmodified E. coli can. Dr. Pollack wanted Ms. Mertz's project halted immediately.

"I said to Janet, 'There's a human experiment I don't want to be part of,'" Dr. Pollack said in a recent interview.

The resulting transcontinental shouting match between Dr. Pollack and Dr. Berg set off a debate among biologists around the world as they contemplated questions that seemed lifted from science fiction. Were genetically modified bacteria superbugs? Would they be more powerful than naturally occurring bacteria? Would scientists who wanted to study them have to move their research to the sort of secure labs used to study diseases like the black plague?

"The notion of being able to move genes between species was an alarming thought," said Alexander Capron, a professor of law and medicine at the University of Southern California in Los Angeles. "People talked about there being species barriers—you're reorganizing nature in some way."

As researchers joined in the debate, they came to the conclusion that strict controls were needed on such experiments until scientists understood the implications. In 1975 the elite of the field gathered at the Asilomar conference center in Pacific Grove, California. There, they recommended that all molecular biologists refrain from doing certain research and abide by stringent regulations for other experiments. To monitor themselves, they set up a committee at the National Institutes of Health to review and approve all research projects.

It took just a few years—and hundreds of experiments—before the most urgent questions had their answers. Over and over again, scientists created bacteria with all manner of added or deleted genes and then mixed them with naturally occurring bacteria.

But rather than creating superbugs, the scientists found themselves struggling to keep the engineered bacteria from dying as the more robust naturally occurring bacteria crowded them out.

It turned out that adding almost any gene to bacteria cells only weakened them. They needed coddling in the laboratory to survive. And the *E. coli* that Ms. Mertz had wanted to use were among the feeblest of all.

By the mid-1980s, the Institutes of Health lifted its restrictions. Even scientists like Dr. Pollack, who sounded the initial alarm, were satisfied that the experiments were safe.

"The answer came out very clearly," he said. "Putting new genes into bacteria did not have the unintended consequence of making the bacteria dangerous."

That decision echoed through industry like the sound of a starter's pistol. First out of the gate were the pharmaceutical companies, with a rapid series of experiments on how the new science could be used in medicines. Hundreds of drugs went into development, including human insulin for diabetes, Activase for the treatment of heart attacks, Epogen for renal disease, and the hepatitis B vaccine.

"It's been huge," said Dr. David Golde, physician in chief at Memorial Sloan-Kettering Cancer Center in New York. "It has changed human health."

The success that modifying living organisms would bring the pharmaceutical industry quickly attracted attention from some of the nation's largest agricultural companies, eager to extend their staid businesses into an arena that Wall Street had endowed with such glamour.

Reaching Out: Monsanto Takes a Soft Approach

In June 1986, Mr. Harbison took control of Monsanto's push into biotechnology, a project snared in mystery and infighting. A nineteen-year veteran of Monsanto who had recently become its president and chief operating officer, he formed a committee to lead the charge.

"There is little more important than this task in our corporation at this time," Mr. Harbison wrote to the thirteen executives selected for the assignment.

"We recognized early on," Mr. Harbison said in a recent interview, "that while developing lifesaving drugs might be greeted with fanfare, monkeying around with plants and food would be greeted with skepticism." And so Mr. Harbison drafted a plan to reach out to affected groups—from environmentalists to farmers —to win their support.

That same month, the company's lobbying effort for regulation began to show its first signs of success. The Environmental Protection Agency (EPA), the Department of Agriculture, and the Food and Drug Administration were given authority over different aspects of the business, from field testing of new ideas to the review of new foods.

In an administration committed to deregulation, the heads of some agencies had been opposed to new rules. At an early meeting, William Ruckelshaus, then the head of the EPA, expressed skepticism that his agency should play any role in regulating field testing, according to people who attended. That was overcome only when Monsanto executives raised the specter of congressional hearings about the use of biotechnology to create crops that contain their own pesticides, these people said.

By fall, Monsanto's strategy committee was developing a plan for introducing biotechnology to the public. A copy of a working draft, dated October 13, 1986, listed what the committee considered the major challenges: organized opposition among environmental groups, political opportunism by elected officials, and lack of knowledge among reporters about biotechnology.

It also highlighted more complex issues, including ethical questions about "tinkering with the human gene pool" and the lack of economic incentives to transfer the technology to the Third World, where it would probably do the most good.

To solve political problems, the document suggested engaging elected officials and regulators around the world, "creating support for biotechnology at the highest U.S. policy levels," and working to gain endorsements for the technology in the presidential platforms of both the Republican and Democratic Parties in the 1988 election.

To deal with opponents, the document said, "Active outreach will encourage public interest, consumer and environmental groups to develop supportive positions on biotechnology, and serve as regular advisers to Monsanto."

Former Monsanto executives said that while they felt confident of the new food's overall safety, they also recognized that bioengineering raised concerns about possible allergens, unknown toxins, or environmental effects. Beyond that, there was a reasonable philosophical anxiety about human manipulation of nature.

"If this business was going to work, one of the things we had to do was engage in a dialogue with all of the stakeholders, including the consumer groups and the more rational environmental organizations," said Mr. Carpenter, who headed the biotechnology strategy group. "It wasn't Nobel Prize thinking."

A Blunder: Decision on Milk Causes a Furor

Even as Monsanto was assembling its outreach strategy, other documents show that it was making strides toward what former executives now acknowledge was a major strategic blunder. The company was preparing to introduce to farmers the first product from its biotechnology program: a growth hormone produced in genetically altered bacteria. Some on the strategy committee pushed for marketing a porcine hormone that would produce leaner and bigger hogs.

But, simply because the product was further along in development, the company decided to go forward with a bovine growth hormone, which improves milk production in cows—despite vociferous objections of executives who feared that tinkering with a product consumed by children would ignite a national outcry.

"It was not a wise choice to go out with that product first," Mr. Harbison acknowledged. "It was a mistake."

Scientists who watched the events remain stunned by Monsanto's decisions.

"I don't think they really thought through the whole darn thing," Dr. Virginia Walbot, a professor of biological sciences at Stanford University, said of Monsanto's decision to market products that benefited farmers rather than general consumers. "The way Thomas Edison demonstrated how great electricity was, was by providing lights for the first nighttime baseball game. People were in awe. What if he had decided to demonstrate the electric chair instead? And what if his second product had been the electric cattle prod? Would we have electricity today?"

The decision touched off a furor. Jeremy Rifkin, director of the Foundation on Economic Trends, an opponent of biotechnology, joined with family-farm groups worried about price declines and other organizations in a national campaign to keep the Monsanto hormone out of the marketplace. Some supermarket chains shunned the idea; several dairy states moved to ban it. The first step toward the shopping cart brought only bad news.

One year later, in 1987, the EPA agreed to allow another company, Advanced Genetic Sciences, to test bioengineered bacteria meant to make plants resistant to frost. But under the agency's guidelines, it had to declare the so-called ice-minus bacteria a new pesticide—classifying frost as the pest.

On April 28 and May 28, strawberry and potato plants were sprayed in two California cities. Photographs of scientists in regulation protective gear—spacesuits with respirators—were broadcast around the world, generating widespread alarm.

"It was surreal," said Dr. Steven Lindow, a professor at the University of California at Berkeley, who helped develop the bacteria.

For the executives at Monsanto, these troubling experiences reinforced their commitment to the strategy of inclusion and persuasion.

The most complex challenge came in Europe, where there was deep distrust of the new foods, particularly among politically powerful farmers. Faced with such resistance, Mr. Harbison said Monsanto began subtly shifting its attention from the lucrative European market to Asia and Africa. The hope was that the economic realities of a global agricultural marketplace would eventually push Europe toward a more conciliatory attitude.

But by the early 1990s, company executives said, everything would change. Mr. Harbison retired. Soon, Monsanto's strategy for biotechnology was being overseen by Robert Shapiro, the former head of Monsanto's Nutrasweet unit, who in 1990 had been named head of the agricultural division.

In no time, former executives said, the strategy inside the company began to change. Mr. Shapiro demonstrated a devout sense of mission about his new responsibilities, these executives said. He repeatedly expressed his belief that Monsanto could help change the world by championing bioengineered agriculture, while simultaneously turning in stellar financial results.

Eager to get going, he shelved the go-slow strategy of consultation and review. Monsanto would now use its influence in Washington to push through a new approach.

Mr. Carpenter, the former head of the company's biotechnology strategy group, recalled going to a meeting with Mr. Shapiro, and cautioning that it seemed risky to tamper with a strategic approach that had worked well for the company in the past. But, he said, Mr. Shapiro dismissed his concerns.

"Shapiro ignored the stakeholders and almost insulted them and proceeded to spend all of his political coin trying to deal directly with the government on a political basis rather than an open basis," Mr. Carpenter said.

Mr. Shapiro, now the nonexecutive chairman of the Pharmacia Corporation, which Monsanto merged with last year, declined to comment. But in an essay published earlier this year by Washington University in St. Louis, he acknowledged that Monsanto had suffered from some of the very faults cited now by critics. "We've learned that there is often a very fine line between scientific confidence on the one hand and corporate arrogance on the other," he wrote. "It was natural for us to see this as a scientific issue. We didn't listen very well to people who insisted that there were relevant ethical, religious, cultural, social, and economic issues as well."

Turning Point: Objections by Scientists

On May 26, 1992, the vice president, Dan Quayle, proclaimed the Bush administration's new policy on bioengineered food. "The reforms we announce today will speed up and simplify the process of bringing better agricultural products, developed through biotech, to consumers, food processors, and farmers," Mr. Quayle told a crowd of executives and reporters in the Indian Treaty Room of the Old Executive Office Building. "We will ensure that biotech products will receive the same oversight as other products, instead of being hampered by unnecessary regulation."

With dozens of new grocery products waiting in the wings, the new policy strictly limited the regulatory reach of the Food and Drug Administration (FDA), which had oversight responsibility for foods headed to market.

The announcement—a salvo in the Bush administration's "regulatory relief" program—was in lock step with the new position of industry that science had proved safety concerns to be baseless.

"We will not compromise safety one bit," Mr. Quayle told his audience.

In the FDA's nearby offices, not everyone was so sure.

Among them was Dr. Louis J. Pribyl, one of seventeen government scientists working on a policy for genetically engineered food. Dr. Pribyl knew from studies that toxins could be unintentionally created when new genes were introduced into a plant's cells. But under the new edict, the government was dismissing that risk and any other possible risk as no different from those of conventionally derived food. That meant biotechnology companies would not need government approval to sell the foods they were developing.

"This is the industry's pet idea, namely that there are no unintended effects that will raise the FDA's level of concern," Dr. Pribyl wrote in a fiery memo to the FDA scientist overseeing the policy's development. "But time and time again, there is no data to back up their contention."

Dr. Pribyl, a microbiologist, was not alone at the agency. Dr. Gerald Guest, director of the FDA Center of Veterinary Medicine, wrote that he and other scientists at the center had concluded there was "ample scientific justification" to require tests and a government review of each genetically engineered food before it was sold.

Three toxicologists wrote, "The possibility of unexpected, accidental changes in genetically engineered plants justifies a limited traditional toxicological study."

The scientists were displaying precisely the concerns that Monsanto executives from the 1980s had anticipated—and indeed had considered reasonable. But now, rather than trying to address those concerns, Monsanto, the industry, and official Washington were dismissing them as the insignificant worries of the uninformed. Under the final FDA policy that the White House helped usher in, the new foods would be tested only if companies did it. Labeling was ruled out as potentially misleading to the consumer, since it might suggest that there was reason for concern.

"Monsanto forgot who their client was," said Thomas N. Urban, retired chairman and chief executive of Pioneer Hi-Bred International, a seed company. "If they had realized their client was the final consumer, they should have embraced labeling. They should have said, 'We're for it.' They should have said, 'We insist that food be labeled.' They should have said, 'I'm the consumer's friend here.' There was some risk. But the risk was a hell of a lot less."

Even some who presumably benefited directly from the new policy remain surprised that it was adopted. "How could you argue against labeling?" said Roger Salquist, the former chief executive of Calgene, whose Flavr Savr tomato, engineered for slower spoilage, was the first fruit of biotechnology to reach the grocery store. "The public trust has not been nurtured," he added.

In fact, the FDA policy was just what the small band of activists opposed to biotechnology needed to rally powerful global support to their cause.

"That was the turning point," said Jeremy Rifkin, the author and activist who in 1992 had already spent more than a decade trying to stop biotechnology experiments. Immediately after Vice President Quayle announced the FDA's new policy, Mr. Rifkin began calling for a global moratorium on biotechnology as part of an effort that he and others named the "pure food campaign."

He quickly began spreading the word to small activist groups around the world that the United States had decided to let the biotechnology industry put the foods on store shelves without tests or labels. Mr. Rifkin said that he got support from dozens of small farming, consumer, and animal rights groups in more than thirty countries. In Europe, these small groups helped turn the public against genetically altered foods, tearing up farm fields and holding protests before television cameras.

If the FDA had required tests and labels, Mr. Rifkin said, "it would have been more difficult for us to mobilize the opposition."

Today, the handful of nonprofit groups that joined Mr. Rifkin's in lobbying the FDA for stronger regulation in 1992 have multiplied to fifty-four. Those groups, including the Sierra Club, Friends of the Earth, the Natural Resources Defense Council, Public Citizen, and the Humane Society of the United States, signed a petition this spring demanding that the government take genetically engineered foods off the market until they are tested and labeled.

"There is absolutely no question that the voluntary nature of the policy was unacceptable to many," said Andrew Kimbrell, one of the early activists to oppose biotechnology and now the executive director of the Center for Food Safety, which filed the petition.

The FDA policy has also helped organizations like Mr. Kimbrell's raise money. In late 1998, groups opposed to biotechnology approached the hundreds of foundations that give regularly to environmental causes and told them about the government's decision to let the companies regulate themselves. Since then, the foundations have given the groups several million dollars out of concern over the policy, said Christina Desser, a lawyer in San Francisco involved in the fund-raising effort.

There was also an about-face in the approach to dealing with overseas markets. As the Clinton administration came to Washington, Monsanto maintained its close ties to policymakers—particularly to trade negotiators. For example, Mr. Shapiro was friends with Mickey Kantor, the United States trade negotiator who would eventually be named a Monsanto director.

Confrontation in trade negotiations became the order of the day. Senior administration officials publicly disparaged the concerns of European consumers as the products of conservative minds unfamiliar with the science.

"You can't put a gun to their head," Mr. Harbison said of the toughened trade strategy with Europe. "It just won't sell."

And it didn't. Protests erupted in Europe, and genetically modified foods became the rallying point of a vast political opposition. Exports of the foods slowed to a stop. With a vocal and powerful opposition growing in both Europe and America, the perceived promise of biotechnology foods began to slip away.

By the end of the decade, the magnitude of Monsanto's error in abandoning its slow, velvet-glove strategy of the 1980s was apparent. Mr. Shapiro himself acknowledged as much. In the fall of 1999, he appeared at a conference sponsored by Greenpeace, the environmental group and major biotechnology critic.

There, while declaring his faith in biotechnology, Mr. Shapiro acknowledged that his company was guilty of "condescension or indeed arrogance" in its efforts to promote the new foods. But it was too late for a recovery. Soon after that speech, with the company's stock price in the doldrums because of its struggles with agricultural biotechnology, Monsanto itself ended its existence as an independent company. It was taken over by Pharmacia, a New Jersey drug company.

In recent months, biotechnology has been struggling with the consequences of its blunders. Leading food companies like Frito-Lay and Gerber have said they will avoid certain bioengineered foods. And grain companies like Archer Daniels Midland and Cargill have asked farmers to separate their genetically modified foods from their traditional ones. That, in turn, creates complex, costly, and—as the Starlink fiasco shows—at times, flawed logistical requirements for farmers.

Efforts have been made by industry and government to assuage public concerns—although critics of the technology maintain that the attempts do not go far enough. Last week, the FDA announced proposed rule changes requiring the submission of certain information that used to be provided voluntarily. But even supporters of the rule change say that it will make little practical difference in the way the business works, since companies have universally submitted all such information in the past, even under the voluntary standard.

And the industry itself has started down a new path, with a multimillion-dollar advertising campaign promoting genetically engineered foods as safe products that provide enormous benefits to populations around the world—an effort that some food industry officials say has come ten years too late.

"For the price of what it would have cost to market a new breakfast cereal, the biotech industry probably could have saved itself a lot of the struggle that it is going through today," said Gene Grabowski, a spokesman with the Grocery Manufacturers of America, a trade group.

And in recent weeks, Monsanto itself has announced plans to chart a new course—one with striking similarity to the course abandoned in 1992—reviving its outside consultations with environmental, consumer, and other groups with concerns or interest in the technology.

For the corporate veterans who set the original strategy, this is scant solace. A dream they had worked so hard to achieve had, at the very least, been set back by years.

"You can't imagine how I have bled over this," said Mr. Carpenter, the former head of biotechnology strategy for Monsanto. "They lost the battle for the public trust."

2

The Green Revolution Strikes Gold

Mary Lou Guerinot

For millennia, breeders have concentrated on modifying the traits of plants to influence their growth performance in the field. The late-twentieth-century version of this effort is the production of transgenic plants. Crops such as Roundup Ready soybeans developed by Monsanto and corn expressing *Bacillus thuringiensis* (Bt) toxin reduce costs to the farmer by minimizing the application of herbicides and insecticides. Other genetically engineered traits increase the cash value of a crop providing us, for example, with canola plants that produce oils high in unsaturated fatty acids. However, the crops that would make the biggest difference for the largest number of people in the world are those that would serve as better sources of essential nutrients. Because extreme poverty continues to limit access of much of the world's population to food, it is important that affordable food be as nutritious as possible. The report in chapter 3, by Ye et al.,[1] who engineered rice grains to produce provitamin A (beta-carotene), exemplifies the best that agricultural biotechnology has to offer a world whose population is predicted to reach 7 billion by 2013.

Although half of the world's population eats rice daily and depends on it as their staple food, rice is a poor source of many essential micronutrients and vitamins. In Southeast Asia, 70 percent of children under the age of five suffer from vitamin A deficiency, leading to vision impairment and increased susceptibility to disease.[2] The United Nations Children's Fund (UNICEF) predicts that improved vitamin A nutrition could prevent 1 to 2 million deaths each year among children aged one to four years.[3]

How, then, can one improve the provitamin A content of rice? Mammals make vitamin A from beta-carotene, which is one of the most abundant carotenoids found in plants. Carotenoids are yellow, orange, and red pigments that are essential components of the photosynthetic membranes of all plants. They serve as accessory light-harvesting pigments and as antioxidants that quench tissue-damaging free radicals such as singlet oxygen species. Rice in its milled form contains neither beta-carotene nor any of its immediate precursors. In their successful bid to engineer rice to produce beta-carotene, Ye and colleagues have been greatly aided by recent progress in dissecting the carotenoid biosynthetic pathway.[4]

Carotenoids, along with a variety of other compounds including gibberellins, sterols, chlorophylls, and tocopherols, are derived from the general isoprenoid biosynthetic pathway. Immature rice endosperm synthesizes the carotenoid precursor geranyl geranyl diphosphate (GGPP).[5] To convert GGPP to beta-carotene, Ye et al. programmed the endosperm to carry out the necessary additional enzymatic steps.

Two molecules of GGPP must first be condensed to form phytoene, which is then desaturated to lycopene and finally cyclized to form beta-carotene. The successful introduction of three genes that encode the additional enzymes is a technical tour de force. Most traits engineered to date have only required the introduction of a single gene. The authors also needed to ensure that the introduced genes would only be expressed in the endosperm as this is the part of the rice grain that remains after polishing.

Most remarkably, the investigators successfully introduced two of the required genes (encoding phytoene synthase and phytoene desaturase) on a construct that did not have a selectable marker. They achieved this by simultaneously introducing another construct, which carried the third gene of interest (lycopene beta-cyclase) as well as a selectable antibiotic resistance gene. This cotransformation strategy should enable Ye et al. to segregate the antibiotic resistance gene away from the phytoene synthase and phytoene desaturase genes, thereby addressing one of the major concerns of opponents of genetically engineered crops. Such plants should still be able to produce beta-carotene because the authors have also shown that plants engineered with standard transformation procedures to express only the phytoene synthase and phytoene desaturase genes do not accumulate lycopene as predicted. Instead these plants produce essentially the same end products (beta-carotene, lutein, and zeaxanthin) as plants engineered to express all three carotenoid genes. The authors speculate that the enzymes necessary to convert lycopene into beta-carotene, lutein, and zeaxanthin are constitutively expressed in normal rice endosperm or are induced when lycopene is formed. In addition, the fact that rice plants normally do make carotenoids should go a long way toward calming fears about "Frankenfoods." Perhaps the only objection to golden rice, in the end, will be its color.

There is plenty of work still to do. Initial calculations suggest that these engineered plants can provide enough provitamin A to satisfy the recommended dietary allowance with a daily ration of rice. But only when true-breeding lines are available will it be possible to accurately determine levels of each type of carotenoid. The production of various carotenoids other than beta-carotene could provide additional health benefits as carotenoids have been implicated in reducing the risk of certain types of cancers, cardiovascular disease, and age-related macular degeneration. Fortunately, excess dietary beta-carotene, in contrast to excess vitamin A, has no harmful effects, so plants with enhanced beta-carotene content should be a safe and effective means of vitamin delivery.

Field-testing will tell us whether production of carotenoids in rice endosperm will entail any metabolic trade-offs. Shunting more of the common precursor GGPP into carotenoid production might result in a decrease in other compounds whose synthesis is dependent on GGPP. For example, tomatoes engineered to produce more phytoene exhibit signs of dwarfism, attributed to a thirtyfold reduction in the plant hormone gibberellic acid, which shares the precursor GGPP with phytoene.[6] However, unlike tomato plants that express phytoene synthase in all their tissues, the rice plants engineered by Ye et al. express the introduced phytoene synthase only in the endosperm, which reduces the potential for metabolic disruption throughout the plant.

Presumably, it should be possible to engineer the pathways for many of the thirteen essential vitamins into plants, once the pathways are known and the corresponding genes have been cloned.[7] Indeed, the model plant *Arabidopsis* has already been successfully engineered to synthesize vitamin E.[8] Improving the mineral content of plants so that they can serve as sources of the fourteen minerals required in the human diet presents researchers with a different set of challenges.[9] Unlike vitamins, which are synthesized by the plants themselves, plants must take up essential minerals from the soil. Iron deficiency is the leading nutritional disorder in the world today, affecting over 2 billion people. As with vitamins, many of the world's staple foods are not good sources of iron. Current efforts are centered on understanding how plants take up and store iron.[10] Rice has been engineered to have higher levels of the iron storage protein ferritin in the grain,[11] but the question remains as to whether these engineered rice plants will be a good source of dietary iron.

The road to better nutrition is not paved with gold and, hence, agribusiness has not centered its efforts on the nutritional value of food. The work that culminated in the production of golden rice was funded by grants from the Rockefeller Foundation, the Swiss Federal Institute of Technology, and the European Community Biotech Program. Like the plant varieties that made the green revolution so successful, the rice engineered to produce provitamin A will be freely available to the farmers who need it most. One can only hope that this application of plant genetic engineering to ameliorate human misery without regard to short-term profit will restore this technology to political acceptability.

The author is in the Department of Biological Sciences, 6044 Gilman, Dartmouth College, Hanover, NH 03755, USA. E-mail: M.L.Guerinot@Dartmouth.edu

Notes

1. Xudong Ye et al., "Engineering the Provitamin A (ß-Carotene) Biosynthetic Pathway into (Carotenoid-free) Rice Endosperm," *Science* 287, no. 5451 (2000): 303.

2. www.who.int/nut.

3. www.unicef.org/vitamina.

4. Francis X. Cunningham and Elisabeth Gantt, "Genes and Enzymes of Carotenoid Biosynthesis in Plants," *Annual Review of Plant Physiology Plant Molecular Biology* 59 (1998): 557; Joseph Hirschberg, "Production of High-value Compounds and Vitamin E," *Current Opinion in Biotechnology* 10, no. 2 (1999): 186.

5. Peter Burkhardt et al., "Transgenic Rice (*Oryza sativa*) Endosperm Expressing Daffodil (*Narcissus pseudonarcissus*) Phytoene Snythase Accumulates Phytoene, a Key Intermediate of Provitamin A Biosynthesis," *Plant Journal* 11, no. 5 (1997): 1071.

6. Rupert Fray et al., "Constitutive Expression of a Fruit Phytoene Synthase Gene in Transgenic Tomatoes Causes Dwarfism by Redirecting Metabolites from the Gibberellin Pathway," *Plant Journal* 8 (1995): 693.

7. Michael A. Grusak and Dean DellaPenna, "Improving the Nutrient Composition of Plants to Enhance Human Nutrition and Health," *Annual Review of Plant Physiology Plant Molecular Biology* 50 (1999): 133.

8. David Shintani and Dean DellaPenna, "Elevating the Vitamin E Content of Plants through Metabolic Engineering," *Science* 282, no. 5396 (1998): 2098.

9. Grusak and DellaPenna, "Improving the Nutrient Composition of Plants to Enhance Human Nutrition and Health."

10. David Eide et al., "A Novel Iron-Regulated Metal Transporter from Plants Identified by Functional Expression in Yeast," *Proceedings of the National Academy of Sciences USA* 93, no. 11 (1996): 5624; Nigel J. Robinson et al., "A Ferric-chelate Reductase for Iron Uptake from Soils," *Nature* 397, no. 6721 (1999): 694.

11. Fumiyaki Goto et al., "Iron Fortification of Rice Seed by the Soybean Ferritin Gene," *Nature Biotechnology* 17, no. 3 (1999): 282.

3

Engineering the Provitamin A (ß-Carotene) Biosynthetic Pathway into (Carotenoid-Free) Rice Endosperm

Xudong Ye, Salim Al-Babili, Andreas Klöti, Jing Zhang, Paola Lucca, Peter Beyer, and Ingo Potrykus

Rice (*Oryza sativa*), a major staple food, is usually milled to remove the oil-rich aleurone layer that turns rancid upon storage, especially in tropical areas. The remaining edible part of rice grains, the endosperm, lacks several essential nutrients, such as provitamin A. Thus, predominant rice consumption promotes vitamin A deficiency, a serious public health problem in at least twenty-six countries, including highly populated areas of Asia, Africa, and Latin America. Recombinant DNA technology was used to improve its nutritional value in this respect. A combination of transgenes enabled biosynthesis of provitamin A in the endosperm.

Vitamin A deficiency causes symptoms ranging from night blindness to those of xerophthalmia and keratomalacia, leading to total blindness. In Southeast Asia, it is estimated that a quarter of a million children go blind each

Reprinted with permission from *Science* 287 (2000): 303–305. Copyright 2000 American Association for the Advancement of Science.

year because of this nutritional deficiency.[1] Furthermore, vitamin A deficiency exacerbates afflictions such as diarrhea, respiratory diseases, and childhood diseases such as measles.[2] It is estimated that 124 million children worldwide are deficient in vitamin A[3] and that improved nutrition could prevent 1 million to 2 million deaths annually among children.[4] Oral delivery of vitamin A is problematic,[5] mainly due to the lack of infrastructure, so alternatives are urgently required. Success might be found in supplementation of a major staple food, rice, with provitamin A. Because no rice cultivars produce this provitamin in the endosperm, recombinant technologies rather than conventional breeding are required.

Immature rice endosperm is capable of synthesizing the early intermediate geranylgeranyl diphosphate, which can be used to produce the uncolored carotene phytoene by expressing the enzyme phytoene synthase in rice endosperm.[6] The synthesis of ß-carotene requires the complementation with three additional plant enzymes: phytoene desaturase and ζ-carotene desaturase, each catalyzing the introduction of two double bonds, and lycopene ß-cyclase, encoded by the *lcy* gene. To reduce the transformation effort, a bacterial carotene desaturase, capable of introducing all four double bonds required, can be used.

We used *Agrobacterium*-mediated transformation to introduce the entire ß-carotene biosynthetic pathway into rice endosperm in a single transformation effort with three vectors (fig. 1).[7] The vector pB19hpc combines the sequences for a plant phytoene synthase (*psy*) originating from daffodil[8] (*Narcissus pseudonarcissus*; GenBank accession number X78814) with the sequence coding for a bacterial phytoene desaturase (*crtI*) originating from *Erwinia uredovora* (GenBank accession number D90087) placed under control of the endosperm-specific glutelin (Gt1) and the constitutive CaMV (cauliflower mosaic virus) 35S promoter, respectively. The phytoene synthase cDNA contained a 5'-sequence coding for a functional transit peptide,[9] and the *crtI* gene contained the transit peptide (*tp*) sequence of the pea *Rubisco* small subunit.[10] This plasmid should direct the formation of lycopene in the endosperm plastids, the site of geranylgeranyl-diphosphate formation.

To complete the ß-carotene biosynthetic pathway, we co-transformed with vectors pZPsC and pZLcyH. Vector pZPsC carries *psy* and *crtI*, as in plasmid pB19hpc, but lacks the selectable marker *aphIV* expression cassette. Vector pZLcyH provides lycopene ß-cyclase from *Narcissus pseudonarcissus*[11] (Gen-Bank accession number X98796) controlled by the rice glutelin promoter and the *aphIV* gene controlled by the CaMV 35S promoter as a selectable marker. Lycopene ß-cyclase carried a functional transit peptide allowing plastid import.[12]

Precultured immature rice embryos (n = 800) were inoculated with *Agrobacterium* LBA4404/pB19hpc. Hygromycin-resistant plants (n = 50) were analyzed for the presence of *psy* and *crtI* genes (fig. 2). Meganuclease I–Sce I digestion released the ~10-kb insertion containing the *aphIV*, *psy*, and *crtI* expression cassettes. Kpn I was used to estimate the insertion copy number. All samples analyzed carried the transgenes and revealed mostly single insertions.

Immature rice embryos (n = 500) were inoculated with a mixture of *Agrobac-*

REPORTS

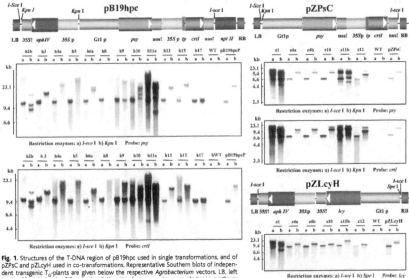

Fig. 1. Structures of the T-DNA region of pB19hpc used in single transformations, and of pZPsC and pZLcyH used in co-transformations. Representative Southern blots of independent transgenic T$_0$-plants are given below the respective *Agrobacterium* vectors. LB, left border; RB, right border; "I", polyadenylation signals; p, promoters; *psy*, phytoene synthase; *crtI*, bacterial phytoene desaturase; *lcy*, lycopene β-cyclase; *tp*, transit peptide.

terium LBA4404/ pZPsC and LBA4404/pZLcyH. Co-transformed plants were identified by Southern hybridization, and the presence of pZPsC was analyzed by restriction digestion. Presence of the pZLcyH expression cassettes was determined by probing I-Sce I– and Spe I– digested genomic DNA with internal *lcy* fragments. Of sixty randomly selected regenerated lines, all were positive for *lcy* and twelve contained pZPsC as shown by the presence of the expected fragments: 6.6 kb for the I-Sce I–excised *psy* and *crtI* expression cassettes from pZPsC and 9.5 kb for the *lcy* and *aphIV* genes from pZCycH (fig. 1). One to three transgene copies were found in co-transformed plants. Ten plants harboring all four introduced genes were transferred into the greenhouse for setting seeds. All transformed plants described here showed a normal vegetative phenotype and were fertile.

Mature seeds from T$_0$ transgenic lines and from control plants were air dried, dehusked, and, in order to isolate the endosperm, polished with emery paper. In most cases, the transformed endosperms were yellow,

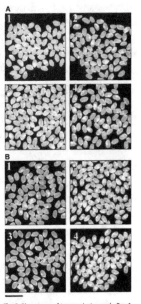

Fig. 2. Phenotypes of transgenic rice seeds. Bar, 1 cm. **(A)** Panel 1, untransformed control; panels 2 through 4, pB19hpc single transformants lines h11a (panel 2), h15b (panel 3), h6 (panel 4). **(B)** pZPsC/pZLcyH co-transformants lines z5 (panel 1), z11b (panel 2), z4a (panel 3), z18 (panel 4).

indicating carotenoid formation. The pB19hpc single transformants (fig. 2A) showed a 3:1 (colored/noncolored) segregation pattern, whereas the pZPsC/pZLcyH co-transformants (fig. 2B) showed variable segregation. The pB19hpc single transformants, engineered to synthesize only lycopene (red), were similar in color to the pZPsC/pZLcyH co-transformants engineered for ß-carotene (yellow) synthesis.

Seeds from individual lines (1 g for each line) were analyzed for carotenoids by photometric and by high-performance liquid chromatography (HPLC) analyses.[13] The carotenoids found in the pB19hpc single transformants accounted for the color; none of these lines accumulated detectable amounts of lycopene. Instead, ß-carotene, and to some extent lutein and zeaxanthin, were formed (fig. 3). Thus, the lycopene α(ε)- and ß-cyclases and the hydroxylase are either constitutively expressed in normal rice endosperm or induced upon lycopene formation.

The pZPsC/pZLcyH co-transformants had a more variable carotenoid pattern ranging from phenotypes similar to those from single transformations to others that contain ß-carotene as almost the only carotenoid. Line z11b is such an example (fig. 3C and fig. 2B, panel 2) with 1.6 µg/g carotenoid in the endosperm. However, reliable quantitations must await homozygous lines with uniformly colored

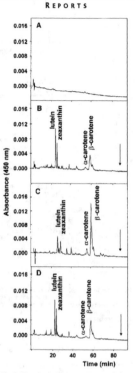

Fig. 3. The carotenoid extracts from seeds (1 g for each line) were subjected quantitatively to HPLC analysis. (A) Control seeds, (B) line h2b (single transformant), (C) line z11b (co-transformant), and (D) z4b (co-transformant). The site of lycopene elution in the chromatogram is indicated by an arrow.

grains. Considering that extracts from the sum of (colored/noncolored) segregating grains were analyzed, the goal of providing at least 2 µg/g provitamin A in homozygous lines (corresponding to 100 µg retinol equivalents at a daily intake of 300 g of rice per day), seems to be realistic.[14] It is not yet clear whether lines producing provitamin A (ß-carotene) or lines possessing additionally zeaxanthin and lutein would be more nutritious, because the latter have been implicated in the maintenance of a healthy macula within the retina.[15]

Notes

The *crtI* gene fused to the transit peptide sequence was kindly provided by N. Misawa (Kirin Co., Ltd., Japan). We thank W. Dong and P. Burkhardt for their valuable contributions, S. Klarer, K. Konja, and U. Schneider-Ziebert for skillful technical assistance, and R. Cassada for correcting the English version of the manuscript. Supported by the Rockefeller Foundation (1993–1996), the European Community Biotech Program (FAIR CT96,

1996–1999) (P.B.), the Swiss Federal Office for Education and Science (I.P.), and by the Swiss Federal Institute of Technology (1993–1996).

1. Alfred Sommer, "New Imperatives for an Old Vitamin (A)," *Journal of Nutrition* 119 (1988): 96.

2. James P. Grant, *The State of the World's Children* (Oxford: Oxford University Press, 1991); Keith P. West Jr., Gene R. Howard, and Alfred Sommer, "Vitamin A and Infection: Public Health Implications," *Annual Review of Nutrition* 9 (1989): 63.

3. Jean H. Humphrey, Keith P. West Jr., and Alfred Sommer, "Vitamin A Deficiency and Attributable Mortality among Under-5-Year-Olds," *World Health Organization Bulletin* 10 (1992): 225.

4. West, Howard, and Sommer, "Vitamin A and Infection: Public Health Implications."

5. Antoinette Pirie, *Proceedings of the Nutrition Society* 42 (1983): 53; Alfred Sommer, "Larges Doses of Vitamin A to Control Vitamin A Deficiency," in *Elevated Dosages of Vitamins: Benefits and Hazards,* eds. Paul Walter, Georg Brubacher, and Hannes Stähelin (Toronto: Hans Huber, 1989), 37–41.

6. Peter Burkhardt et al., "Transgenic Rice (*Oryza sativa*) Endosperm Expressing Daffodil (*Narcissus pseudonarcissus*) Phytoene Synthase Accumulates Phytoene, a Key Intermediate of Provitamin A Biosynthesis," *Plant Journal* 11, no. 5 (1997): 1071.

7. Three vectors—pUC18, pPZP100, and pBin19[16]—were digested with Eco RI and Hind III and a synthetic linker flanking by meganuclease I-Sce I including Kpn I, Not I, and Sma I (5'-AATTCATTACCCTGTTATCCCTACCCGGGCGGCCGCGGTAC-CATTACCCTGTTATCCCTAA-3') and (5'-AGCTTTAGGGATAACAGGGTAATG-GTACCGCGGCCGCCCGGGTAGGGATAACGGGTAATG-3') were introduced, forming pUC18M, pPZP100M, and pBin19M, respectively. An intermediate vector was made by insertion of the *crtI* expression cassette excised from Hind III/Eco RI-digested pUCET4, originally derived from pYPIET4,[17] into pBluescriptKS with Hind III/Eco RI digestion, followed by insertion of *psy* expression cassette from Sac II-blunted/Kpn I-digested pGt1psyH[17] into the Kpn I/Xho I-blunted previous vector. Finally, *crtI* and *psy* expression cassettes were isolated with Kpn I/Not I digestion and inserted into Kpn I/Not I-digested pUC18M and designated as pBaal3. pBin19hpc was made by insertion of a Kpn I fragment originally from pCIB900[18] containing *aphIV* selectable marker gene into pBaal3, followed by digestion of the I-Sce I fragment of the resulting plasmid and insertion into I-Sce I–digested pBin19M. pZPsC was obtained by insertion of the I-SceI fragment of pBaal3 bearing the *psy* and *crtI* genes into I-Sce I–digested pPZP100M. pZLcyH was constructed by digestion of pGt1LcyH with I-Sce I and insertion of the resulting fragment, carrying *lcy* and *aphIV*, into I-Sce I–restricted pPZP100M. The three vectors were separately electroporated into *Agrobacterium tumefaciens* LBA4404[19] with corresponding antibiotic selection. Callus induction: Immature seeds of japonica rice cultivar TP 309 at milk stage were collected from greenhouse-grown plants, surface-sterilized in 70 percent ethanol (v/v) for 1 min, incubated in 6 percent calcium hypochloride for one hour on a shaker, and rinsed three to five times with sterile distilled water. Immature embryos were then isolated from the sterilized seeds and cultured onto NB medium [N6 salts and B5 vitamins, supplemented with 30 g/l maltose, 500 mg/l proline, 300 mg/l casein hydrolate, 500 mg/l glutamine, and 2 mg/l 2,4-D (pH 5.8)]. After four to five days, the coleoptiles were removed, and the swelled scutella were subcultured onto fresh NB medium for three to five days until

inoculation of *Agrobacterium*. Transformation: one-week-old precultured immature embryos were immersed in *Agrobacterium tumefaciens* LBA 4404 cell suspension as described.[20] For co-transformation, LBA4404/pZPsC [optical density at 600 nm (OD_{600}) = 2.0] mixed with an equal volume of LBA4404/pZLcyH (OD_{600} = 1.0) was used for inoculation after acetonsyrigone induction. The inoculated precultured embryos were co-cultivated onto NB medium supplemented with 200 mM acetonsyringone for three days, subcultured on recovery medium (NB with 250 mg/l cefotaxime) for one week and then transferred onto NB selection medium in the presence of 30 mg/l hygromycin and 250 mg/l cefotaxime for four to six weeks. Transgenic plants were regenerated from recovered resistant calli on NB medium supplemented with 0.5 mg/l NAA and 3 mg/l BAP in four weeks, rooted and transferred into the greenhouse.

8. Michael Schledz et al., "Phytoene Synthase from *Narcissus pseudonarcissus*: Functional Expression, Galactolipid Requirement, Topological Distribution in Chromoplasts and Induction during Flowering," *Plant Journal* 10, no. 5 (1996): 781.

9. Michael Bonk et al., "Chloroplast Import of Four Carotenoid Biosynthetic Enzymes in Vitro Reveals Differential Fates Prior to Membrane Binding and Oligomeric Assembly," *European Journal of Biochemistry* 247, no. 3 (1997): 942.

10. Norihiko Misawa et al., "Functional Expression of the *Erwina uredovora* Carotenoid Biosynthesis Gene crtl in Transgenic Plants of ß-carotene Biosynthesis Activity and Resistance to the Bleaching Herbicide Norflurazon," *Plant Journal* 4, no. 5 (1993): 833.

11. Salim Al-Babili, Elias Hobeika, and Peter Beyer, "A cDNA Encoding Lycopene Cyclase (GenBank X98796) from *Narcissus pseudonarcissus* L. (PGR96-107)," *Plant Physiolgy* 112 (1996): 1398.

12. Bonk, "Chloroplast Import of Four Carotenoid Biosynthetic Enzymes in Vitro Reveals Differential Fates Prior to Membrane Binding and Oligomeric Assembly."

13. Dehusked seeds were polished for six hours with emery paper on a shaker. The endosperm obtained was ground to a fine powder and 1 g was extracted repeatedly with acetone. Combined extracts were used to record the ultraviolet-visible spectrum, allowing quantification using ε_{450nm} 134,000 1^{-1} mol $^{-1}$ cm $^{-1}$ for ß-carotene. The samples were dried and the residue quantitatively applied in 30 µl chloroform to HPLC for analysis using a photodiode array detector (Waters) and a C_{30} reversed-phase column (YMC Europe GmbH) with the solvent system A [methanol: *tert*-butylmethyl ether (1: 1, v/v)] and system B [methanol:*tert*-butylmethyl ether:H_2O (6: 1.2 : 1.2, v/v/v)], using a gradient of 100 percent B to 43 percent B within twenty-five minutes, then to 0 percent B within a further seventy-five minutes. Final conditions were maintained for an additional ten minutes. Photometric quantifications were reexamined by HPLC using synthetic all-*trans* lycopene as an external standard.

14. Burkhardt et al., "Transgenic Rice (*Oryza sativa*) Endosperm Expressing Daffodil (*Narcissus pseudonarcissus*) Phytoene Synthase Accumulates Phytoene, a Key Intermediate of Provitamin A Biosynthesis."

15. John T. Landrum et al., "A One Year Study of the Macular Pigment: The Effect of 140 Days of a Lutein Supplement," *Experimental Eye Research* 65, no. 1 (1997): 57.

16. Celeste Yanish-Perron, Jeffrey Vieria, and Joachim Messing, "Improved M13 Phage Cloning Vectors and Host Strains: Nucleotide Sequences of the M13mp18 and pUC19 Vectors," *Gene* 33 (1985): 103–109; Peter T. J. Hajdukiewicz et al., "The Small, Versatile pPZP Family of Agrobacterium Binary Vectors for Plant Transformation," *Plant Molecualr Biology* 25 (1994): 989; Michael Bevan, "Binary Agrobacterium Vectors for Plant Transformation," *Nucleic Acids Research* 12 (1984): 8711.

17. Misawa et al., "Functional Expression of the *Erwina uredovora* Carotenoid Biosynthesis Gene crtl in Transgenic Plants of ß-carotene Biosynthesis Activity and Resistance to the Bleaching Herbicide Norflurazon."

18. Joachim Wuenn et al., "Transgenic Indica Rice Breeding *thurigiensis* Line IR58 Expressing Synthetic cryIA(b) Gene from Bacillus Provides Effective Insect Pest Control," *Bio/Technology* 14 (1996): 171.

19. André Hoekema et al., "Binary Vector Strategy Based on Separation of Vir- and T-region of the *Agrobacterium tumefaciens* Ti-plasmid," *Nature* 303 (1983): 179–80.

20. Murielle Uzé et al., "Plasmolysis of precultured Immature Embryos Improves *Agrobacterium*-mediated Gene Transfer to Rice (*Oryza sativa* L.)," *Plant Science* 130 (1997): 87.

4

Genetically Engineered "Golden Rice" Is Fool's Gold

Greenpeace Statement

Manila/Amsterdam: Genetically engineered "Golden Rice" containing provitamin A will not solve the problem of malnutrition in developing countries according to Greenpeace. The genetic engineering (GE) industry claims vitamin A rice could save thousands of children from blindness and millions of malnourished people from vitamin A deficiency (VAD)–related diseases. But a simple calculation based on the product developers' own figures show an adult would have to eat at least twelve times the normal intake of 300 grams to get the daily recommended amount of provitamin A.[1]

Syngenta, one of the world's leading genetic engineering companies and pesticide producers, which owns many patents on the "Golden Rice," claims a single month of marketing delay of "Golden Rice" would cause 50,000 children to go blind.[2]

Greenpeace calculations show however, that an adult would have to eat at least 3.7 kilos of dry weight rice, that is, around 9 kilos of cooked rice, to satisfy his/her daily need of vitamin A from "Golden Rice." In other words, a normal

---

Greenpeace, http://www.greenpeace.org/extra/?item_id=5441.

daily intake of 300 grams of rice would, at best, provide 8 percent percent of the vitamin A needed daily. A breast-feeding woman would have to eat at least 6.3 kilos in dry weight, converting to nearly 18 kilos of cooked rice per day.[3]

"It is clear from these calculations that the GE industry is making false promises about "Golden Rice." It is a nonsense to think anyone would or could eat this much rice, and there is still no proof that it can provide any significant vitamin benefits anyway," said Greenpeace Campaigner Von Hernandez in the Philippines, where the first grains of the genetically engineered rice had been delivered to the International Rice Research Institute last month for breeding into local rice varieties. "This whole project is actually based on what can only be characterised as intentional deception. We recalculated their figures again and again, we just could not believe serious scientists and companies would do this."

In addition, one of the main sponsors of "Golden Rice," the Rockefeller Foundation, has told Greenpeace the GE industry has "gone too far" in its promotion of the product. While upholding its principal support for the project, Rockefeller Foundation President Gordon Conway, wrote to Greenpeace: "the public relations uses of Golden Rice have gone too far. The industry's advertisements and the media in general seem to forget that it is a research product that needs considerable further development before it will be available to farmers and consumers."[4]

"The European markets have resoundingly rejected GE products, consumers worldwide don't want them in their food, and the industry is desperate for alternative markets. "Golden Rice" has been presented as a quick fix for a global problem. It isn't, and the cash-driven propaganda about the product is swamping attempts to enforce existing effective solutions, and carry out further work on other sustainable, reliable methods to address the problem," added Hernandez.

Genetically engineered rice does not address the underlying causes of vitamin A deficiency (VAD), which are mainly poverty and lack of access to a more diverse diet. For the short-term, measures such as supplementation (that is, pills) and food fortification are cheap and effective. Promoting the use and the access to food naturally rich in provitamin A, such as red palm oil, will also help addressing the VAD-related sufferings. The only long-term solution is to work on the root causes of poverty and to ensure access to a diverse and healthy diet.[5]

Notes

1. Notes to Editors United Nations' World Health Organization/Food and Agriculture Organisation and the U.S. National Academy of Science recommendations on daily vitamin A intake.

2. Dr. Adrian C. Dubock, of Zeneca Plant Science (now Syngenta): "The levels of expression of provitamin A that the inventors were aiming at, and have achieved, are sufficient to provide the minimum level of provitamin A to prevent the development of irre-

versible blindness affecting 500,000 children annually, and to significantly alleviate Vitamin A deficiency affecting 124,000,000 children in twenty-six countries." "One month delay = 50,000 blind children month," at a conference on "Sustainable Agriculture in the New Millennium" in Brussels, May 28–31, 2000.

3. Greenpeace briefing paper "Vitamin A: Natural Sources vs Golden Rice" [online], www.greenpeace.org/.

4. "The False Promise of GE Rice" and the letter to Greenpeace U.K., January 22, 2001 [online], www.greenpeace.org/~geneng/.

5. Nutritionists have pointed out that numerous problems converge to cause vitamin A deficiency. In a recent letter to the *New York Times*, Dr. Marion Nestle noted that "conversion of beta-carotene to vitamin A, and transport in the body to the tissues that use vitamin A, require diets adequate in fat and protein. People whose diets lack these nutrients or who have intestinal diarrheal diseases—common in developing countries—cannot obtain Vitamin A from Golden Rice."

5

Golden Rice and the Greenpeace Dilemma

Ingo Potrykus

My offer to Greenpeace for a dialogue over golden rice was honest and a response to the reaction of Benedikt Haerlin (campaign leader of Greenpeace International), who accepted a moral obligation in this discussion. His reaction was the only way out of the problem that Greenpeace would otherwise have lost credibility in front of the press, which insisted in a clear answer on this matter, and Benedikt Haerlin was clever enough to realize this.

I respect his statement and I am looking forward to further discussions with him. The "hysteric" reactions of other Greenpeace activists to this step of normalization show that not everybody there realizes that if Greenpeace is continuing with its unqualified attacks against the responsible assessment of golden rice, Greenpeace will soon have a credibility problem far more severe than that coming up in context with the Brent Spar case. At least part of the media have realized that there is not much substantiation behind the routine arguments as far as golden rice is concerned.

Source: Greenpeace (http://www.greenpeace.org), "Golden Rice and the Greenpeace Dilemma": Second Response to Greenpeace from Prof. Ingo Potrykus, 15 February 2001.

As Greenpeace activists come again and again with the argument that release into the environment is too dangerous, I invite them to construct a realistic, concrete case. I have not found, in three years of discussions with numerous environmentalists, a scenario which could justify banning the field testing of golden rice. As the pathway is already in rice (and in every green plant), and the difference is only in its activity in the endosperm, it is very hard to construct any selective advantage for golden rice in any environment, and, therefore, any environmental hazard. The same holds true for all the other standard arguments, and I refer to "The Golden Rice Tale," available on the internet, and published in March in the journal "In-Vitro" [*In Vitro Cellular and Developmental Biology—Plant* 37, no. 2 (March 2001): 93–100].

It was very educating to see how selective Greenpeace was when citing from my statement—leaving everything out which did not fit into their view, and emphazizing selectively what they could use against me. Where is the difference to the PR campaigns Greenpeace likes to complain about? The "information" from Greenpeace was so distorted that I received compaints that I was ignoring the fact that daily allowance values did not mean much and that far lower provitamin values could already be expected to have beneficial effects (the point I was making in my response!).

This shows how Greenpeace has been able to transmit a completely wrong message by citing me. Here follows a citation from one of the responses to my "wrong" view:

"As I would assume you know, there is vast difference in the amount of vitamin A needed to reduce mortality, vs that needed to prevent blindness, vs that needed to prevent night-blindness and other like symptoms, vs that which satisfies actual metabolic needs, vs that which is equal to the recommended allowance, vs that which might be considered for optimal intake, vs that which might trigger toxicity symptoms. The vastness of those quantitative differences is further exaggerated in individuals whose metabolic need for this essential nutrient has been modified by an extended period of deprivation. Clearly in individuals whose diet is almost devoid of vitamin A dietary intake at levels representing only a small fraction of the 'recommended allowance' offers the potential to have a significant impact on both morbidity and mortality."

When I stated that I acknowleged that Greenpeace had identified a weak point in our strategy, I referred to the fact that only experimental data gained from nutrition studies with golden rice varieties could clarify how much provitamin A we would need to offer per gram of rice. This data will be available only after one to two years from now. With this data in hand, the optimal lines can then be determined for the final breeding adjustment.

I invite Greenpeace activists to specify in which area they see potential problems so that we can take care of these concerns in the process of the needs assessment and the extended safety assessments. But I expect concrete proposals, not blunt statements like "it is too dangerous to release transgenic plants into the environment." Please take the trouble to think about the case of golden rice.

To those who feel that they must prevent golden rice under all circumstances (for whatever political, ethical, religious reason) I would like to repeat: Golden rice will be used to complement traditional interventions to fight vitamin A deficiency. We need complementation because of the 500,000 blind children per year we have on the background of traditional interventions.

If you plan to destroy test fields to prevent responsible testing and development of golden rice for humanitarian purposes, you will be accused of contributing to a crime against humanity. Your actions will be carefully registered and you will, hopefully, have the opportunity to defend your illegal and immoral actions in front of an international court.

Golden Rice Hoax

When Public Relations Replace Science

Vandana Shiva

While the complicated technology transfer package of "golden rice" will not solve vitamin A problems in India, it is a very effective strategy for corporate take-over of rice production, using the public sector as a Trojan horse.

"Golden Rice": A Technology for Creating Vitamin A Deficiency

Golden rice has been heralded as the miracle cure for malnutrition and hunger of which 800 million members of the human community suffer.

Herbicide-resistant and toxin-producing genetically engineered plants can be objectionable because of their ecological and social costs. But who could possibly object

Dr. Vandana Shiva, "The 'Golden Rice' Hoax—When Public Relations Replace Science," 15 February 2001, http://www.online.sfsu.edu/~rone/GEssays/goldenricehoax.html.

to rice engineered to produce vitamin A, a deficiency found in nearly 3 million children, largely in the Third World?

As remarked by Mary Lou Guerinot, the author of the commentary on vitamin A rice in *Science*, one can only hope that this application of plant genetic engineering to ameliorate human misery without regard to short-term profit will restore this technology to political acceptability.

Unfortunately, vitamin A rice is a hoax, and will bring further dispute to plant genetic engineering where public relations exercises seem to have replaced science in promotion of untested, unproven, and unnecessary technology.

The problem is that vitamin A rice will not remove vitamin A deficiency (VAD). It will seriously aggravate it. It is a technology that fails in its promise.

Currently, it is not even known how much vitamin A the genetically engineered rice will produce. The goal is 33.3 µg (micrograms)/100g of rice. Even if this goal is reached after a few years, it will be totally ineffective in removing VAD.

Since the daily average requirement of vitamin A is 750 µg of vitamin A and one serving contains 30g of rice according to dry weight basis, vitamin A rice would only provide 9.9 µg which is 1.32 percent of the required allowance. Even taking the 100g figure of daily consumption of rice used in the technology-transfer paper would only provide 4.4 percent of the RDA.

In order to meet the full needs of 750 micrograms of vitamin A from rice, an adult would have to consume 2.272 kg of rice per day. This implies that one family member would consume the entire family ration of 10 kg from the Public Distribution System (PDS) in four days to meet vitamin A needs through "golden rice."

This is a recipe for creating hunger and malnutrition, not solving it.

Besides creating vitamin A deficiency, vitamin A rice will also create deficiency in other micronutrients and nutrients. Raw milled rice has a low content of fat (0.5g/100g). Since fat is necessary for vitamin A uptake, this will aggravate vitamin A deficiency. It also has only 6.8g/100g of protein, which means less carrier molecules. It has only 0.7g/100g of iron, which plays a vital role in the conversion of beta-carotene (precursor of vitamin A found in plant sources) to vitamin A.

Superior Alternatives Exist and Are Effective

A far more efficient route to removing vitamin A deficiency is biodiversity conservation and propagation of naturally vitamin A–rich plants in agriculture and diets. Table 1 gives sources rich in vitamin A used commonly in Indian foods.

In spite of the diversity of plants evolved and bred for their rich vitamin A content, a report of the Major Science Academies of the World—the Royal Society, U.K., the National Academy of Sciences of the USA, the Third World Academy of Science, the Indian National Science Academy, the Mexican Academy of Sciences, the Chinese Academy of Sciences, and the Brazilian Academy of Sciences—on transgenic plants and world agriculture has stated, vitamin A deficiency

Source Hindi name/ Content (µg/100g)
(Amaranth leaves) Chauli saag=266–1,166
(Coriander leaves)—Dhania=1,166–1,333
(Cabbage) Bandh gobi=217
(Curry leaves)-Curry patta=1,333
(Drumstick leaves)-Saijan patta1=283
(Fenugreek leaves)-Methi-ka-saag=450
(Radish leaves)-Mooli-ka-saag=750
(Mint)-Pudhina=300
(Spinach)-Palak saag=600
(Carrot)-Gajar=217–434
(Pumpkin [yellow])-Kaddu=100–120
(Mango [ripe])-Aam=500
(Jackfruit)-Kathal=54
(Orange)-Santra=35
(Tomato [ripe])-Tamatar=32
(Milk [cow, buffalo])-Doodh=50–60
(Butter)-Makkhan=720–1,200
(Egg [hen])-Anda=300–400
(Liver [Goat, sheep])-Kalegi=6,600–10,000
Cod liver oil=10,000–100,000

causes half a million children to become partially or totally blind each year.

Traditional breeding methods have been unsuccessful in producing crops containing a high vitamin A concentration, and most national authorities rely on expensive and complicated supplementation programs to address the problem. Researchers have introduced three new genes into rice, two from daffodils and one from a microorganism. The transgenic rice exhibits an increased production of beta-carotene as a precursor to vitamin A and the seed is yellow in color. Such yellow, or golden rice, may be a useful tool to help treat the problem of vitamin A deficiency in young children living in the tropics.

It appears as if the world's top scientists suffer a more severe form of blindness than children in poor countries. The statement that "traditional breeding has been unsuccessful in producing crops high in vitamin A" is not true given the diversity of plants and crops that Third World farmers, especially women, have bred and used, which are rich sources of vitamin A, such as coriander, amaranth, carrot, pumpkin, mango, and jackfruit.

It is also untrue that vitamin A rice will lead to increased production of beta-carotene. Even if the target of 33.3 µg of vitamin A in 100g of rice is achieved, it will be only 2.8 percent of the beta-carotene we can obtain from amaranth leaves, and 2.4 percent of the beta-carotene obtained from coriander leaves, curry leaves, and drumstick leaves. Even the World Bank has admitted that the rediscovery and use of local plants and the conservation of vitamin A–rich green leafy vegetables and fruits have dramatically reduced VAD-threatened children over the past twenty years in very cheap and efficient ways. Women in Bengal use more than two hundred varieties of field greens. Over 3 million people have benefited greatly from a food-based project for removing VAD by increasing vitamin A availability through home gardens. The higher the diversity of crops the better the uptake of pro-vitamin A.

The reason there is vitamin A deficiency in India in spite of the rich biodiversity base and indigenous knowledge base in India is because the green revolution technologies wiped out biodiversity by converting mixed cropping systems to monocultures of wheat and rice and by spreading the use of herbicides which destroy field greens.

In spite of effective and proven alternatives, a technology-transfer agreement has been signed between the Swiss government and the government of India for the transfer of genetically engineered vitamin A rice to India.

The ICAR [Indian Council of Agricultural Research], the ICMR [Indian Council of Medical Research], ICDS [Integrated Child Development Services], USAIUD [U.S. Agency for International Development], UNICEF [the United Nations Children's Fund], and the WHO [World Health Organization] have been identified as potential partners. The breeding and transformation is to be carried out at Tamil Nadu Agricultural University, Coimbatore, Central Rice Research Institute, Cuttack and Punjab Agricultural University, and Ludhiana and University of Delhi, South Campus.

The Indian varieties in which the vitamin A traits are expected to be engineered have been identified as IR 64, Pusa Basmati, PR 114, and ASD 16.

Dr. M. S. Swaminathan has been identified as "Godfather" to ensuring public acceptance of genetically engineered rice. The Department of Biotechnology (DBT) and ICAR are also potential partners for guaranteeing public acceptance and steady progress of the project.

Genetically engineered vitamin A rice will aggravate this destruction since it is part of an industrial-agriculture, intensive-input package. It will also lead to major water scarcity since it is a water-intensive crop and displaces water-prudent sources of vitamin A.

Transferring an Illusion to India

The first step in the technology transfer of vitamin A rice requires a needs assessment and an assessment of technology availability. One assessment shows that vitamin A rice fails to pass the needs test.

The technology-availability issue is related to whether the various elements and methods used for the construction of transgenic crop plants are covered by intellectual property rights. Licenses for these rights need to be obtained before a product can be commercialized. The Cornell-based ISAAA (International Service for the Acquisition of Agri-biotech Application) has been identified as the partner for ensuring technology availability by having material transfer agreements signed between the representative authority of the ICAR and the "owners" of the technology, Dr. Ingo Potrykus and Dr. Peter Beyer. In addition, Novartis and Kerin Breweries have patents on the genes used as constructs for the vitamin A rice.

At a public hearing on biotechnology at the U.S. Congress on 29 June 2000, Astra-Zeneca stated they would be giving away royalty-free licenses for the development of "golden rice."

At a workshop organized by the M. S. Swaminathan Research Foundation,

Dr. Barry of Monsanto's Rice Genome Initiative announced that it will provide royalty-free licenses for all its technologies that can help the further development of "golden rice."

Hence these gene giants—Novartis, Astra-Zeneca, and Monsanto—are claiming exclusive ownership to the basic patents related to rice research. Further, neither Monsanto nor Astra-Zeneca said they will give up their patents on rice—they are merely giving royalty-free licenses to public sector scientists for development of "golden rice." This is an arrangement for a public subsidy to corporate giants for research and development since they do not have the expertise or experience with rice breeding which public institutions have.

Not giving up the patents, but merely giving royalty-free licenses implies that the corporations like Monsanto would ultimately like to collect royalties from farmers for rice varieties developed by public-sector research systems. Monsanto has stated that it expects long-term gains from these IPR arrangements, which implies markets in rice as "intellectual property" which cannot be saved or exchanged for seed. The real test for Monsanto would be its declaration of giving up any patent claims to rice now and in the future and joining the call to remove plants and biodiversity out of TRIPS (Trade-Related Aspects of Intellectual Property Rights). Failing such an undertaking by Monsanto, the announcement that Monsanto giving royalty-free licenses for development of vitamin A rice like the rice itself can only be taken as a hoax to establish monopoly over rice production, and reduce rice farmers of India into bio-serfs.

While the complicated technology-transfer package of "golden rice" will not solve vitamin A problems in India, it is a very effective strategy for corporate takeover of rice production, using the public sector as a Trojan horse.

7

Open Letter to Greenpeace

Gordon Conway

THE ROCKEFELLER FOUNDATION

January 22, 2001

Dr. Doug Parr
Greenpeace
Canonbury Villas
London N1 2PN
England

Peter Melchett wrote suggesting that it would be
useful if I responded to the report by Dr. Vandana
Shiva entitled "The Golden Rice Hoax." I am pleased to
do so, and I am also enclosing background information on
Vitamin A deficiency disorders and the foundation's role
in the development of golden rice that you may find
informative.

First, it should be stated that we do not consider
golden rice *the* solution to the vitamin A deficiency
problem. Rather, it provides an excellent complement to

From www.rockfound.org by Gordon Conway, "Open Letter to Greenpeace," 22 January 2001. Reprinted with permission.

fruits, vegetables, and animal products in the diet, and to various fortified foods and vitamin supplements.

Complete balanced diets are the best solution, but the poorer families are, the less likely it is that their children will receive a balanced diet and the more likely they will be dependent on cheap food staples such as rice. This is particularly true in the dry seasons when fruits and vegetables are in short supply and expensive. Animal products are good sources of vitamin A, but will be unavailable to people on vegetarian diets. Unfortunately, in many cases, the bioavailability of sources of vitamin A is very low in green vegetables. And, the more rural these families are, the less likely their children will be reached regularly and effectively by vitamin A fortification and supplementation programs. Still, all these sources can and do make important contributions.

In her comments, Vandana Shiva ignores the fact that vitamin A–deficiency disorders result from a deficiency of vitamin A, not a complete absence of vitamin A in the diet. Vitamin A–deficient individuals are lacking 10 percent, 20 percent, or 50 percent of their daily requirements, not 100 percent. Hence, any additional contribution toward daily requirements would be useful. We calculated that the best golden rice lines reported in *Science* could contribute 15 to 20 percent of the daily requirements.

It should also be noted that the paper published in *Science* reported on the very first set of rice plants producing beta-carotene in the grain. The inventors have since made further improvements both in the level of beta-carotene production and with the elimination of the antibiotic-resistance gene.

Note also that if women consume this added source of vitamin A, it will improve their status, thereby increasing the concentration of vitamin A in the breast milk, a secondary but important source of vitamin A intake for young infants. The fact that hundreds of millions of children remain vitamin A deficient indicates that more needs to be done, complementary strategies need to be tried, and that golden rice also has the potential to make important contributions.

I do not know exactly what point Dr. Shiva is trying to make in the section on transferring the technology to India. Yes, needs assessments are being conducted so that the trait can be transferred to varieties grown where the beta-carotene is most needed. She seems to think that Indians are too dependent on rice. That may or may not be true, but even so, I do not see why we should not try to make it a more nutritious food. Where possible, scientists do use conventional breeding to improve the nutrition of crops, including increased beta-carotene production, but this was not possible with rice.

Finally, I agree with Dr. Shiva that the public relations uses of golden rice have gone too far. The industry's advertisements and the media in general seem to forget that it is a research product that needs considerable further development before it will be available to farmers and consumers.

I hope these comments and the enclosures are useful.

Gordon Conway

Part 2

Ethics in Agriculture

Introduction

Food produced behind the farm gate is consumed at the kitchen plate. If we raise ethical issues about food, are we not just unnecessarily complicating an otherwise routine part of our lives? Can we not leave the production of food to farmers, the science of food to agriculturalists, the processing of food to food scientists, the packaging and labeling to industries and regulators, and the stocking and pricing to the grocery boards and supermarkets? Moreover, what is it about genetically modified foods in particular that warrants ethical outcry about modern food systems?

Concrete reasons for the ethical controversy surrounding genetically modified foods are discussed in other sections of this collection. Some worry that genetically modified foods will have some kind of negative consequences, or that we should take precautions against the introduction of genetically modified foods because we cannot foresee what the outcomes will be. How we address these worries depends heavily on knowing the facts about the nature and magnitude of the threats posed by GM foods. Knowing these facts is not enough, however. Facts don't tell us what to do—for that we need to consider how those facts bear on the values that we hold. For instance, if a genetically modified plant is known to be harmful to nontarget organisms, that alone doesn't prescribe a specific course of action. We need to consider whether we value the crop more than we wish to avoid the harm caused by it.

Advocates and opponents of genetically modified foods typically express themselves with powerful ethical arguments. Prince Charles thinks it is immoral to play God with nature. Dawkins thinks it is immoral to turn one's back on scientific innovation. Entrenched positions such as these abound: It is immoral to allow monopoly control of agriculture; it is immoral to deny developing countries new technologies; it is immoral to mix the DNA of unrelated species; it is immoral to halt technologies that may benefit the environment. And so forth. We could fill several volumes cataloguing the variety and nuance in the ethical positions for and against GM foods.

Here we wish to avoid a laundry list of ethical positions as much as we wish to avoid any single ethical position. We aim for something more general; the articles in this section try to establish some balance and ethical rigor in the controversy surrounding GM foods. Paul Thompson notes that agricultural ethics are often driven by specific issues, but argues that there is a deeper level at which the problems with agricultural ethics are fundamentally about public trust in science and the advancement of biotechnology in developed countries. Thompson makes an eloquent call for a practical ethical discourse that goes beyond the emotion and opinion that dominates single-issue perspectives. This view is echoed by Marc Saner, who points out that ethical objections about the *process* by which GM foods are created often get confused with ethical objections about the *products* themselves. Like Thompson, Saner sees the debate on GM foods being stalled by the lack of clarity about what issues—"If we want fairness, respect, and progress in this debate, then we all have to say what we really mean."

Saner's prescription makes obvious sense, but it is difficult to put in practice. Ethical arguments about biotechnology are complicated, difficult to put clearly, and do not always result in clear consensus, a point made by Wolfgang van den Daele later in this volume. Magnus and Caplan observe that much of the public debate on GM foods attends to the potential risks associated with the technology, and transacts the moral debate in terms of a cost-benefit analysis. Right action is that which, in the face of risk, leads to the fewest harmful consequences. Yet what counts as harm is often a technical matter, not a moral matter. In fact, when morals are introduced—particularly morals Saner calls "technology transcending"—they are often dismissed because they lie outside of a technical assessment of risk. This pattern of debate shortchanges the moral dimension of GM foods. Magnus and Caplan try to show that when someone like Prince Charles says that GM foods are objectionable because we are playing God, he could be making any one of a number of serious moral claims. Magnus and Caplan think that we should have fairness, respect, and progress in the debate on GM foods, and help others to say what they really mean.

Gary Comstock captures many of the considerations outlined in the other papers, and tries to rationalize them into a framework for resolving debates about GM foods that seeks convergence on a moral decision from three standard moral systems. At one point in time, Comstock believed that GM foods were unethical, but careful reflection has now changed his mind. He argues that no reasonable objection stands in the way of GM foods if developing countries have equal access to agri-food biotechnology, if GM foods are not demonstrably harmful in any way, and if the scientific pursuit of GM foods encourages scientific innovation and wisdom in the natural sciences. Comstock discusses a study that shows how the persuasive rhetoric of single-issue advocacy blocks agreement on the ethical standing of GM food.

8

Bioethics Issues in a Biobased Economy

Paul B. Thompson

There is now a twenty-five-year history of debate over ethical issues associated with recombinant DNA, beginning with the 1975 Asilomar conference to consider the risks and advisability of basic gene-transfer research. Early in that history, issues associated with medicine or the manipulation of human DNA quickly came to be treated as wholly distinct from those associated with every other application of biotechnology. Today, virtually all nonhuman, nonmedical applications of biotechnology are classed as agricultural biotechnology. The only significant exception to this generalization is genetic engineering and cloning for xenotransplantation, which overlaps both medical and agricultural categories.

Controversy over specifically agricultural biotechnology really began in about 1984, when Jeremy Rifkin's lawsuit forced the National Institutes of Health's Recombinant DNA Advisory Committee (the RAC) to consider

NABC Report 12 in *The Biobased Economy of the Twenty-first Century: Agriculture Expanding into Health, Energy, Chemicals, and Materials* (Eds. Allan Eaglesham, William F. Brown, and Ralph W. F. Hardy) Published by the National Agricultural Biotechnology Council, Boyce Thompson Institute, Tower Road, Ithaca, NY 14853. (Single copies of NABC Report 12 are available free of charge.)

the possible environmental impact of ice-nucleating bacteria proposed to protect crops such as strawberries from frost damage. Social scientists had, by that time, already begun to speculate on the possible impact of a new generation of technologies on the structure of American agriculture, the lot of developing countries, and on the organization and funding of agricultural research worldwide. The National Agricultural Biotechnology Council was founded in 1989 in part as a forum in which to air these controversies. As we gather for the twelfth annual conference to discuss a new generation of technologies not geared to food production, I cannot but sense a hope that these new technologies are so full of prospect that the ethical controversies of the past will now fall by the wayside. As I myself argued in my 1996 remarks to this group,[1] there are reasons why that hope need not be vain. But I believe it is best to begin with a look to the past.

Most ethical issues that were tied to agricultural biotechnology over the last fifteen years fall into one of four categories: food safety, environmental impact, animal ethics, and social consequences. I will speak very briefly about each, then will focus the balance of my remarks on a fifth type of issue that spills over from the category of social consequences to encompass the entire debate over agricultural biotechnology. This fifth class concerns how bioethical issues are addressed within advanced industrial democracies, and takes up the question of public trust in science. One could say that comportment with respect to ethical issues is itself the most significant ethical issue facing the scientists, administrators, and public servants charged with the development of agricultural biotechnology. However, virtually all of the issues that have been tied to agricultural biotechnology in the last twenty-five years could have also been raised with respect to other technologies, both within agriculture and for society at large. Debate over agricultural biotechnology is, in this sense, a surrogate for debate over technological progress itself.

Ethical Issues for Agricultural Biotechnology: A Quick Survey

If one's cue were taken from the newspapers or from industry spokespersons, the hottest issue associated with agricultural biotechnology would be food safety. The core issue of ethics associated with the safety of genetic transformation applied to foods (or so-called GMOs) concerns the comparative emphasis on science-based food-safety risk assessment as opposed to a policy of informed consumer consent.[2] Some argue that individual consumers must not be put in a position where they are unable to apply their own values in choosing whether to eat GMOs. Others argue that the matter of whether genetic transformation has been used is immaterial to the underlying values (such as safety and healthfulness) that are the basis of consumer choice. They argue that the very act of informing consumers about GMO foods would mislead consumers into making choices that are not consistent with the underlying purposes that are sought through the purchase and consumption of food. This is an ethical issue rather than a simple dispute over facts because one viewpoint stresses individual autonomy and consent, while the other stresses rational optimization. The tension between these two ways of stating the most basic norms of decision making has been endemic to some of the most protracted ethical debates of the last two hundred years. Needless to say, it is possible for reasonable people to disagree.

After food safety, the environmental impact of agricultural biotechnology has received a great deal of play in the media. Some critics of agricultural biotechnology argue that we cannot even imagine the possible environmental consequences of genetic transformation. Other critics note some of the specific environmental consequences that have in fact been imagined with respect to products such as herbicide-tolerant or Bt crops, and argue that the risks are unacceptable. Defenders note the procedures for environmental risk assessment that are in place. They argue that these present adequate safeguards for the environment, and note that agricultural biotechnology may well have environmentally beneficial effects that outweigh any risks.

These environmental debates involve far more controversy over factual issues than do debates over food safety, but they still involve ethics. Like debates over food safety, they involve disputes over the validity and wisdom of relying on off-setting cost-risk-benefit optimization to conceptualize the issues. Even among those who accept the risk-benefit approach, the issues involve value judgements about the relative importance of food production as opposed to the preservation of wildlife and genetic diversity. They involve value judgments about how to proceed in the face of uncertainty, and indeed, about the very nature of uncertainty. The issues involve value judgments even about the nature of nature, as some believe that preserving wildlife and a certain aesthetic character on farms is part of nature conservation, whereas others see agriculture as inimical to wild nature. Again, it is possible for reasonable people to disagree.

Perhaps we should class the potential for biotechnology's impact on animal welfare as a subheading of environmental effects. It has seemed like a different class of issue to most observers because the focus has been on domesticated rather than wild animals, and because the ethical issues themselves are quite different from those listed above. Here, what is contentious, is the possibility of using gene transfer in a way that eventuates in an increase in suffering for domesticated livestock, or ironically, using gene transfer to relieve suffering by creating animals that are more tolerant of conditions that animal advocates currently find intolerable. Animal welfare is an ethical issue because the moral status of animals is itself one of the most fiercely contested ethical issues of the late twentieth century. Reasonable people disagree.

Finally, there are people who have framed the debate over agricultural biotechnology in terms of its social consequences. Indeed, many of the arguments for the deployment of agricultural biotechnology note its capacity to feed the poor and benefit farmers while keeping the cost of food low for all. Critics, on the other hand, fear that biotechnology will only turn the crank of the technological treadmill that has caused many farm bankruptcies and depleted the population of rural communities for one hundred years. Some critics fear that biotechnology will be the instrument for a similar kind of consolidation of land holdings in the developing world. Others argue that the transformation in the international system of intellectual property rights, which has accompanied the advent of agricultural biotechnology, may not be in the interests of poor farmers in the developing world. Still others have argued that agricultural biotechnology has precipitated a change in the nature of science itself, particularly at public institutions such as universities, resulting in a skewed allocation of resources and corporate control over research priorities. The social consequences of agricultural biotechnology are controversial in part because all of them—those that note biotechnology's putative benefits as well as those that call attention to its social costs—make disputable causal claims about the link between technological innovation and its eventual social impact. With all due respect to my colleagues in the social sciences, the models for social causation in economics, sociology, anthropology, geography, and political science continue to be beset with gaps and ambiguities that render them vulnerable to protracted methodological disputes and ideological influence. For this reason, disputes over social consequences take on a character that is more often political than ethical. Much of what divides disputants over the social consequences of agricultural biotechnology concerns different opinions about the capacity for various forms of social organization, notably private markets and government agencies, to reliably produce desired social outcomes.

Yet, there is still an explicitly ethical dimension to these debates. For example, when someone says that genetic engineering will benefit the poor, he is at least implicitly suggesting that not only is benefiting the poor a good thing, but that it is relatively better than benefiting someone else. There are, thus, ethically grounded notions of fairness and distributive justice lurking in debates over social consequences. Reasonable people disagree about what they and others deserve, what is fair, and how the resources of our society should be distributed. Such disagreements are inherently philosophical, and have been the very stuff of ethics and of social and political philosophy ever since Plato.

So, there are four large issues raised by, or associated with, agricultural biotechnology on which reasonable people disagree. This conference is dedicated to emerging applications of biotechnology that do not involve foods. We may reasonably conclude that issues associated with food safety and individual consent will not be associated with these new agricultural technologies. But this is only one of the four types of issue that have dogged agricultural biotechnology for over fifteen years, and it is arguably the simplest and least intrinsically contentious of the four. I must, therefore, conclude that the hopes for a new day and an ethical pass with respect to the biobased economy are probably not in the cards.

Agricultural Bioethics: Public Discourse and Public Trust

So, finally, we come to that fifth comprehensive issue, which we might call the "trust" issue. Does biotechnology—understood not merely as the laboratory techniques or the products themselves but as the consortium of industry and academic researchers, government regulators, and research administrators that has shepherded recombinant DNA techniques from basic research through product launch—merit the public's trust? Notice that the question of whether biotechnology merits public trust differs from whether biotechnology is, in fact, trusted. When the matter of trust is framed as a question of merit, of trustworthiness, it becomes an ethical issue in itself.

Even in an explicitly ethical mode, the question of trust inevitably connects with the broader public's attitudes and perceptions of biotechnology. My suggestion today is that the way that researchers, regulators, and administrators comport themselves with respect to the ethical issues I have already reviewed, albeit briefly, is the largest single factor in determining whether they are trustworthy. I will make some speculative remarks about public skepticism regarding agricultural biotechnology, but I must stress that I will not try to explain why agricultural biotechnology is mistrusted in fact. Nor do I believe that the relationship between being trustworthy and being trusted in fact is a simple or straightforward one.

First, a simple observation: none of the ethical issues listed above—issues on which reasonable people disagree—depends on active political opposition to biotechnology for their definition or significance. Each would be an ethical issue even if virtually no one was sufficiently concerned about agricultural biotechnology to carry placards, write angry letters, or construct Web pages that espouse a given analysis of each issue, while recruiting fellow travelers. An issue does not become "ethical" simply by virtue of its popularity, but because deep and systematic differences in values and interpretations open up the possibility for incompatible prescriptions for action. Throughout human history, it has often been the case that a small minority, sometimes a single individual, seizes on a vital difference and opposes a strong majority point of view. These minority viewpoints need not, and historically often have not, represented anything even remotely like widespread public doubt or opposition to the mainstream point of view. So we should not equate a response to ethical issues and a response to public concerns.

In some cases, the proper response to public concerns is a public relations campaign designed to sway citizens in the mainstream to a point of view more consistent with one's own interests. Such a campaign may eschew serious discussion of issues, choosing instead to associate a product or person with favorable images, or to associate opponents with unfavorable images. In such cases, the issue that has given rise to public concern is handled strategically. I shall use the term "strategic discourse" for any form of communication that tries to bolster public support for an objective (or mute public opposition) in an effective and efficient manner. Characteristically, a form of communication is strategic whenever the alteration or manipulation of audience attitudes and behavior is the dominant criterion for success.

I hope it is evident to everyone that strategic discourse is never an appropriate response to an ethical issue. In having too little concern with mutual understanding, strategic discourse disrespects those with differing values and differing points of view. Discourse ethics is a program in philosophy that prescribes a general approach for ethical issues.[3] We might summarize it in commonsense terms by saying that ethical issues must not be treated simply as obstacles to be overcome in the pursuit of other goals. They must instead be addressed seriously and in their own terms. When one is presented with an ethical objection to an opinion or course of action, one has a responsibility to ensure that one has first understood the force of that objection. Second, one must either alter the opinion or course of action to accommodate the objection, or offer a response that explains why the objection has been rejected. This means that those who offer an ethical objection are owed a reply. The reply should restate the objection in terms that the person who offered the objection can accept. If the terms are not accepted, one must conclude that one has not understood the objection, and try again.

If your reply to an objection involves a rejection of it, you owe the person who offered the objection an opportunity to reply to your reply, which, of course, may occasion further objections and replies. Obviously, this is a process that can go on at some length, so we must regard this characterization of discourse ethics as an idealization, and we must recognize that time and resource constraints limit the extent to which ideal discourse can be realized in practice. There is the further problem that the back and forth process of objection and reply can itself be deployed not in pursuit of seriousness and mutual respect, but as a delaying tactic. Anyone who has ever attended a public meeting on biotechnology within fifty miles of the Washington beltway knows exactly what I am talking about. Despite these shortcomings, I believe that it is still possible to conduct practical ethical discourse. While falling short of the unrealizable ideal case in which all objections are fully answered, practical discourse does treat ethical issues with the seriousness that they demand.

I believe that serious practical discourse is possible because I believe that I do it all the time myself. It is the standard to which I have aspired in all my research and writing on agricultural biotechnology. I have seen it at the "bioethics workshops" sponsored by Iowa State University, and even on occasion at the annual meetings of the National Agricultural Biotechnology Council (NABC). As further evidence, I would submit that that the Ethical, Legal, and Social Implications (ELSI) program of the Human Genome Initiative has supported a great deal of serious practical discourse on the goals and implications of human genetics. I believe that there should be a similar program in agricultural biotechnology, but here I get ahead of my main message.

Strategic and practical discourse are analogous to some criteria we rely upon when we determine whether an individual person is trustworthy. Trustworthy people display thoughtfulness of purpose and a clear capacity to be mindful of the interests of those by whom they are trusted. We do not trust people who seem to be making reference to their own immediate goals and self-interest at every moment. Similarly, I think that we can say that groups or associations of people who always seem to be engaged in strategic discourse, and never in serious practical discourse, are manifestly not trustworthy. This is not necessarily a judgment that reflects on the moral character of the individuals involved. People who are fine, upstanding, and virtuous citizens in their own right may well be involved in groups or associations that are untrustworthy in virtue of the fact that serious discourse about ethical issues occurs infrequently in these groups and associations. We should not expect groups and associations to avoid strategic discourse on every occasion. That would be like asking someone to be a saint, always putting others' interests before her own. But just as we mistrust the person who seems unable to even contemplate a situation with respect to others' interests, we mistrust the group or association that displays no evident interest in, or experience with, serious practical discourse.

We can bring this observation to the point at hand by considering the three key technologies of the postwar era as described by Martin Bauer: nuclear power, information technology (IT), and biotechnology. According to Bauer,[4] these three are particularly relevant to the problem of public acceptance and trust, because each has been presented to the public as a technology that would revolutionize the way we live. While the scientists, engineers, regulators, and power-company officials who developed and promoted nuclear energy displayed seriousness with respect to the safety of their technology, they have never been particularly willing to engage in practical discourse about the social, legal, and ethical issues posed by nuclear power generation. In contrast, computer professionals have carried on robust debates about a host of ethical issues from privacy rights to intellectual property and the impact of a wired society on interpersonal relations. Early on, they formed the Computer Professionals for Social Responsibility to promote debate over the risks of inadvertent nuclear war due to computer failure, and this group went on to promote both discussion and activism about access to computers for the poor. Even as the public seemed willing to embrace information technology uncritically, the critical voices emerged from within the culture of the computer industry, and demanded that ethical issues be taken seriously.

How then does biotechnology fare in the comparison? My answer is, better than nuclear power, but worse than IT. On the plus side, molecular biologists got off to an admirable start with Asilomar, and the previously mentioned ELSI program has ensured that medical bioethicists are deeply involved in discussions of the future human applications of gene technology. On the agriculture side, there are a few programs here and there, including my own Center for Biotechnology Policy and Ethics, which operated at Texas A&M University from 1991 to 1998. The executive council of the National Agricultural Biotechnology Council adopted a comprehensive endorsement of the need for universities to create a climate hospitable to debate and learning about the ethical dimensions of biotechnology in 1997. The National Agricultural Biotechnology Council annual conferences, which began in 1989, are themselves the most visible and substantive vehicle for nonstrategic discourse on ethical issues in North America.

On the minus side, I must say that these activities have gone along in fits and starts. Scientists and administrators have been far more interested in talking about ethical issues when agricultural biotechnology was getting negative publicity in the press than when things were going smoothly. Furthermore, many substantive criticisms of biotechnology have not been treated as concerns deserving respect and reply, much less a change in direction. When environmental or social issues are raised, defenders of biotechnology too often shift the subject to food safety or attack the sincerity and motives of their critics. This tendency to change the subject reveals a preoccupation with strategic thinking, and undermines an observer's confidence that serious issues are being treated seriously. I cannot help but draw the same conclusion that a casual observer of the debate would draw. Commitment to serious practical discourse and a critical consciousness among the individuals and organizations who have been involved in the application of molecular biology to agriculture and in the development of new agricultural biotechnologies has not been particularly deep.

Conclusion

Among those who have thought and written about the ethical issues that arise in connection with agricultural biotechnology, I have never been one who thought the use of recombinant techniques posed unique risks or exceptional ethical issues. I do think that the organization and culture of agricultural R&D is insufficiently attentive to a wide range of social, environmental, legal, and ethical issues that ride along with any significant technological innovation. I thus think that biotechnology provides an important case study and object lesson for some of the questions that we should be debating with respect to the ecological meaning of agriculture, and the impact of technical change on our social institutions, not to mention the poor. In one sense, I regret that I have not taken this opportunity to address some of those questions directly, but there is only so much that can be done at any given time or place.

It will not suffice to leave these issues to the final-stage regulators or adopters of technology. Scientists, educators, and administrators must institutionalize continuous critical reflection on their activities, and they must find some way to make that reflection effective in shaping the agenda for research and the deployment of technology. I am not of the opinion that the present status quo is wholly inadequate with respect to its capacity for ethical reflection and serious practical discourse. Indeed, remarkable strides have been made during the last twenty-five years. Nevertheless, we may not be sanguine about the status quo, either.

There is the distinct prospect that the specific technologies being discussed at this conference will be described and promoted in a manner that will only perpetuate the tendency to avoid seriousness with respect to ethical issues, and will provide even greater opportunity to deploy the strategy of changing the subject. How often I have heard the phrase "All we need is a product with consumer benefits!" For the life of me, I cannot find a way to interpret such language as anything other than a thoroughly strategic preoccupation with the manipulation of biotechnology's public image. I can appreciate that not everyone involved with biotechnology needs to be engaged in serious discussion of the issues I have raised in my talk. There is room for people who are concerned with its public image, and who are preoccupied with selling a product. I just hope there is room for something else, too.

Notes

1. Paul B. Thompson, "Tying It All Together," in *Agricultural Biotechnology: Novel Products and New Partnerships, NABC Report 8,* eds. Ralph W. F. Hardy and Jane B. Segelken (Ithaca, N.Y.: National Agricultural Biotechnology Council, 1996).

2. Paul B. Thompson, *Food Biotechnology in Ethical Perspective* (London : Blackie Academic for Chapman and Hall [distributed by Aspen Publishing], 1997).

3. Jürgen Habermas, "Discourse Ethics: Notes on a Program of Philosophical Justification," in *The Communicative Ethics Controversy,* eds. Seyla Benabib and Fred Dallmayr (Cambridge: MIT Press, 1990.)

4. Martin Bauer, *Resistance to New Technology: Nuclear Power, Information Technology, and Biotechnology* (Cambridge: Cambridge University Press, 1995).

9

Real and Metaphorical Moral Limits in the Biotech Debate

Marc A. Saner

From the beginning, the regulatory and public debate over biotechnology has been closely tied to the question of whether it matters what process is used to develop a product. Generally speaking, critics of genetic engineering argue that, yes, it does matter, while proponents argue that, no, only the features of the product matter.

I argue here that this question is at the root of the ethical debate over modern biotechnology. I argue further that fully understanding this question is of critical importance in moving the ethical debate surrounding modern biotechnology ahead.

The process-versus-product view is important to the ethical analysis because the two views neatly map onto the distinction between nonconsequentialist and consequentialist ethics. Simply put, nonconsequentialists formulate ethical prescriptions that stand regardless of the consequences, whereas consequentialists consider consequences in ethical decision making. For example, a nonconsequentialist may hold that killing an innocent human being is wrong under all circumstances. In contrast, a consequentialist would not prohibit such killing absolutely, but would attempt to judge it in the context of predicted consequences (for example, one can imagine a lifeboat situation in which the choice is between a single act of murder and the probable death of all passengers). The critical point is that nonconsequentialists may formulate absolute moral limits, whereas consequentialists will prefer to formulate ethical prescriptions contingent on the forecasting of consequences.

Within this classification, the process view is nonconsequentialist and the product view is consequentialist. Nonconsequentialists may argue that some or all types of genetic engineering are wrong, because these methods lie beyond a moral limit. An expression of this view would be, for example, "The genetic engineering of humans violates the basic dignity that all humans possess." In support of this line of argument, one can point out that metaphysical concepts, such as "dignity" and the prescription of moral limits, are common ingredients of existing and widely accepted legal and moral frameworks; for example, in human rights declarations. The categorical refusal to consider the use of human embryos as a source of stem cells is based on such an approach. Arguing that some nonhuman animals also have "dignity" that could be violated can extend this line of argument. Finally, one may single out genetic engineering as the worst offender within biological technologies, all of which may be considered a threat to the "intrinsic value" of nature (the existence of such value is implied in the United Nations World Charter for Nature of 1982) or to the "integrity of ecological systems" (the existence of such integrity is explicit in the new Earth Charter Initiative).[1] I would call such moral arguments "the prescription of a 'real' moral limit."

In contrast, consequentialists would argue that no method is intrinsically wrong, morally speaking. What really matters is the harm that may result, and such harm should be forecasted with risk-assessment methodology. Very broad categories of goods may be considered within this approach. As a result, risk assessment may have to be conducted not just with human health and the economy in mind, but also to assess environmental, aesthetic, social, and political change. Still, what matters is the risk of harm; all decisions are contingent on the prediction and consideration of risks and benefits. This line of argument does not support the view that a type of research is intrinsically immoral.

The differentiation between nonconsequentialists and consequentialists suitably characterizes two extreme approaches to the evaluation of a new technology. In practice, however, advocates and opponents of modern biotechnology often combine consequentialist and nonconsequentialist elements.

For example, Greenpeace's slogan "no genetic manipulation of nature"[2] appears to describe a moral limit. In reality, however, Greenpeace debates the issue using science and (consequentialist) risk language. Greenpeace is not alone. In the public debate, all opponents are pressed to provide a whole list of arguments that often have the structure: first, genetic engineering is absolutely wrong; and second, the projected risks are too high considering the projected benefits. This prompts the question: Why do we need to add a risk argument after stating the moral argument?

The absolutist, nonconsequentialist moral prescription would trump the contingent risk argument in any case—even if the balance of benefits and risk would call for the use of biotechnology on consequentialist moral grounds, as the industry keeps arguing. Perhaps the moral language is just a metaphor to strongly suggest a conclusion reached on the basis of (consequentialist) risk. I would call this position "the prescription of a metaphorical moral limit."

The problem with this approach is that it lacks clarity. Is a metaphorical moral limit specified to illustrate that the consequences are thought to be so severe that only an absolute prohibition will do? Or, is risk language used to convince science-minded individuals who may not be inclined to accept the true reason, the real moral limit specified first? Lack of clarity on this point fuels the rhetoric in the debate.

It is perhaps helpful here to consider an ethical prescription of the second order—a prescription for the way ethical prescriptions should be used in this debate. I believe it is, in principle, defensible to argue for the prohibition of a technological method on moral grounds, even when the argument is based on an extension of traditional moral limits (for example, an extension from a human-centered approach to one not centered on humans). I note in this context that religious freedom is a human right. I further believe it is defensible to scrutinize closely the control structure over vital resources, such as food and health care, or to scrutinize closely the conditions for release of persistent technologies that may be hard to trace or manage, and in cases in which it is difficult to assign liabilities. However, I do not believe it defensible to call for prohibition when tight regulation is in order, or to argue for tight regulation as a tactic toward achieving the real goal of prohibition.

Similarly, if the primary goal is profit, then one should avoid the argument that "we have to feed the world." The use of imprecise language or rhetoric entails a very real cost: when it becomes necessary to alter one's stance over time, then credibility and trust are at risk. A loss of credibility and trust hurts advocates in both camps—and most of all, the public.

In a nutshell, a second-order viewpoint of the ethical debate leads to a straightforward prescription. If we want fairness, respect, and progress in this debate, then we all have to say what we really mean.

Notes

1. See http://www.earthcharter.org/draft/charter/htm.
2. http://www.greenpeace.org/~geneng/.

10

Food for Thought

The Primacy of the Moral in the GMO Debate

David Magnus and Arthur Caplan

The Limits of Risk/Benefit Analysis

Debate about the acceptability of genetically modified organisms (GMOs) in agriculture and other settings has been framed almost entirely as a question of risk and benefit.[1] Proponents of GMOs point to the many benefits that their application will bring. Critics contend that the case for their safety has not been adequately addressed and thus the risks of GMOs outweigh any benefit that this new technology might bring forward. One consequence of this way of framing the debate is that it circumscribes the role of values and transforms the debate into what seems mainly to be a matter of getting the facts straight. At the extreme, some have claimed that the acceptability of GMOs is entirely a scientific matter and one in which nonexperts should play no role. Scientists and only scientists can tell us what the benefits and risks of GMOs are and thus whether GMOs are worthwhile.

Critics often try to respond to the narrowing of the debate into risk/benefit terms by noting that science is not the last word when it comes to food. This places the critics in the position of constantly appearing antiscientific or indifferent to the findings of science. The GMO debate takes on the appearance of being a pro- and antiscience battle.

But, values play at least some role even when the question is risk versus benefit. How much risk is too much, what sorts of benefits are worthwhile and to what extent are they valued, and above all, how much uncertainty is acceptable where risk is at issue are all issues that allow for values to be introduced. Even so, the role of values is fairly limited in this way of framing the debate.

In this paper, we will try to focus on this debate from the perspective of the moral and even religious values that are at stake in the debate, the sort of issues that scientific and regulatory bodies (for example, the National Research Council report and genetically modified plants) that have been addressing this issue often acknowledge and then choose to ignore. We start by summarizing the risks and benefits that are the locus of most of the current discussion.

Real Benefit and Real Risk

In one sense, a core problem in the GMO debate is that there are obviously benefits to be had from their use. Current GMOs include crops that are of benefit to farmers. Future GMOs will include foods that have far greater nutritional and even pharmaceutical benefit, crops that can grow in regions that currently cannot provide enough food for subsistence, cattle whose milk can be modified to offer medicinal benefit, and foods that are more desirable in terms of traits that the public wants or microorganisms that can battle against pollutants or dangerous microbes. It is worth noting that the market forces that largely determine which of these products are developed are complicated and that there are important trade-offs— the traits that may be needed to feed a starving world are different than the traits that farmers in the United States want, and both may differ from the characteristics that the paying public supports. But, the fact that benefit is likely to be obtained from GMOs seems incontrovertible. The same is true of risk.

Most of the criticisms of GMOs focus on two kinds of risks. First, there is concern about food safety. The second is what risk do GMOs pose to the environment?

A potential major risk of GMOs is an adverse impact on the health of those who eat them. Will some individuals develop allergic reactions? The new technology makes it possible to cross species barriers with impunity. Will a shellfish gene placed in a tomato lead to never-before-seen types of allergic reaction? Consider a detailed case. "Bt corn," a common GMO, includes a gene from *Bacillus thuingienis*, which produces a pesticide that kills the European corn borer. One of the problems with Bt corn is that while the pesticide given off by these plants is safe for human consumption, it is likely that insects will soon develop a resistance.

StarLink is a new variation of Bt that includes a protein (Cry9C) that does not break down as easily in the body. It may therefore postpone resistance. However, it also has some of the characteristics of food allergens. The fact that it will remain in the body longer increases the risk of allergic reactions in some people (though there are no verified cases of this at the present time). StarLink corn was approved for animal feed, but not for human consumption. Unfortunately, it is difficult if not impossible to keep the corn supply for animals and humans separate. The feed is often in the same silos and at least some of the corn from one field can send seed to another. The result has been the discovery of small amounts of StarLink corn throughout the food supply. Granted, a very small trace of Cry9C in a fast-food taco may not be the greatest health problem involved in such a meal. But still there is an additional risk of allergic reaction associated with this form of GM corn.

In addition to allergic reactions, some critics worry that each new generation of pesticide-producing GMOs could lead to a buildup of harmful poisons in the body of those who ingest them. After the corn borer develops resistance to the Bt crops, will there be a temptation to attempt to develop plants that produce more and more toxic substances?

Richard Lewontin has raised an additional concern. Because there is little control over where in a genome a gene is placed, it is possible for an inserted gene to disrupt a regulatory region. This might result in, for example, an increase in the amount of an unregulated protein in the organism, which could lead to unintended harmful consequences. A vegetable that produces a toxic protein that under normal circumstances is circumscribed either in the amount of the protein produced, or in the part of the plant where it is produced (for example, parts that are not ingested) could become toxic if a new gene is introduced into the genes that regulate the genes that produce that protein.

A second set of concerns arises over the environmental impact of GMOs. There are several different risks. First, there are worries about gene flow and the escape of genes. The same genes that may one day make it possible for plants to grow in poor, salty soil or in relatively arid regions could create an ecological nightmare if those same genes should be introduced to other plants. This can happen through outcrossing between the GMOs and closely related plants. For example, GM wheat could cross with native grasses in South America to alter the makeup of the ecosystem and potentially create "superweeds." Even in the absence of gene flow, the GMOs themselves could become "superweeds" as a result of the traits that make them better suited to new habitats. The environmental trade-off for technology that makes it possible to produce sustainable agriculture in regions where it cannot "naturally" flourish is the significant risk of loss of biodiversity and the unchecked spread of plants into unintended regions.

In addition to these concerns over the ecosystem and the creation of super-weeds, there is a worry over the potential impact of some GMOs on nontarget organisms. Cornell researchers found that pollen from Bt corn could kill the larvae of monarch butterflies who ingested it. This raised the fear that these engineered crops could kill butterflies and other nontarget organisms in addition to the corn borer. The results of subsequent field research to determine whether the Bt corn really represents a threat outside of the lab have been mixed.

Similarly, creating genetically modified animals and fish could lead to problems. Genetically engineered salmon could lead to the widespread introduction of new genes into wild fish runs. If the genes spread sufficiently, they could undermine the wild types. Current aquaculture methods that use nets to contain GM fish result in relatively large numbers of fish escaping into the wild. The problem of "killer bees" was a result of laboratory organisms that escaped into the wild. Genetically modified mice and other mammals could create pests that will be much more difficult to eliminate.

Genetically engineered microorganisms present even greater environmental and health concerns. It will soon be possible to engineer bacteria and viruses to produce deadly pathogens. This could well open a new era in biological weapons in addition to the environmental problems that could result from the release of organisms into the environment. It has been a number of years since the Supreme Court allowed the patenting of a genetically modified bacterium that could eat oil. The environmental assessment of the widespread introduction of engineered microorganisms has only barely begun to receive attention.[2]

Utilitarianism as a Framework

There is at least one major philosophical tradition that embodies the risk/benefit approach. Consequentialists would argue that weighing these risks and benefits could tell us which outcome will produce the best outcome, that is, maximize utility. On this view, we must do our best to assess and weigh the trade-offs between the benefits to be gained and the risks to human health and the environmental impact for each usage of the technology. This will require assessing the probabilities of each potential outcome. On this basis, some products may be found to be too dangerous to be developed, while others will be worthwhile.

The utilitarian perspective would also yield a way of assessing any regulatory framework. From this perspective, no current system is adequate to ensure that only those products that ought to be developed are developed. The present system in the United States, which is widely acknowledged as having one of the strongest regulatory systems, is quite complex and disjointed. Organisms that have been modified to give off a pesticide fall under the jurisdiction of the Environmental Protection Agency (EPA) while a GMO produced in the same way to express a gene to provide a nutritional benefit fall under the Food and Drug Administration (FDA). Some of these agencies have no experience with environmental-impact assessment and this may be left entirely out of the review process. There is a growing sense that the FDA, the U.S. Department of Agriculture, and the EPA are not sufficiently rigorous or consistent in how they regulate GMOs and that there should be a single set of standards, which more closely resembles the way the FDA handles drug development.

It is worth noting that the single most important factor for the differences between European and American attitudes toward GMOs (there is much more widespread opposition in Europe) is the amount of confidence in the regulatory institutions that protect the food supply. After "mad cow disease," foot-and-mouth disease outbreaks, poisoned cola, and dioxin exposure in poultry, Europeans do not trust their governments to provide safe food. A similar loss in confidence among U.S. consumers could have a similar effect.

Although a utilitarian framework captures much of the debate (especially with respect to regulatory systems), we feel that this way of framing the issues is far too narrow. There is no agreement regarding what consequences are deemed desirable, how to weigh them against various risks, how the burdens and benefits of taking risk to achieve benefit should be borne and by whom, who should be involved in exposure to risk to get benefit (consenting adults, children, the public as a whole, and so on), or how to weigh short-term and long-term consequences. The move toward risk/benefit, whether by proponents or critics, represents more of a values consensus than is warranted. Risk/benefit analyses tend to obscure important value questions hiding them amid a thicket of complicated models, complex quantitative analysis, or rhetorical appeals to unknown risks. There are at least three major problems with the consequentialist approach.

First, this framework requires the weighing and calculating of very different kinds of outcomes in commensurable measures. But the difficulty of weighing monetary savings versus the value of health is difficult if not impossible. Trying to find a way to assess the relative value of an ecosystem is doomed.

Second, many of the things people care about when thinking about risks and benefits are not really about the relative weighting of the factors, but rather their distribution: who is exposed to risk, who benefits, and who gets to decide? People object to being exposed to risk if it is not of their own choosing and if it is done for the benefit of others (even if the amount of risk is smaller than risks they are willing to expose themselves to on a routine basis). These features point to the fact that considerations of justice and fairness may well matter more than utility in most people's assessment of the technology.

This points to a third inadequacy in the consequentialist framework. It fails to capture many of the values at stake in the debate.

Playing God and Other Values

When it was announced that a sheep named Dolly was a clone from an adult cell, the widespread hysteria over cloning had very little to do with any analysis of risks and benefits of the technology. Rather, there was a clear sense of "moral repugnance" that was expressed in a variety of ways—talk of the immorality of attempts to prolong life eternally or images of clones acting as automatons subject to the will and purpose of others and, above all, a sense of the "unnaturalness" of the act.

Similarly, there is evidence that the fundamental framing of the GMO issue is in similar terms (and in fact is a related response). The term "Frankenfoods" suggests the modern Promethean myth—human hubris in the pursuit of knowledge can lead to our own unwitting destruction. Moreover, it is simply arrogant for humans to usurp the role of nature and natural design. Talk of "playing God" permeates and underlies the GMO debate just as it does the cloning controversy.[3] Indeed, there is at least some empirical evidence to back up the claim that people are primarily motivated by moral concerns when they oppose GMOs (those who think in terms of risks and benefits tend to be at least cautious supporters of the technology).[4] Prince Charles's well-known criticisms are couched in religious and value terms, rather than in terms of risks and benefits. Many of the most prominent advertisements that activists have used to attack biotechnology focus on "playing God" or the "unnaturalness" of GMOs.

What are we to make of these concerns? Leon Kass, Charles Krauthammer, George Will, and others have argued (in the context of cloning) that we should give great weight to the signpost of moral repugnance.[5] Others simply say they find GMOs disgusting, creepy, or to use a term introduced by Arthur Caplan into bioethics debates, "yucky." On this view, the visceral response is either significant for its own sake or at a minimum a way of capturing a whole set of objections that are difficult to articulate but clearly perceived. On the other hand, there are those who object to this type of concern as sentimental or irrational or both. This type of visceral response tends to be ephemeral—as people become used to the technology, the concern dissipates. On this view, concern over "playing God" is nothing more than prejudice against the new and unfamiliar.[6]

It is a mistake to see the concern over "playing God" or statements about the "yuck" factor of particular foods or creatures as the end of conversation. Moral sensibilities are clearly relevant to a moral assessment. But contrary to Kass, they are merely a starting point not an end point.

To explore the concept of moral repugnance in detail, we gathered a group of representatives from various Western and non-Western religions who all have experience in thinking about biotechnology, the environment, or food issues to provide input into the meaning of "playing God." Based upon our discussions as well as analysis of media images, and other literature, we argue that the concept has multiple overlapping meanings and represents a number of concerns.

One concern that is expressed is the concept of "hubris." There is some concern that there are unintended consequences to biotechnology, and we lack humility as we transform nature. There are many stories in many different traditions that demonstrate the disaster that can ensue when our hubris leads us to ignore the limitations in our knowledge. This is made particularly acute as many traditions see nature infused with value. From this way of understanding the "playing God" objection, it follows that the "precautionary principle" is an important and legitimate concept—one that can be grounded not only in risk/benefit terms, but also in terms of moral meaning.

A second way of understanding playing God is through the concept of stewardship. In every tradition, there is a tension between the need to make use of what we are given, to "improve the garden," and the need to "preserve the garden," to maintain and protect the status quo. This tension reflects two different views of nature. On the one hand, it is something to be utilized for the improvement of life. On the other hand, there is a view that rejects this more instrumental understanding of nature for a view which prefers acceptance and reverence for the world as it is given to us. While some critics and some supporters tend toward one or the other of these worldviews, most of the religious traditions strive for a proper balance between the two. Playing God is a function of an inappropriate balance, which would make us poor stewards.

Another key set of concepts in understanding the meaning of playing God is the concepts of *power* and *accountability*. An important part of the unease that many people feel is the question of not just what the risks are but who controls them, who is exposed to them, and who are the intended beneficiaries of the technologies. The scientific community and the large biotech industries are not inherently democratic institutions. In this case, *power* is associated both with knowledge and economic resources. These are institutions that do not seem at all *accountable* to the general public in any way, and this is a source of great anxiety. The initial GM foods certainly play into these concerns when they are products such as Roundup Ready or even Bt crops. These input-trait modifications do not obviously benefit the general public—the same public that is exposed to any risk associated with these crops. It is also worth noting that there seems to be a stronger anti-science bent to the recent opposition to GMOs and other biotechnologies compared to earlier environmental opposition to new technologies. Whereas Rachel Carson saw science as the ultimate arbiter of the problems with pesticides like DDT, many recent critics of biotech question the authority of science to address issues of risk assessment. The scientific community is seen by some as creating a new secular priesthood—an authoritarian structure in which science has replaced God. This sense of playing God yields a mistrust of the scientific community and underscores the importance of seeking ways to both democratize and incorporate values into the process of decision making in technological matters.

Finally, there is a sense of playing God that refers to violating in some significant sense the "purity" of organisms as they exist. Many religious traditions have dietary restrictions that may or may not be jeopardized by the ability to introduce genes from one species into a quite unrelated species. There is a further spiritual concern. While many traditions see the power to create as one way in which we are created in God's image, there is the risk that the arrogant pursuit of recreating nature to suit our needs leads us to usurp God's power. It suggests that God did not properly create the world and requires human action. Again, there is a tension between the need to improve the world that we find and to preserve and protect it. These concerns suggest the need for caution and possibly a need to make sure that consumers are informed about changes that may be relevant (for example, many Muslims would find it objectionable to eat foods that had porcine genes).

Each of these senses of playing God represents a challenge. It is important to see that these represent a starting point in dialogue about the technology, not an end point. It is also worth noting that the explicitly religious concerns all mirror more general moral considerations about justice, benevolence, and the moral significance of nature. Any argument that takes as its starting point that nature should not be altered significantly has failed to recognize the tension and the need for balance between instrumental and preservationist aspects of our obligations. While it is important to proceed with humility, to be aware of the limitations of our knowledge, and to be cognizant of the importance of questions of justice and power, the hope remains that public discussion of values in these debates can allow for the appropriate progress of science and technology.

Accepting a consequentialist framework does nothing to settle the GMO debate and is not the only moral response to GMOs. There is ample evidence that framing the issues in this way does not accurately capture the nature of public concern. Nor does it capture the values that are in play when the question is GMOs. There are core values, even religious values, that are relevant but that are not amenable to analysis in a simple risk/benefit framework or consequentialist mindset. There are deontological and even spiritual concerns that many people articulate that cannot be blunted by pointing to any amount of relative benefit that would be created by GMOs. At its core, the debate about GMOs is a debate about what values count and what framework they should be counted in.

Notes

This research was made possible by a grant from the Rockefeller Foundation. Some of this material is printed in David Magnus, "Genetically Modified Organisms," *Lahey Clinic, Medical Ethics,* spring 2001.

1. Richard Lewontin, "Genes in the Food!" *New York Review of Books* 48, no. 10 (21 June 2001): 81–84; NRC report, *Genetically Modified Pest-Protected Plants: Science and Regulation* (Washington D.C.: National Academy Press, 2000).

2. Mildred Cho et al., "Ethical Considerations in Synthesizing a Minimal Genome," *Science* 286, no. 5447 (1999): 2087–90.

3. Michael Pollan, "Playing God," *New York Times Magazine,* 25 October 1998.

4. George Gaskell et al., "Worlds Apart? The Reception of Genetically Modified Foods in Europe and the United States," *Science* 285, no. 5426 (1999): 384–88.

5. Leon Kass and James Q. Wilson, *The Ethics of Human Cloning* (Washington D.C.: AEI Press, 1998).

6. Greg Pence, *Whose Afraid of Human Cloning?* (Lanham, Md.: Rowman and Littlefield, 1998).

11

Ethics and Genetically Modified Foods

Gary Comstock

Much of the food consumed in the United States is genetically modified. Genetically modified (GM) food derives from microorganisms, plants, or animals manipulated at the molecular level to have traits that farmers or consumers desire. These foods often have been produced using techniques in which "foreign" genes are inserted into the microorganisms, plants, or animals. Foreign genes are genes taken from sources other than the organism's natural parents. In other words, genetically modified plants contain genes they would not have contained if researchers had only used traditional plant-breeding methods.

Some consumer advocates object to GM foods, and sometimes they object on ethical grounds. When someone opposes GM foods on ethical grounds, he typically has some reason or other for his opposition. We can scrutinize his reasons and, when we do so, we are doing applied ethics. Applied ethics involves identifying people's arguments for various conclusions and then analyzing those arguments to determine whether the arguments support the conclusions. A critical goal here is to decide whether an argument is sound. A sound argument is one in which all of the premises are true and no mistakes have been made in reasoning.

Ethically justifiable conclusions inevitably rest on two kinds of claims: (a) empirical claims, or factual assertions about how the world *is,* claims ideally based on the best available scientific observations, principles, and theories, and (b) normative claims, or value-laden assertions about how the world *ought to be,* claims ideally based on the best available moral judgments, principles, and theories.

Is it ethically justifiable to pursue genetically modified crops and foods? There is an objective answer to this question, and we will try here to figure out what it is. But we must begin with a proper, heavy, dose of epistemic humility, acknowledging that few ethicists at the moment seem to think that they know the final answer.

Should the law allow GM foods to be grown and marketed? The answer to this, and every, public policy question rests ultimately with us, citizens who will in the voting booth and shopping market decide the answer. To make up our minds, we will use feelings, intuitions, conscience, and reason. However, as we citizens are, by and large, not scientists, we must, to one degree or other, rest our factual understanding of the matter on the opinions of scientific experts. Therefore, ethical responsibility in the decision devolves heavily upon scientists engaged in the new GM technology.

Ethical Responsibilities of Scientists

Science is a communal process devoted to the discovery of knowledge, and to open and honest communication of knowledge. Its success, therefore, rests on two different kinds of values.

Epistemological values are values by which scientists determine which knowledge claims are better than others. The values include clarity, objectivity, capacity to explain a range of observations, and ability to generate accurate predictions. Claims that are internally inconsistent are jettisoned in favor of claims that are consistent, and fit with established theories. (At times, anomalous claims turn out to be justifiable, and an established theory is overthrown, but these occasions are rare in the history of science.) Epistemological values in science also include fecundity, the ability to generate useful new hypotheses; simplicity, the ability to explain observations with the fewest number of additional assumptions or qualifications; and elegance.

Personal values, including honesty and responsibility, are a second class of values, values that allow scientists to trust their peers' knowledge claims. If scientists are dishonest, untruthful, fraudulent, or excessively self-interested, the free flow of accurate information so essential to science will be thwarted. If a scientist plagiarizes the work of others or uses fabricated data, that scientist's work will become shrouded in suspicion and otherwise reliable data will not be trusted. If scientists exploit those who work under them, or discriminate on the basis of gender, race, class, or age, then the mechanisms of trust and collegiality undergirding science will be eroded.

The very institution of scientific discovery is supported, indeed, permeated with values. Scientists have a variety of goals and functions in society, so it should be no surprise that they face different challenges.

University scientists must be scrupulous in giving credit for their research to all who deserve credit; careful not to divulge proprietary information; and painstaking in maintaining objectivity, especially when funded by industry. Industry scientists must also maintain the highest standards of scientific objectivity, a particular challenge since their work may not be subject to peer-review procedures as strict as those faced by university scientists. Industry scientists must also be willing to defend results of their research that are not favorable to their employers' interests. Scientists employed by nongovernmental organizations face challenges, as well. Their objectivity must be maintained in the face of an organization's explicit-advocacy agenda, and in spite of the fact that their research might provide results that might seriously undermine the organization's fund-raising attempts. All scientists face the challenges of communicating complex issues to a public that receives them through media channels that often are not equipped to communicate the qualifications and uncertainties attaching to much scientific information.

At its core, science is an expression of some of our most cherished values. The public largely trusts scientists, and scientists must in turn act as good stewards of this trust.

A Method for Addressing Ethical Issues

Ethical objections to GM foods typically center on the possibility of harm to persons or other living things. Harm may or may not be justified by outweighing benefits. Whether harms are justified is a question that ethicists try to answer by working methodically through a series of questions:[1]

1. What is the harm envisaged? To provide an adequate answer to this question, we must pay attention to how significant the harm or potential harm may be (will be it severe or trivial?); who the "stakeholders" are (that is, who are the persons, animals, even ecosystems, who may be harmed?); the extent to which various stakeholders might be harmed; and the distribution of harms. The last question directs attention to a critical issue, the issue of justice and fairness: Are those who are at risk of being harmed by the action in question different from those who may benefit from the action in question?

2. What information do we have? Sound ethical judgments go hand in

hand with thorough understanding of the scientific facts. In a given case, we may need to ask two questions: Is the scientific information about harm being presented reliable, or is it fact, hearsay, or opinion? and, What information do we not know that we should know before making the decision?

3.   What are the options? In assessing the various courses of action, emphasize creative problem solving, seeking to find "win-win" alternatives in which everyone's interests are protected. Here we must identify what objectives each stakeholder wants to obtain; how many methods are available by which to achieve those objectives; and what advantages and disadvantages attach to each alternative.

4.   What ethical principles should guide us? There are at least three secular ethical traditions:

   • Rights theory holds that we ought always to act so that we treat human beings as autonomous individuals, and not as mere means to an end.
   • Utilitarian theory holds that we ought always to act so that we maximize good consequences and minimize harmful consequences.
   • Virtue theory holds that we ought always to act so that we act the way a just, fair, good person would act.

Ethical theorists are divided about which of these three theories is best. We manage this uncertainty through the following procedure. Pick one of the three principles. Using it as a basis, determine its implications for the decision at hand. Then, adopt a second principle. Determine what it implies for the decision at hand. Repeat the procedure with the third principle. Should all three principles converge on the same conclusion, then we have good reasons for thinking our conclusion morally justifiable.

5.   How do we reach moral closure? Does the decision we have reached allow all stakeholders either to participate in the decision or to have their views represented? If a compromise solution is deemed necessary in order to manage otherwise intractable differences, has the compromise been reached in way that has allowed all interested parties to have their interests articulated, understood, and considered? If so, then the decision may be justifiable on ethical grounds.

   There is a difference between *consensus* and *compromise. Consensus* means that the vast majority of people agree about the right answer to a question. If the group cannot reach a consensus but must, nevertheless, take some decision or other, then a *compromise* position may be necessary. But neither consensus nor compromise should be confused with the right answer to an ethical question. It is possible that a society might reach a consensus position that is unjust. For example, some societies have held that women should not be allowed to own property. That may be a consensus position, or even a compromise position, but it should not be confused with the truth of the matter. Moral closure is a sad fact of life;

we sometimes must decide to undertake some course of action even though we know that it may not be, ethically, the right decision, all things considered.

Ethical Issues Involved in the Use of Genetic Technology in Agriculture

Discussions of the ethical dimensions of agricultural biotechnology are sometimes confused by a conflation of two quite different sorts of objections to GM technology: intrinsic and extrinsic. It is critical not only that we distinguish these two classes, but keep them distinct throughout the ensuing discussion of ethics.

Extrinsic objections focus on the potential harms consequent upon the adoption of GMOs. Extrinsic objections hold that GM technology should not be pursued because of its anticipated results. Briefly stated, the extrinsic objections go as follows. GMOs may have disastrous effects on animals, ecosystems, and humans. Possible harms to humans include perpetuation of social inequities in modern agriculture, decreased food security for women and children on subsistence farms in developing countries, a growing gap between well-capitalized economies in the Northern hemisphere and less capitalized peasant economies in the South, risks to the food security of future generations, and the promotion of reductionistic and exploitative science. Potential harms to ecosystems include possible environmental catastrophe, inevitable narrowing of germplasm diversity, and irreversible loss or degradation of air, soils, and waters. Potential harms to animals include unjustified pain to individuals used in research and production.

These are valid concerns, and nation-states must have in place testing mechanisms and regulatory agencies to assess the likelihood, scope, and distribution of potential harms through a rigorous and well-funded risk-assessment procedure. It is for this reason that I have said, above, that GM technology must be developed responsibly and with appropriate caution. However, these extrinsic objections cannot by themselves justify a moratorium, much less a permanent ban, on GM technology, because they admit the possibility that the harms may be minimal and outweighed by the benefits. How can one decide whether the potential harms outweigh potential benefits unless one conducts the research, field tests, and data analysis necessary to make a scientifically informed assessment?

In sum, extrinsic objections to GMOs raise important questions about GMOs, and each country using GMOs ought to have in place the organizations and research structures necessary to insure their safe use.

There is, however, an entirely different sort of objection to GM technology, a sort of objection that, if it is sound, would indeed justify a permanent ban.

Intrinsic objections allege that the process of making GMOs is objectionable *in itself.* This belief is defended in several ways, but almost all of the formulations are related to one central claim, the unnaturalness objection:

It is unnatural to genetically engineer plants, animals, and foods (**UE**).

If **UE** is true, then we ought not to engage in bioengineering, however unfortunate may be the consequences of halting the technology. Were a nation to accept **UE** as the conclusion of a sound argument, then much agricultural research would have to be terminated and potentially significant benefits from the technology sacrificed. A great deal is at stake.

In *Vexing Nature? On the Ethical Case Against Agricultural Biotechnology,* I discuss fourteen ways in which **UE** has been defended.[2] For present purposes, those fourteen objections can be summarized as follows:

(1) **To engage in ag biotech is to** *play God.*
(2) **To engage in ag biotech is to** *invent world-changing technology.*
(3) **To engage in ag biotech is** *illegitimately to cross species boundaries.*
(4) **To engage in ag biotech is to** *commodify life.*

Let us consider each claim in turn.

(1) **To engage in ag biotech is to** *play God.*

In a Western theological framework, humans are creatures, subjects of the Lord of the Universe, and it would be impious for them to arrogate to themselves roles and powers appropriate only for the Creator. Shifting genes around between individuals and species is taking on a task not appropriate for us, subordinate beings. Therefore, to engage in bioengineering is to play God.

There are several problems with this argument. First, there are different interpretations of God. Absent the guidance of any specific religious tradition, it is logically possible that God could be a being who wants to turn over to us all divine prerogatives; or explicitly wants to turn over to us at least the prerogative of engineering plants; or who doesn't care what we do. If God is any of these beings, then the argument fails because playing God in this instance is not a bad thing.

The argument seems to assume, however, that God is not like any of the gods just described. Assume that the orthodox Jewish and Christian view of God is correct, that God is the only personal, perfect, necessarily existing, all-loving, all-knowing, and all-powerful being. On this traditional Western theistic view, finite humans should not aspire to infinite knowledge and power. To the extent that bioengineering is an attempt to control nature itself, the argument would go, bioengineering would be an unacceptable attempt to usurp God's dominion.

The problem with this argument is that not all traditional Jews and Christians think that this God would rule out genetic engineering. I am a practicing evangelical Christian and the chair of my local church's council. In my tradition, God is thought to endorse creativity, and scientific and technological development, including genetic improvement. Other traditions have similar views. In the mystical writings of the Jewish Kabbalah, God is understood as one who expects humans to be cocreators, technicians working with God to improve the world. At least one Jewish philosopher, Baruch Brody, has suggested that biotechnology may be a vehicle ordained by God for the perfection of nature.[3]

I personally hesitate to think that humans can "perfect" nature. However, I have become convinced that genetic modification might help humans to rectify some of the damage we have already done to nature. And I believe God may endorse such an aim. For humans are made in the divine image. God desires that we exercise the spark of divinity within us. Inquisitiveness in science is part of our nature. Creative impulses are not found only in the literary, musical, and plastic arts. They are part of molecular biology, cellular theory, ecology, and evolutionary genetics, too. It is unclear why the desire to investigate and manipulate the chemical bases of life should not be considered as much a manifestation of our godlike nature as the writing of poetry and the composition of sonatas. As a way of providing theological content for **UE**, then, argument (1) is unsatisfactory because it is ambiguous and contentious.

(2) **To engage in ag biotech is to** *invent world-changing technology*.

Let us consider (2) in conjunction with a similar objection (2a).

(2a) **To engage in ag biotech is to** *arrogate historically unprecedented power* **to ourselves.**

The argument here is not the strong one, that biotech gives us divine power, but the more modest one, that it gives us a power we have not had previously. But it would be counterintuitive to judge an action wrong simply because it has never been performed. On this view, it would have been wrong to prescribe a new herbal remedy for menstrual cramp, or to administer a new anaesthetic. But that seems absurd. More argumentation is needed to call historically unprecedented actions morally wrong. What is needed is to know *to what extent* our new powers will transform society, whether we have witnessed prior transformations of this sort, and whether those transitions are morally acceptable.

We do not know how extensive the ag biotech revolution will be, but let us assume that it will be as dramatic as its greatest proponents assert. Have we ever witnessed comparable transitions? The change from hunting and gathering to agriculture was an astonishing transformation. With agriculture came not only an increase in the number of humans on the globe, but the first appearance of complex cultural activities: writing, philosophy, government, music, the arts, and architecture. What sort of power did people arrogate to themselves when they moved from hunting and gathering to agriculture? The power of civilization itself.[4]

Ag biotech is often oversold by its proponents. But suppose that they are right, that ag biotech brings us historically unprecedented powers. Is this a reason to oppose it? Not if we accept agriculture and its accompanying advances, for when we accepted agriculture, we arrogated to ourselves historically unprecedented powers.

In sum, the objections stated in (2) and (2a) are not convincing.

(3) **To engage in ag biotech is** *illegitimately to cross species boundaries.*

The problems with this argument are both theological and scientific. I will leave it to others to argue the scientific case that nature gives ample evidence of generally fluid boundaries between species. The argument assumes that species boundaries are distinct, rigid, and unchanging while, in fact, species now appear to be messy, plastic, and mutable. To proscribe the crossing of species borders on the grounds that it is unnatural seems scientifically indefensible. It is also difficult to see how (3) could be defended on theological grounds. None of the scriptural writings of the Western religions proscribe genetic engineering, of course, because genetic engineering was undreamed of at the time the holy books were written. Now, one might argue that such a proscription may be derived from Jewish or Christian traditions of scriptural interpretation. Talmudic laws against mixing "kinds," for example, might be taken to ground a general prohibition against inserting genes from "unclean" species into clean species. Here's one way the argument might go: For an observant Jew to do what scripture proscribes is morally wrong; Jewish oral and written law proscribe the mixing of kinds (for example, eating milk and meat from the same plate; yoking donkeys and oxen together); bioengineering is the mixing of kinds; therefore, for a Jew to engage in bioengineering is morally wrong.

But this argument fails to show that bioengineering is intrinsically objectionable in all of its forms for everyone. The argument might prohibit *Jews* from engaging in certain *kinds* of biotechnical activity but not all; it would not prohibit, for example, the transferring of genes *within* a species, nor, apparently, the transfer of genes from one clean species to another clean species. Incidentally, it is worth noting that the Orthodox community has accepted transgenesis in its food supply. Seventy percent of cheese produced in the United States is made using a GM product, chymosin. This cheese has been accepted as kosher by Orthodox rabbis.[5]

In conclusion, it is difficult to find a persuasive defense of (3) either on scientific or religious grounds.

(4) **To engage in ag biotech is to** *commodify life.*

The argument here is that genetic engineering treats life in a reductionistic manner, reducing living organisms to little more than machines. Life is sacred and not to be treated as a good of commercial value only, to be bought and sold to the highest bidder.

Could we apply this principle uniformly? Would not objecting to the products of GM technology on these grounds also require that we object to the products of ordinary agriculture on the same grounds? Is not the very act of bartering or exchanging crops and animals for cash vivid testimony to the fact that every culture on earth has engaged in the commodification of life for centuries? If one accepts commercial trafficking in non-GM wheat and pigs, then why should we object to commercial trafficking in GM wheat and GM pigs? Why should it be wrong for us to treat DNA the way we have previously treated animals, plants, and viruses?[6]

While (4) may be true, it is not a sufficient reason to object to GM technology because our values and economic institutions have long accepted the commodification of life. Now, one might object that various religious traditions have never accepted commodification, and that genetic engineering presents us with an opportunity to resist, to reverse course. Leon Kass,[7] for example, has argued that we have gone too far down the road of dehumanizing ourselves and treating nature as a machine, and that we should pay attention to our emotional reactions against practices such as human cloning. Even if we cannot defend these feelings in rational terms, our revulsion at the very idea of cloning humans should carry great weight. Mary Midgley[8] has argued that moving genes across species boundaries is not only "yucky" but, perhaps, a monstrous idea, a form of playing God.

Kass and Midgley have eloquently defended the relevance of our emotional reactions to genetic engineering but, as both admit, we cannot simply allow our emotions to carry the day. As Midgley writes, "Attention to . . . sympathetic feelings [can stir] up reasoning that [alters] people's whole world view."[9] But as much hinges on the reasoning as on the emotions.

Are the intrinsic objections sound? Are they clear, consistent, and logical? Do they rely on principles we are willing to apply uniformly to other parts of our lives? Might they lead to counterintuitive results?

Counterintuitive results are results we strongly hesitate to accept because they run counter to widely shared, considered moral intuitions. If a moral rule or principle leads to counterintuitive results, then we have a strong reason to reject it. For example, consider the following moral principle, which we might call the doctrine of naïve consequentialism (NC):

Always improve the welfare of the most people (NC).

Were we to adopt NC, then we would be not only permitted but required to sacrifice one healthy person if by doing so we could save many others. If six people need organ transplants (two need kidneys, one needs a liver, one needs a heart, and two need lungs), then NC instructs us to sacrifice the life of the healthy person so as to transplant that person's six organs to the other six. But this result, that we are *obliged* to sacrifice innocent people to save strangers, is wildly counterintuitive. This result gives us a strong reason to reject NC.

I have argued that the four formulations of the unnaturalness objection considered above are unsound insofar as they lead to counterintuitive results. I do not take this position lightly. Twelve years ago, I wrote "The Case against bGH," an article, I have been told, that "was one of the first papers by a philosopher to object to ag biotech on explicitly ethical grounds." I then wrote a series of other articles objecting to GM herbicide-resistant crops, transgenic animals, and, indeed, all of agricultural biotechnology.[10] I am acquainted with worries about GM foods. But, for reasons that include the weakness of the intrinsic objections, I have come to change my mind. The sympathetic feelings on which my anti-GMO worldview was based did not survive the stirring up of reasoning.

Why Are We Careful with GM Foods?

I do not pretend to know anything like the full answer to this question, but I would like to be permitted the luxury of a brief speculation about it. The reason may have to do with a natural, completely understandable, and wholly rational tendency to take precautions with what goes into our mouths. When we are in good health and happy with the foods available to us, we have little to gain from experimenting with a new food, and no reason to take a chance on a potentially unsafe food. We may think of this disposition as the precautionary response.

When faced with two contrasting opinions about issues related to food safety, consumers place great emphasis on negative information. The precautionary response is particularly strong when a consumer sees little to gain from a new food technology. When a given food is plentiful, it is rational to place extra weight on negative information about any particular piece of that food. It is rational to do so, as Dermot Hayes points out, even when the source of the negative information is known to be biased.

There are several reasons for us to take a precautionary approach to new foods. First, under conditions in which nutritious, tasty food is plentiful, we have nothing to gain from trying a new food if, from our perspective, it is in other respects identical to our current foods. Suppose on a rack in front of me there are eighteen dozen maple-frosted Krispy Kreme doughnuts, all baked to a golden brown, all weighing three ounces. If I am invited to take one of them, I have no reason to favor one over the other.

Suppose, however, that a naked man runs into the room with wild hair flying behind him yelling that the sky is falling. He approaches the rack and points at the third doughnut from the left on the fourth shelf from the bottom. He exclaims, "This doughnut will cause cancer! Avoid it at all costs, or die!" There is no reason to believe this man's claim and yet, since there are so many doughnuts freely available, why should we take any chances? It is rational to select other doughnuts, since all are alike. Now, perhaps one of us is a mountain climber who loves taking risks. He might be tempted to say, "Heck, I'll try that doughnut." In order to focus on the right question here, the risk takers should ask themselves whether they would select the tainted doughtnut to take home to feed to their two-year-old daughter. Why impose any risk on your loved ones when there is no reason to do so?

The Krispy Kreme example is meant to suggest that food tainting is both a powerful and an extraordinarily easy social act. It is powerful because it virtually determines consumer behavior. It is easy because the tainter does not have to offer any evidence of the food's danger at all. Under conditions of food plenty, rational consumers do and should take precautions, avoiding tainted food no matter how untrustworthy the tainter.

Our tendency to take precautions with our food suggests that a single person with a negative view about GM foods will be much more influential than many people with a positive view. The following experiment lends credibility to this hypothesis. In a willingness-to-pay experiment, Hayes and colleagues paid eighty-seven primary food shoppers $40 each.[11] Each participant was assigned to a group ranging in size from a half-dozen to a dozen members. Each group was then seated at a table at lunchtime and given one pork sandwich. In the middle of each table was one additional food item, an irradiated pork sandwich. Each group of participants was given one of three different treatments: (a) the *pro-irradation* treatment, (b) the *anti-irradition* treatment, or (c) the *balanced* treatment.

Each treatment began with all of the participants at a table receiving the same, so-called neutral description of an irradiated pork sandwich. The description read, in part, like this:

> The U.S. FDA has recently approved the use of ionizing radiation to control Trichinella in pork products. This process results in a ten-thousand-fold reduction in Trichinella organisms in meat. The process does not induce measurable radioactivity in food.

After the participants read this description, they would proceed to conduct a silent bid in order to purchase the right to exchange their nonirradiated sandwich for the irradiated sandwich. Whoever bid the highest price would be able to buy the sandwich for the price bid by the second-highest bidder. In order to provide participants with information about the opinions of the others at their table so that they could factor this information into their future bids, the lowest and highest bids of each round were announced before the next round of bidding began. At the end of the experiment, one of the ten bidding rounds would be selected at random, and the person bidding the highest amount in that round would have to pay the second-highest price bid during that round for the sandwich.

After five rounds of bidding, the second-highest bids in all three groups settled rather quickly at an equilibrium point, roughly, twenty cents. That is, someone at every table was willing to pay twenty cents for the irradiated pork sandwich, but no one in any group would pay more than twenty cents. The bidding was repeated five times in order to give participants the opportunity to respond to information they were getting from others at the table, and to insure the robustness of the price.

After five rounds of bidding, each group was given additional information.

Group (a), the so-called Pro group, was provided with a description of the sandwich that read, in part:

> Each year, 9,000 people die in the United States from food-borne illness. Some die from Trichinella in pork. Millions of others suffer short-term illness. Irradiated pork is a safe and reliable way to eliminate this pathogen. The process has been used successfully in twenty countries since 1950."

The pro-group participants were informed that the author of this positive description was a pro-irradiation food-industry group. After the description was read, five more rounds of bidding began. The price of the irradiated sandwich quickly shot upward, reaching eighty cents by the end of round ten. A ceiling price was not reached, however, as the bids in every round, including the last, were significantly higher than the preceding round. The price, that is, was still going up when the experiment was stopped (see table 1).

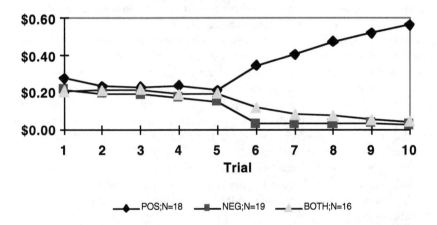

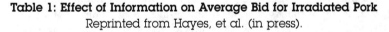

Table 1: Effect of Information on Average Bid for Irradiated Pork
Reprinted from Hayes, et al. (in press).

After its first five rounds of bidding, group (b) was provided with a different description. It read, in part:

> In food irradiation, pork is exposed to radioactive materials. It receives 300,000 rads of radiation—the equivalent of 30 million chest x rays. This process results in radiolytic products in food. Some radiolytic products are carcinogens, and linked to birth defects. The process was developed in the 1950s by the Atomic Energy Commission.

This source of this description was identified to the bidders as "Food and Water," an anti-irradiation activist group in England. After group (b) read this description, it began five more rounds of bidding. The bid went down, quickly reaching zero. After the first five rounds produced a value of twenty cents in group (b) for the pork sandwich described in a "neutral" way, *no one* in this group would pay a penny for the irradiated sandwich described in a "negative" way. This result was obtained even though the description was clearly identified as coming from an activist, nonscientific group.

After five rounds of bidding on the neutral description, the third group, group (c), received *both* the positive and the negative descriptions. One might expect that this group's response would be highly variable, with some participants scared off by the negative description and others discounting it for its unscientific source. Some participants might be expected to bid nothing while others would continue to bid highly.

However, the price of the sandwich in the third, so-called balanced group, also fell quickly. Indeed, the price reached zero almost as quickly as it did in group (b), the negative group. That is, even though the third group had both the neutral and the positive description in front of them, no one exposed to the negative description would pay two cents for the irradiated sandwich.

Hayes's study illuminates the precautionary response, and carries implications for the GM debate. These implications are that, given neutral or positive descriptions of GM foods, consumers initially will *pay more* for them. Given negative descriptions of GM foods, consumers initially will *not* pay more for them. Finally, and this is the surprising result, given *both* positive and negative descriptions of GM foods, consumers initially will *not* pay more for them. Both sides in the GM food debate should be scrupulous in providing reasons for all of their claims. But especially for their negative claims.

In a worldwide context, the precautionary response of those facing food abundance in developed countries may lead us to be insensitive to the conditions of those in less fortunate situations. Indeed, we may find ourselves in the following ethical dilemma.

For purposes of argument, make the following three assumptions. (I do not believe that any of the assumptions is implausible.) First, assume that GM food is safe. Second, assume that some GM "orphan" foods, such as rice enhanced with iron or vitamin A, or virus-resistant cassavas, or aluminum-tolerant sweet potatoes, may be of great potential benefit to millions of poor children. Third, assume that widespread anti-GM information and sentiment, no matter how unreliable on scientific grounds, could shut down the GM infrastructure in the developed world.

Under these assumptions, consider the possibility that by tainting GM foods in the countries best suited to conduct GM research safely, anti-GM activists could bring to a halt the range of money-making GM foods marketed by multinational corporations. This result might be a good or a bad thing. However, an unintended side effect of this consequence would be that the new GM orphan crops mentioned above might not be forthcoming, assuming that the development and commercialization of these orphan crops is dependent upon the answering of fundamental questions in plant science and molecular biology that will only be answered if the research agendas of private industry are allowed to go forward along with the research agendas of public research institutions.

Our precautionary response to new food may put us in an uncomfortable position. On the one hand, we want to tell "both sides" of the GM story, letting people know both about the benefits and the risks of the technology. On the other hand, some of the people touting the benefits of the technology make outlandish claims that it will feed the world while some of the people decrying the technology make unsupported claims that it will ruin the world. In that situation, however, those with unsupported negative stories to tell carry greater weight than those with unsupported positive stories. Our precautionary response, then, may well lead in the short term, at least, to the rejection of GM technology. Yet, the rejection of GM technology could indirectly harm those children most in need, those who need what I have called the orphan crops.

Are we being forced to choose between two fundamental values, the value of free speech versus the value of children's lives?

On the one hand, open conversation and transparent decision-making processes are critical to the foundations of a liberal democratic society. We must reach out to include everyone in the debate, and allow people to state their opinions about GM foods, whatever their opinion happens to be, whatever their level of acquaintance with the science and technology happens to be. Free speech is a value not to be compromised lightly.

On the other hand, stating some opinions about GM food can clearly have a tainting effect, a powerful and extraordinarily easy consequence of free speech. Tainting the technology might result in the loss of this potentially useful tool. Should we, then, draw some boundaries around the conversation, insisting that each contributor bring some measure of scientific data to the table, especially when negative claims are being made? Or are we collectively prepared to leave the conversation wide open? That is, in the name of protecting free speech, are we prepared to risk losing an opportunity to help some of the world's most vulnerable?

The Precautionary Principle

As a thirteen-year-old, I won my dream job, wrangling horses at Honey Rock Camp in northern Wisconsin. The image I cultivated for myself was the weathered cowboy astride Chief or Big Red, dispensing nuggets to awestruck young rider wanna-bes. But I was, as they say in Texas, all hat.

"Be careful!" was the best advice I could muster.

Only after years of experience in a western saddle would I have the skills to size up various riders and advise them properly on a case-by-case basis. You should slouch more against the cantle and get the balls of your feet onto the stirrups. You need to thrust your heels in front of your knees and down toward the animal's front hooves. You! Roll your hips in rhythm with the animal, and stay away from the horn. You, stay alert for sudden changes of direction.

Only after years of experience with hundreds of different riders would I realize that my earlier generic advice, well-intentioned though it was, had been of absolutely no use to anyone. As an older cowboy once remarked, I might as well have been saying, "Go crazy!" Both pieces of advice were equally useless in making good decisions about how to behave on a horse.

Now, as mad cow disease grips the European imagination, concerned observers transfer fears to genetically modified foods, advising: "Take precaution!" Is this a valuable observation that can guide specific public-policy decisions, or well-intentioned but ultimately unhelpful advice?

As formulated in the 1992 Rio Declaration on Environment and Development, the precautionary principle states that "lack of full scientific certainty shall not be used as a reason for postponing cost-effective measures to prevent environmental degradation." The precautionary approach has led many countries to declare a moratorium on GM crops on the supposition that developing GM crops might lead to environmental degradation. The countries are correct that this is an implication of the principle. But is it the only implication?

Suppose global warming intensifies and comes, as some now darkly predict, to interfere dramatically with food production and distribution. Massive dislocations in international trade and corresponding political power follow global food shortages, affecting all regions and nations. In desperate attempts to feed themselves, billions begin to pillage game animals, clear-cut forests to plant crops, cultivate previously nonproductive lands, apply fertilizers and pesticides at higher-than-recommended rates, and kill and eat endangered and previously non-endangered species.

Perhaps not a likely scenario, but not entirely implausible, either. GM crops could help to prevent it, by providing hardier versions of traditional lines capable of growing in drought conditions, or in saline soils, or under unusual climactic stresses in previously temperate zones, or in zones in which we have no prior agronomic experience. On the supposition that we might need the tools of genetic engineering to avert future episodes of crushing human attacks on what Aldo Leopold called "the land," the precautionary principle requires that we develop GM crops. Yes, we lack full scientific certainty that developing GM crops will prevent environmental degradation. True, we do not know what the final financial price of GM research and development will be. But if GM technology were to help save the land, few would not deem that price cost-effective. So, according to the precautionary principle, lack of full scientific certainty that GM crops will prevent environmental degradation shall not be used as a reason for postponing this potentially cost-effective measure.

The precautionary principle commits us to each of the following propositions:

(1) We must not develop GM crops.
(2) We must develop GM crops.

As (1) and (2) are plainly contradictory, however, defenders of the principle should explain why its implications are not incoherent.

Much more helpful than the precautionary principle would be detailed case-by-case recommendations crafted upon the basis of a wide review of nonindustry-sponsored field tests conducted by objective scientists expert in the construction and interpretation of ecological and medical data. Without such a basis for judging this use acceptable and that use unacceptable, we may as well advise people in the GM area to go crazy. It would be just as helpful as "Take precaution!"

Religion and Ethics

Religious traditions provide an answer to the question "How, overall, should I live my life?" Secular ethical traditions provide an answer to the question "What is the right thing to do?" When in a pluralistic society a particular religion's answers come into genuine conflict with the answers arrived at through secular ethical deliberation, we must ask how deep is the conflict. If the conflict is so deep that honoring the religion's views would entail dishonoring another religion's views, then we have a difficult decision to make. In such cases, the conclusions of secular ethical deliberation must override the answers of the religion in question.

The reason is that granting privileged status to one religion will inevitably discriminate against another religion. Individuals must be allowed to follow their consciences in matters theological. But if one religion is allowed to enforce its values on others in a way that restricts the others' ability to pursue their values, then individual religious freedom has not been protected.

Moral theorists refer to this feature of nonreligious ethical deliberation as the *overridingness* of ethics. If a parent refuses a lifesaving medical procedure for a minor child on religious grounds, the state is justified in overriding the parent's religious beliefs in order to protect what secular ethics regards as a value higher than religious freedom: the life of a child.

The overridingness of ethics applies to our discussion only if a religious group claims the right to halt GM technology on purely religious grounds. The problem here is the confessional problem, of one group attempting to enforce its beliefs on others. I mean no disrespect to religion; as I have noted, I am a religious person, and I value religious traditions other than my own. Religious traditions have been the repositories and incubators of virtuous behavior. Yet each of our traditions must in a global society learn to coexist peacefully with competing religions, and with nonreligious traditions and institutions.

If someone objects to GM technology on purely religious grounds, we must ask on what authority she speaks for her tradition, whether there are other, conflicting, views within her tradition, and whether acting on her views will entail disrespecting the views of people from other religions. It is, of course, the right of each tradition to decide its attitude about genetic engineering. But in the absence of other good reasons, we must not allow someone to ban GM technology for narrowly sectarian reasons alone. To allow such an action would be to disrespect the views of people who believe, on equally sincere religious grounds, that GM technology is not necessarily inconsistent with God's desires for us.

Minority Views

When in a pluralistic society the views of a particular minority come into genuine conflict with the views of the majority, we must ask a number of questions: How deep is the conflict? How has the minority been treated in the past? If the minority has been exploited, have reparations been made? If the conflict is so deep that honoring the minority's views would entail overriding the majority's views, then we have a difficult decision to make. In such cases, the conclusions of the state must be just, taking into account the question of past exploitation and subsequent reparations, or lack thereof. This is a question of justice.

The question of justice would arise in the discussion of GM technology if the majority favored GM technology, while the minority claimed the right to halt GM technology. If the minority cited religious arguments to halt GMOs, yet the majority believed that halting GMOs would result in loss of human life, then the state faces a decision very similar to the one discussed in the prior section. In this case, secular policy decisions may be justified in overriding the minority's religious arguments insofar as society deems the value of human life higher than the value of religious freedom.

However, should the minority cite past oppression as the reason that their values ought to predominate over the majority's, then a different question must be addressed. Here, the relevant issues have to do with the nature of past exploitation; its scope and depth; and the sufficiency of efforts, have there been any, to rectify the injustice and compensate victims. If the problem is long-standing and has not been addressed, then imposing the will of the majority would seem a sign of an unjust society insensitive to its past misdeeds. If, on the other hand, the problem has been carefully addressed by both sides and, for example, just treaties arrived at through fair procedures have been put in place, are being enforced, are rectifying past wrongs, and are preventing new forms of exploitation, then the minority's arguments would seem to be far weaker. This conclusion would be especially compelling if it could be shown that the lives of *other* disadvantaged peoples might be put at risk by honoring a particular minority's wish to ban GMOs.

Conclusion

Earlier I described a method for reaching ethically sound judgments. It was on the basis of that method that I personally came to change my mind about the moral acceptability of GM crops. My opinion changed as I took full account of three considerations: (a) the rights of people in various countries to choose to adopt GM technology (a consideration falling under the human rights principle); (b) the balance of likely benefits over harms to consumers and the environment from GM technology (a utilitarian consideration); and (c) the wisdom of encouraging discovery, innovation, and careful regulation of GM technology (a consideration related to virtue theory).

Is it ethically justifiable to pursue genetically modified crops and foods? I have come to believe that three of our most influential ethical traditions converge on a common answer. Assuming we proceed responsibly and with appropriate caution, the answer is yes.

Notes

1. In describing this method, I have drawn on an ethics assessment tool devised by Dr. Courtney Campbell, Philosophy Department, Oregon State University, and presented at the Oregon State University Bioethics Institute in Corvallis, Oregon, summer 1998.

2. Gary Comstock, *Vexing Nature? On the Ethical Case against Agricultural Biotechnology* (Boston and Dordrecht: Kluwer Academic Publishers, 2000).

3. Brody Baruch, private communication.

4. William McNeill, "Gains and Losses: A Historical Perspective on Farming," 1989 Iowa Humanities Lecture, National Endowment for the Humanities and Iowa Humanities Board, Oakdale Campus, Iowa City, Iowa, 1989.

5. Jonathan Gressel, observation at the Annual Meeting of the Weed Science Society of America, Chicago, 10 February 1998; See also, Alan Ryan et al., *Genetically Modified Crops: The Ethical and Social Issues* (London: The Royal Society, 1999), sec. 1.38.

6. Dorothy Nelkin and M. Susan Lindee, *The DNA Mystique: The Gene as Cultural Icon* (New York: Freeman, 1995).

7. Leon R. Kass, *Toward a More Nature Science: Biology and Human Affairs* (New York: Free Press, 1988); Kass, "Beyond Biology: Will Advances in Genetic Technology Threaten to Dehumanize Us All?" *New York Times*, 23 August 1998 [online], http://www.nytimes.com/books/98/08/23/reviews/980823.23kassct.html.

8. Mary Midgley, "Biotechnology and Monstrosity: Why Should We Pay Attention to the 'Yuk Factor,'" *Hastings Center Report* 30, no. 5 (2000): 7–15.

9. Ibid.

10. See also, Gary Comstock, "The Case against bGH," *Agriculture Human Values* 5 (1988): 26–52. The other essays are reprinted in Comstock, *Vexing Nature?* chaps. 1–4.

11. Dermot Hayes, John A. Fox, and Jason F. Shogren, "Consumer Preferences for Food Irradiation: How Favorable and Unfavorable descriptions Affect Preferences for Irradiated Pork in Experimental Auctions," *Journal of Risk and Uncertainty* 24, no. 1 (2002): 75–95.

Acknowledgments

I learned much from discussing these ideas with colleagues, especially Gary Varner, Tony Smith, Ned Hettinger, Marc Saner, Rob Streiffer, Dermot Hayes, Kristen Hessler, Fred Kirschenmann, and C. S. Prakash. I was also fortunate to participate in several conversations on the topic during the past few months, and would like to express gratitude to my hosts, including:

- Three local chapters of the American Chemical Society at Eastern Oregon University (Richard Hermens); Washington State University (Roger Willett); and Seattle University (Susan Jackels), in October 2000.
- The "New Zealand Royal Commission on Genetic Modification"; a public audience in Wellington, New Zealand (sponsored by the New Zealand Life Sciences Network, and Francis Wevers); and St. John's College, Auckland, NZ (Graham Redding), November 2000.
- The "Plant Sciences Institute Colloquium," Iowa State University, February 2001 (Stephen Howell).
- "Biotech Issues 2001," an Extension In-Service conference at Colorado State University (Bob Zimdahl and Pat Kendall); and a seminar in the CSU Philosophy Department (Phil Cafaro and Holmes Rolston); both in February.
- The 2001 Annual Meeting of the American Association for the Advancement of Science, San Francisco, in February (Katherine R. Smith and Nicole Ballenger).
- A seminar at the University del Pais Vasco/Euskal Herriko Unib., Vitoria, Spain, in March (Marta Salona, Mertxe de Renobales).
- A seminar in the Depto. de Microbiologia e Instituto de Biotecnologia, Universidad de Granada, Spain, in March (Enrique Ianez).
- A colloquium on environmental ethics, "Colóquio Ética Ambiental: uma ética para o futuro," at Faculdade de Letras da Universidade de Lisboa, March 2000 (Cristina Beckert).
- A seminar at the Center for International Development and Science, the Technology and Public Policy Program, and the Belfer Center for Science and International Affairs, Harvard University, March 2001 (Calestous Juma and Derya Honca).
- The National Agricultural Biotechnology Council, annual meeting, May 2001 (Diane Birt, Colin Scanes, and Lynn Westgren).
- The Center for Judaism and the Environment, and Center for Business Ethics, Jerusalem College for Technology, Israel (Akiva Wolff, Pinchas Rosenstein, and Jacqueline Rose); and "Symposium 2001: Plant Biotechnology, Its Benefits Versus Its Risks," Tel Aviv University, Israel, May 2001 (Bernie Epel and Roger Beachy).

Section 3 ("Ethical Issues Involved in the Use of Genetic Technology") is reprinted from Comstock, *Vexing Nature? On the Ethical Case against Agricultural Biotechnology* (Boston/Dordrecht: Kluwer Academic Publishers, 2000), pp. 182–95, with the kind permission of the publisher.

Other portions of the paper were written with support of the Cooperative State Research, Education, and Extension Service, U.S. Department of Agriculture, under Agreement No. 00-52100-9617.

Part 3

Religion

Introduction

Jacob had no hesitation about changing the nature of Laban's animals. Does this therefore mean that, in the light of Judaism and Christianity, it is legitimate for us to do what we will with animals and plants? Did God give us animals and plants to do whatever we want to do, so long as it suits our own ends? Not according to the Prince of Wales, who would have us respect the end results of God's creation and (apparently) leave them untouched. But as Richard Dawkins—atheist though he may be—points out, things cannot be this simple. Surely it is not God's will that we turn our backs on the possibilities of organic change when these possibilities are directed for the good of all? Does God really want us to eschew using genes inserted into bacteria to make something like insulin? Clearly not!

The pieces in this section address some of the religious issues involved in the genetically modified foods debate. First, Pope John Paul II—a man who although doctrinally conservative is very sympathetic to science—makes it clear that he is, in principle, in favor of even something as innovative and drastic as genetic modification of organisms. He urges us to open ourselves "to all of the developments of the technological era." He warns that we must safeguard jealously, "the perennial values that characterize you." But, endorsing our science-based efforts, the Pope quotes King David's Psalm 65: "You visit the earth and water it, you greatly enrich it."

Anglican plant scientist Joe Perry offers us a very careful discussion of the religious issues surrounding the use of genetically modified crops. He is far from willing to endorse uncritically any possible intervention into natural processes, and draws back strongly from all attempts to apply modern molecular biology to such enterprises as the cloning of human beings. However, in principle, he sees no barriers to the Christian who would produce genetically modified foods, and Perry is especially critical of Prince Charles for (as he sees it) urging us to "dichotomize science and faith." Indeed, as a Christian, Perry believes that there is much biblical support for the practice, especially inasmuch as it is seen as a tool to feed the hungry. However, we must beware against "harming the environment unduly."

Carl Feit is both a respected molecular biologist and an orthodox rabbi. In a piece especially commissioned for this volume, he considers how the Jewish laws on behavior—permissible foods, permitted breeding practices, and the like—impinge on the GM food issue. He recognizes that he is dealing with issues that are far beyond the imagination of the ancient Jews who formulated the laws, but that is nothing new. There is a long tradition of modifying and interpreting scripture in the light of modern cultural and scientific advances. Although outsiders often think of the Jewish laws as restrictive, Feit argues that they promote a "full flowering of human creative genius within a spiritually enriching environment" and he argues, with the Christian thinkers in this section, that GM foods are permissible—even that it can be morally obligatory to work on them to further human welfare. God did not give us our talents and expect us to ignore or reject them.

12

Jubilee of the Agricultural World

Pope John Paul II

Saturday, 11 November 2000

Ladies and Gentlemen,
Dear Brothers and Sisters,

I am pleased to be able to meet you on the occasion of the Jubilee of the Agricultural World, for this moment of celebration and reflection on the present state of this important sector of life and the economy, as well as on the ethical and social perspectives that concern it.

I thank Cardinal Angelo Sodano, secretary of state, for his kind words expressing the sentiments and expectations of all those present. I respectfully greet the dignitaries, including those of different religious backgrounds who are representing various organizations and are present this evening to offer us the contribution of their testimony.

The Jubilee of farmers coincides with the traditional "Thanksgiving Day" promoted in Italy by the praiseworthy Confederation of Farmers, to whom I extend my most cordial greetings. This "Day" makes a strong appeal

Reprinted with permission from the Internet site of the Holy See, www.vatican.va.

to the perennial values cherished by the agricultural world, particularly to its marked religious sense. To give thanks is to glorify God who created the land and its produce, to God who saw that it was "good" (Gn 1: 12) and entrusted it to man for wise and industrious safekeeping.

Dear men and women of the agricultural world, you are entrusted with the task of making the earth fruitful. A most important task, whose urgent need today is becoming ever more apparent. The area where you work is usually called the "primary sector" by economic science. On the world economic scene, your sector varies considerably, in comparison to others, according to continent and nation. But whatever the cost in economic terms, plain good sense is enough to highlight its real "primacy" with respect to vital human needs. When this sector is underappreciated or mistreated, the consequences for life, health, and ecological balance are always serious and usually difficult to remedy, at least in the short term.

The Church has always had special regard for this area of work, which has also been expressed in important magisterial documents. How could we forget, in this respect, Blessed John XXIII's *Mater et Magistra*? At the time he put his "finger on the wound," so to speak, denouncing the problems that were unfortunately making agriculture a "depressed sector" in those years, regarding both "labor productivity" and "the standard of living of farm populations."[1]

In the time between *Mater et Magistra* and our day, it certainly cannot be said that these problems have been solved. Rather it should be noted that there are others in addition, in the framework of new problems stemming from the globalization of the economy and the worsening of the "ecological question."

The Church obviously has no "technical" solutions to offer. Her contribution is at the level of Gospel witness and is expressed in proposing the spiritual values that give meaning to life and guidance for practical decisions, including at the level of work and the economy.

Without doubt, the most important value at stake when we look at the earth and at those who work is the principle that brings the earth back to her Creator: *The earth belongs to God!* It must therefore be treated according to his law. If, with regard to natural resources, especially under the pressure of industrialization, an irresponsible culture of "dominion" has been reinforced with devastating ecological consequences, this certainly does not correspond to God's plan. "Fill the earth and subdue it; and have dominion over the fish of the sea and over the birds of the air" (Gn 1: 28). These famous words of Genesis entrust the earth to man's *use*, not *abuse*. They do not make man the absolute arbiter of the earth's governance, but the Creator's "coworker": a stupendous mission, but one which is also marked by precise boundaries that can never be transgressed with impunity.

This is a principle to be remembered in agricultural production itself, whenever there is a question of its advance through the application of biotechnologies, which cannot be evaluated solely on the basis of immediate economic interests. They must be submitted beforehand to rigorous scientific and ethical examination, to prevent them from becoming disastrous for human health and the future of the earth.

The fact that the earth belongs constitutively to God is also the basis of the principle, so dear to the Church's social teaching, of the universal destination of the earth's goods.[2] What God has given man, he has given with the heart of a father who cares for his children, no one excluded. God's earth is therefore also man's earth and that of *all* mankind! This certainly does not imply the illegitimacy of the right to property, but demands a conception of it and its consequent regulation which will safeguard and further its intrinsic "social function."[3]

Every person, every people, has the right to live off the fruits of the earth. At the beginning of the new millennium, it is an intolerable scandal that so many people are still reduced to hunger and live in conditions unworthy of man. We can no longer limit ourselves to academic reflections: we must rid humanity of this disgrace through appropriate political and economic decisions with a global scope. As I wrote in my Message to the Director-General of the FAO on the occasion of World Food Day, it is necessary "to uproot the causes of hunger and malnutrition."[4] As is widely known, this situation has a variety of causes. Among the most absurd are the frequent conflicts within states, which are often true wars of the poor. And there remains the burdensome legacy of an often unjust distribution of wealth in individual nations and at the world level.

This is an aspect which the celebration of the Jubilee brings precisely to our special attention. For the original institution of the Jubilee, as it is formulated in the Bible, was aimed at reestablishing equality among the children of Israel also by restoring property, so that the poorest people could pick themselves up again and everyone could experience, including at the level of a dignified life, the joy of belonging to the one people of God.

Our Jubilee, 2,000 years after Christ's birth, must also bear this sign of universal brotherhood. It represents a message that is addressed not only to believers, but to all people of good will, so that they will be resolved, in their economic decisions, to abandon the logic of sheer advantage and combine legitimate "profit" with the value and practice of solidarity. As I have said on other occasions, we need a globalization of solidarity, which in turn presupposes a "culture of solidarity" that must flourish in every heart.

Thus, while we never cease to urge the public authorities, the great economic powers, and the most influential institutions to move in this direction, we must be convinced that there is a "conversion" that involves us all personally. We must start with ourselves. For this reason, in the encyclical *Centesimus annus*, along with the discussions of the ecological question, I pointed to the urgent need for a "human ecology." This concept is meant to recall that "not only has God given the earth to man, who must use it with respect for the original good purpose for which it was given to him, but man, too, is God's gift to man. He must therefore respect the natural and moral structure with which he has been endowed."[5] If man loses his sense of life and the security of moral standards, wandering aimlessly in the fog of indifferentism, no policy will be effective for safeguarding both the concerns of nature and those of society. Indeed, it is man who can build or destroy, respect or despise, share or reject. The great problems posed by the agricultural sector, in which you are directly involved, should be faced not only as "technical" or "political" problems, but at their root as "moral problems."

It is therefore the inescapable responsibility of those who work with the name of Christians to give a credible witness in this area. Unfortunately, in the countries of the so-called developed world, an irrational consumerism is spreading, a sort of "culture of waste," which is becoming a widespread lifestyle. This tendency must be opposed. To teach a use of goods which never forgets either the limits of available resources or the poverty of so many human beings, and which consequently tempers one's lifestyle with the duty of fraternal sharing, is a true pedagogical challenge and a very far-sighted decision. In this task, the world of those who work the land with its tradition of moderation and heritage of wisdom accumulated amid much suffering, can make an incomparable contribution.

I am therefore very grateful for this "Jubilee" witness, which holds up the great values of the agricultural world to the attention of the whole Christian community and all society. Follow in the footsteps of your best tradition, opening yourselves to all the developments of the technological era, but jealously safeguarding the perennial values that characterize you. This is also the way to give a hope-filled future to the world of agriculture. A hope that is based on God's work, of which the Psalmist sings: "You visit the earth and water it, you greatly enrich it (Ps. 65: 10).

As I implore this visit from God, source of prosperity and peace for the countless families who work in the rural world, I would like to impart an Apostolic Blessing to everyone at the end of this meeting.

Notes

1. Pope John XXII, *Mater et Magistra,* no. 123–24.
2. Pope John Paul II, *Centesimus annus,* no. 6.
3. *Mater et Maistra,* no. 111; Pope Paul VI, *Populorum progressio,* 23.
4. *L'Osservatore Romano,* English ed., 1 November 2000, 3.
5. *Centesimus annus,* no. 38.

13

Genetically Modified Crops

Joe N. Perry

But, my son, be warned: there is no end of opinions ready to be expressed. Studying them can go on forever, and become very exhausting. (Eccles. 12: 12)

Theological, Ethical, and Socioeconomic Issues

Wisdom shouts in the streets for a hearing . . . "You simpletons!" she cries. "How long will you go on being fools? How long will you scoff at wisdom and fight the facts?" (Prov. 1: 20–22)

GM Crops—Outside God's Will?

I will address the most important issue raised by GM crops: Can we discern whether their development is or is not in accordance with God's will? The verses above

Lecture given by Joe N. Perry at the 2001 Consultation of The Christ and the Cosmos Initiative. Published in "Genetic Engineering: Theological, Ethical, and Socio-Economic Issues," ed. Dr. Brenda Beamond, August 2001. Copyright © 2001 Rothamsted Experimental Station. Reprinted with permission.

from Proverbs, while true, do not help us; both sides of the argument would claim to have true wisdom on their side. Prince Charles,[1] in his response to the year 2000 Reith Lectures, claimed that discernment in this area could be achieved by using our hearts and our minds, that the instinctive, heart-felt awareness buried deep within each one of us would provide the most reliable guide. Although this phraseology is deliberately inclusive of many faiths, for the Christian the sentiment is especially attractive because of its allusion to the need to be guided by the Holy Spirit. While it is important to pray for guidance, I believe that with such a complex issue, we need to base our judgement on a full knowledge of the evidence for and against. I give a greater weight than does the prince to the measured, rational approach.

The prince's article ignored much of the extensive and balanced work done by Christian organizations to meet the challenge of GMOs. There is no need to restate all that here, but readers are encouraged to look at the output of bodies such as the Environmental Information Network of Churches Together in Britain and Ireland, Dr. Donald Bruce's Science, Religion & Technology Project of the Church of Scotland,[2] the John Ray Initiative, the Eco-Congregation Project (launched September 2000 at St Pauls's Cathedral), and Christian Ecology Link. Christian ecologists such as the Rev. Dr. Michael Reiss have written extensively on the issue.[3] Other Christian bodies who have considered the problem are the Church of England Ethical Investment Advisory Group and Christian Aid. The report of the Nuffield Council for Bioethics,[4] while not explicitly Christian in standpoint, is highly relevant. Of course, in this volume readers will also find Professor Derek Burke's lecture *The Ethics of Genetic Engineering*. Indeed, so much has already been written that I will try to focus on new aspects of the debate.

For me, there are three main arguments. Are the processes of manufacture of GM crops explicitly forbidden in the Bible? If not, is the whole concept tainted through the unwarranted usurping of the Creator's function in having created life? Third, if neither of these apply, do GM crops have consequences that must of necessity be outside God's will.

Regarding explicit commandments, this is one of many challenges that could not have been conceived at the time the Bible was written. Perhaps the closest the issue is to being addressed specifically is in Lev. 19: 19: "Do not mate your cattle with a different kind; don't sow your field with two kinds of seed." Such laws in Leviticus, addressing the need to avoid "boundary-crossing" and to keep things separate, often for reasons of health, purity, or cleanliness, are not kept by Christians but are stringently observed by Jews. Yet the issue of GM crops seems not to present a problem for the Office of the Chief Rabbi.[5] An article from autumn 2000 by Rabbi Rashi Simon and Professor Edward Simon, in the *Jewish Chronicle*, gives considerable detail as to why GM food *is* considered kosher. (One interesting point raised is that even if the act of genetic modification of the original DNA were forbidden, which they argue it is not, the resulting plants would still be kosher because, as is often unappreciated, the genetic transformation event happens only once. After the initial GM plant is generated, all subsequent seed used is manufactured through the perfectly standard crossing techniques of plant breeding.)

The second question cannot be considered without consideration of whether God created plant species as some immutable set. Such a view would be challenged immediately by evolution, which I regard as fact not theory. On the other hand, the divisions between species, which in general do not interbreed, appear sharp— but are they really? Biologists would contend that species boundaries are actually indistinct and difficult to define easily.[6] Davies[7] explains in detail how Darwin's concept of a population stresses the uniqueness of every living thing in the world (something no Christian would challenge, at least in the human context) and how Darwin viewed the species as a statistical abstraction. By contrast, Davies traces back the view of a species as an unchangeable type with a 'defining essence' to Plato and Aristotle. He believes the Platonic view of 'eternal and ideal forms,' doctrinal for over 2000 years, is deeply ingrained within our collective psyche. He argues that because of this "it is not surprising that many people today find the mere thought of taking a gene from one species and placing it another as abhorrent." Perhaps Christians have another inner battle to fight before being able to resolve this further implication of Darwinian evolution for our theology? Antisthenes' response to Plato that "I can see a horse, but I cannot see horseness" is perhaps too flippant, but Popper's question, "why cannot there be as many 'essences' in things as there are things?" challenges us to extend further our belief in human uniqueness, toward animals, plants, and all life-forms. Note again that the contrast is between the reductionist and individualistic (Darwinian) and the holistic (Platonic).

Often, this argument is restated in GM debates in terms of whether the technology is "natural" or not.[8] Of course, some of the most telling ripostes to Prince Charles's lecture were to point out that the landscape over which he loves to hunt is completely unnatural and that we have been "tampering with nature" by practising agriculture for over 5,000 years. Examples abound from within "conventional" plant breeding of successive techniques being developed that have pulled at the boundaries of species and forced reproduction to occur between two usually separate species. Chronologically, they include the first cereal hybrid, formed in 1799; the creation in 1876 of the Triticale hybrid between wheat and rye, now the world's principal cereal variety, grown on over 2 million hectares; protoplast fusion in 1906; mutagenesis via x rays in 1927, which has yielded the United Kingdom's favorite barley variety for brewing; and embryo hybrid rescue in 1960. Additionally, many very similar genes are shared in common between unrelated species, such as those that promote resistance to fungal infection.[9] When we realize to what extent this applies to humans,[10] we can perhaps begin to understand why some biologists are puzzled as to *why* we place so much emphasis on these questions of ethics, morals, and theology. I refute any suggestion that these questions should be avoided, but I note that Beringer[11] and Ryan et al.[12] have reported scientists as being "genuinely baffled" why GM crops are deemed intrinsically more likely to cause more environmental problems or be more damaging to health than those bred traditionally. The argument is often offered that GM technology moves a single gene with confidence in the outcome, whereas conventional breeding shuffles tens of thousands of genes with little idea, prescreening, of possible results.

The third question, to some degree linked to the second, concerns to what degree we are permitted or encouraged to use our creative and intellectual gifts to alter our environment and to intervene to affect life and enhance its quality. Very many commentators in both the Judaic and Christian tradition point to Gen. 1:28-30, and Ps. 8:6-8, as providing God's blessing for intervention, with the strict proviso that "to have dominion" should be properly interpreted as to be custodian or steward, not master. Many remark on Gen. 2:15, where God places Adam in the Garden of Eden as gardener, to tend, care, and work. Again the image often quoted[13] is of God giving freedom to humans to help to actively mold the Creation to their needs, *so long as the Creation is respected by not harming the environment unduly*. Given this, permission to grow GM crops would seem to receive plenty of biblical support from both traditions, *so long as they are used to help feed the hungry* (Isa. 3:14–15, Amos 2:6, all of Matthew 5 & 6, Matt. 25:14-25, and so on). Arguments against such use of GMOs in agriculture are not helped by the fact that neither the public at large nor even most activist organizations seem to be against the use of GM technology in medicine. To take an example from within this diocese, a debate at Christ Church, Chorleywood, on 3 November 2000, "that we believe that Genetic Modification represents a positive step forward for humanity" was passed overwhelmingly, mainly on grounds of the benefits of medical research.[14]

To summarize the above, I have answered my own question by concluding that, for me, GM crops are not, *of necessity*, outside God's will. However, there are three important provisos. First, GM issues for humans are completely different from GM issues in animals and those in plants. Our belief in the human soul and that we are created in God's image rules out, for me, human cloning and certain related technologies. Second, the only GM crops considered in depth in this article are genetically modified herbicide-tolerant (GMHT). As is repeated often above, each separate construct must be considered separately. Third, I have not dwelt on the socioeconomic aspects of GM crops. I have focused on the technology itself because that is what I am best qualified to do, but that does not mean that we should not be equally thorough in questioning the socioeconomic aspects, that is, the use to which that technology is put.

Third World Issues

Others have discussed at length the socioeconomic aspects of GM crops on Third World agriculture, particularly subsistence farmers.[15]

Views are extremely polarized, and indeed, much of the impetus driving activist organizations such as Greenpeace and Friends of the Earth's mistrust of GM crops are clearly now derived less from scientific concerns about risk and more from a skepticism of the motives of the very large, transnational, biotechnology corporations. For example, the Christian Aid[16] position was very strongly against GM crops, expressing concern over the degree of ownership and control over the food chain by a handful of large corporations, the threat of biotechnology to small farmers through "terminator technology," the loss of agricultural bio-diversity, and a widening of the gap between rich and poor. Unfortunately that position, expressed forcibly against "golden" vitamin-A rice, strayed too far outside its competence, and made unwise and unsubstantiated comments on the scientific issues. I believe it is being modified. Oxfam appears to have had a consistently better balanced and well-informed policy. An early document[17] points to a need to reconcile the rigid, individualistic patenting system of the developed world with the community-held knowledge systems of poorer countries. This conflict is epit-omized by the contradictions between the 1993 Convention on Biological Diver-sity (which broadly promotes the conservation of agricultural biodiversity) and the 1995 Trade Related Aspects of Intellectual Property Rights agreement of the World Trade Organization (which set up a global system of intellectual property rights [IPR] on plant genetic resources). The need to ensure wide crop biodiver-sity is in any event well recognized *within* conventional agriculture, because of the need to avoid catastrophes such as the United States maize leaf blight epidemic caused by *Helminthosporium maydis* during the 1970s. Biopiracy is already a real danger in pharmaceutical research; the ability of biotechnology to transform indigenous Third World crop plants transmits that danger from medicine to agri-culture. (Indeed, the problems caused by IPR is not limited to the Third World, as witnessed by the latest display of gross ineptitude in public relations by the Mon-santo company, which has just successfully sued small farmer Percy Schmeiser, from Bruno, Saskatchewan, Canada, for using seeds allegedly blown unasked on to his farm from a neighbor's GM crop. Monsanto's position was reportedly that whether or not Schmeiser knew that his canola field was tainted with their GMHT seed, he must still pay their technology fee.)

There is another issue related to GM crops that seems to elicit completely polarized views: The green revolution. On the one hand, organizations such as Greenpeace and Christian Aid claim that millions of subsistence farmers were forced from the land by the adoption of intensive agricultural practices into the cities.[18] On the other hand, it is claimed[19] that the lives of millions of city dwellers were saved by a large and sustainable increase in food production, which would otherwise have involved ploughing up of millions of hectares of wilderness and forest.[20] It is often not appreciated that the varieties responsible for the green rev-olution were developed at international PSREs within the CGIAR group of insti-tutes, such as CIMMYT and IRRI, and distributed from there, and not by large, multinational companies.[21] As world population is set to rise to 8 billion by 2020, food production will no longer keep pace with population growth under sce-narios of conventional agriculture.[22] In the future, there will be a difficult choice, between the status quo and the attempt to increase production in the Third World through the use of GM crops; this is bound to cause intense debate.

Campaigning Organizations

> They won't believe me. They won't do what *I* tell them to. They'll say, "The Lord never appeared to you!" (Exod. 4:1)

There is a chasm of thought between the scientists and the campaigning organizations that is deeply worrying. It would be doubly depressing were the gap to be unbridgeable.

Campaigning organizations such as Greenpeace and Friends of the Earth, by definition, have an agenda. In their debate with scientists over GM crops, both sides are frustrated because they seek to argue with different rules. The campaigning organizations believe passionately in their cause and, to a certain extent, believe that the ends justify the means in a debate. By contrast, scientists have no agenda except to get to the truth, which may be complex; judgements in their terms may require them to balance conflicting evidence. Representatives of the campaigning organizations agree with that approach privately but feel they cannot concede any point in public debate because they fear that their members will not be able to accept a mixed message. In this, they are similar to politicians; every point must be argued, and none conceded. Again by contrast, the scientists want to take each issue separately, agreeing with this, disagreeing with that, and accepting that further evidence needs to be gathered on others. The campaigners' tactic of raising very many issues and attacking vigorously on each one has been termed the "blender" method by those who advise on public relations. The idea is that even if all of them are refutable, the impression is left with the public that there is some truth in the overall argument; this is not the way science proceeds. Campaigning organizations argue in black-and-white terms; in debate they will state conjecture as if it were fact and occasionally selectively misquote where this is advantageous in debate. Scientists are much more cautious, often preferring to admit that they do not fully know an area rather than trying to express an answer that might only be 80 percent correct. Few scientists will ever admit that there is absolutely no risk of some hazard, even if the probability of its occurrence is vanishingly small; this leaves them wide open to the accusation of advising that it is sensible to proceed with a course of action that might conceivably harm the public or the environment. In large areas of the GM debate, such as economic, political or religious aspects, scientists will often deliberately express no opinion whatever.

It is important to understand the different approaches of these groups, and how their different rules of engagement affect how the public perceives them in open debate. Getting both groups to communicate is *the* major challenge for the future. Anyone who doubts that my portrayal is accurate should spend a couple of hours sharing a platform of a public meeting with activist organizations while attempting to explain to the public that the FSE trial site planned near their village will do them no harm. What I worry most about is the level of the debate—that the rational approach will be lost under the ferocity of the attack and the appeal to emotion rather than reason. This is the worrying obverse of the coin which Prince Charles advocated. Nor is my concern about rationalism a lone opinion; it is shared by many completely neutral and independent scientists. In addition, Dr. Patrick Moore, a Canadian ecologist and one of the founding members of Greenpeace in Vancouver in the early 1970s, believes[23] that Greenpeace got it wrong in the mid-1980s when it "abandoned science and logic, just when mainstream society was adopting all the more reasonable items on the environmental agenda. Many environmentalists couldn't make the transition from confrontation to consensus, and their agendas had more to do with class warfare and anticapitalism than the actual science of the environment." On GM crops, Moore fears that we are "entering an era when pagan beliefs and junk science are influencing public policy" and denigrates activists for "preying on people's fear of the unknown." He asks if they are so worried about human health why more of Greenpeace's funding isn't used to tackle tobacco. In the House of Commons debate on GMOs and Biotechnology, 13 January 2000, the MP for Bexleyheath & Crayford, Nigel Beard,[24] described Greenpeace and Friends of the Earth as unjustifiably and unscrupulously attacking the objectivity and competence of those who sit on government committees with "the vehemence of fundamentalist cults" and described likely Greenpeace vandalism of the FSE fields that will produce evidence on whether GM crops would damage the environment as "abandonment of legality and of science and rationality as a foundation of public policy."

Notes

I thank all my colleagues, family, and friends, too many to list here, who have provided help and material for this paper. Ian Crute made many valuable suggestions for improvements. IACR-Rothamsted receives grant-aided support from the Biotechnology and Biological Sciences Research Council of the United Kingdom.

1. Charles, Prince of Wales, "A Reflection on the 2000 Reith Lectures" [online], www.princeofwales.gov.uk/speeches/environment_18052000.html.

2. Donald Bruce and Ann Bruce, eds. *Engineering Genesis* (London: Earthscan, 1998).

3. Roger Straughan and Michael Reiss, *Ethics, Morality, and Crop Biotechnology* (Swindon, U.K.: BBSRC, 1996).

4. Alan Ryan et al., *Genetically Modified Crops: The Ethical and Social Issues* (London: Neuffield Council on Bioethics, 1999).

5. Ibid., sec. 1.38.

6. For example, Straughan and Reiss, *Ethics, Morality, and Crop Biotechnology.*

7. Eric L. Davies et al., "Nematode Parasitism Genes," *Annual Review of Phytopathology* 38 (2000): 365–96; Keith Davies, "What Makes Genetically Modified Organisms so Distasteful?" *Trends in Biotechnology* 19, no. 10 (2001): 424–27.

8. Ryan et al., *Genetically Modified Crops: The Ethical and Social Issues,* sec. 1.32–1.40.

9. http://www.ncbi.nlm.gov/HomoloGene.

10. Eric S. Lander et al., "Initial Sequencing and Analysis of the Human Genome," *Nature* 409, no. 6822 (2001): 860–921.

11. John Beringer, "Reply to Comments in BES Lecture," *Bulletin of the British Ecological Society* 31, no. 2 (2000): 16.

12. Ryan et al., *Genetically Modified Crops: The Ethical and Social Issues.*

13. David Berry et al., *Sustainable Development: Can It Be Made to Work in the Real World* (Cheltenham, U.K.: John Ray Initiative, 2000).

14. Henry McLeish, "GM Debate on the Record," *SEEROUND* (December 2000): 8.

15. Ryan et al., *Genetically Modified Crops: The Ethical and Social Issues*; The Royal Society and other national academies, *Transgenic Plants and World Agriculture* (London: The Royal Society, 2000).

16. Christian Aid, policy briefing on Biotechnology and Genetically Modified Organisms, January 2000, Christian Aid, London.

17. Oxfam, "Biotechnology in Crops: Issues for the Developing World," Oxfam, Oxford, 1998.

18. Vandana Shiva. *The Violence of the Green Revolution* (Penang: Third World Network, 1991).

19. Matt Ridley, "Comment: The Poverty of a Reith Lecturer's Thinking," *London Daily Telegraph,* 16 May 2000.

20. Anthony J. Trewavas, "The Population/Biodiversity Paradox: Agricultural Efficiency to Save Wilderness," *Plant Physiology* 125, no. 1 (2001): 174–79.

21. Norman E. Borlaug, "The Green Revolution: Peace and Humanity" (a speech on the occasion of the awarding of the 1970 Nobel Peace Prize in Oslo, Norway, 11 December 1970); http://www2.iastate.edu/~rjsa/vad/revolution.html.

22. Per Pinstrup-Anderson et al., *World Food Prospects: Critical Issues for the Early Twenty-first Century* (Washington, D.C.: IFPRI, 1999).

23. Patrick Moore, see interview in *New Scientist,* 25 December 2000, reprinted on http://www.agbioworld.org/pr/moore.html, or http://www.greenspirit.com/newscientist.htm, or e-mail patrickmoore@home.com.

24. Nigel Beard, Debate on GMOs and Biotechnology, Westminster Hall, London, 13 January 2000.

14

Genetically Modified Food and Jewish Law (Halakhah)

Carl Feit

Introduction

The pros and cons of the use of genetically modified (GM) foods are being amply covered in other chapters in this book. This chapter will address the issue of the production and use of GM products from the point of view of the Jewish religion.

Jewish religious law (Halakhah) derives from biblical texts that are viewed as divinely mandated. Traditional Judaism also recognizes an ancient oral tradition that consists of interpretations and clarifications of the biblical law. In the course of time, rabbinic enactments and extensions of the biblical law were legislated. This corpus of legal discussion is contained in the Talmud. Post-talmudic Jewish law is found in various codifications of the law, most notably and authoritatively that of R. Yosef Karo called the Shulkhan Arukh (mid-sixteenth century). In addition, there is a tremendous body of literature known as Responsa, which represents the attempts of rabbinic scholars of each generation to apply the principles of Halakhah to the ever changing and developing realia of society. In this article, I will try to present a broad outline of Jewish thinking in this area, but due to the breadth of both the topic and the source material, as well as the technical/legal complexity of some of the concerns raised, this should not be viewed as a complete and definitive presentation.

General Concerns

In general, Judaism takes a very supportive attitude toward scientific and techno-logic advancement. This derives from the view of the rabbis that God created the universe, but left it in an incomplete state. The existence of disease, famine, wars, and pestilence is reflective of the lack of perfection in the world. With the creation of humankind in the form of Adam and Eve, God created a partner, whom he has charged with the task of working to complete the creation, by finding the cure for disease, producing enough food for the hungry, and establishing peace among nations, that all may reap the benefits of God's plentiful bounty. "Be fruitful and multiply, fill the universe and achieve dominion over it" is God's challenge and endorsement of the boldest scientific and cultural undertakings. Nevertheless, all scientific activity, like the rest of human conduct, must also conform to the highest of God's moral and ethical teachings.

Thus, when confronted with new technologies, Judaism looks at two areas. The first is whether or not the undertaking is in consonance with the general mandate to work toward the benefit of humanity and the enhancement of the physical uni-verse. If this answer is affirmative, then the second question is whether any other divine rules are violated or threatened in the process of carrying out the project.

Regarding the first question as to whether or not the production and distri-bution of GM crops can be viewed as working toward the benefit of humanity and the enhancement of the physical universe, that is precisely the subject of the current scientific debate. Certainly the goals of producing GM crops that are more productive, resistant to pests and herbicides, able to grow in arid regions, more resistant to spoilage, and more nutritious, are in keeping with our divine mandate. On the other hand, some of the ominous and dire warnings about the production of chimeric monster organisms would mitigate against such undertakings. In Jewish law we find the concept of *ushmartem et nafshoteichem*, an obligation to pro-tect the physical welfare of both the individual and the community. The Jewish law would seek out the best scientific evidence available to determine if the benefits far outweigh the risks. Since to date, many millions of hectares of GM crops have been grown, and there is no or scant evidence of any significant human health or environmental problems associated specifically with these crops or their products, cautious optimism is warranted. Due care must be made to avoid the introduction of known allergens into food crops and appropriate public health regulatory over-sight should be maintained. The possibility of long-term adverse affects should be kept in mind when establishing such monitoring systems. The divine mandate to humanity regarding the use of the land is *l'avdah u'lshomrah*, to work it but (also) to protect it. Thus an obligation to show concern for, and to refrain from, activities that are detrimental to the environment is a cornerstone of Jewish law.

There is certainly no general Jewish concern about "interfering" with God's plan of creation, by altering certain generic "kinds" created by God. Although some religious thinkers have raised this issue, Judaism's positive endorsement of the domestication of plants and animals as well as Judaism's view of the obligatory nature of healing the sick would amply prove the contrary. Rather than fearing that society is overstepping its boundary and "playing God," Judaism has always stressed that the advancement of individuals and civilization is brought about through a process of Divine imitation or *imitatio Dei*. As God is depicted in Genesis as a creator of worlds and species, this represents a legitimate undertaking of the scientific community.

Having answered with a tentative yes to the first query, it would appear that there are at least two areas of potential concern, where either the production or consumption of GM foods might involve possible violations of Jewish law. One would be the area of Kilayim or forbidden mixtures, and the other would be on the area of Kashrus or forbidden foods.

Forbidden Mixtures—Kilayim

The Bible states:

> "You shall keep my statutes: You shall not mate your animal with a diverse kind; you shall not sow your fields with diverse kinds of seed; and a garment containing a shaatnez mixture shall not come upon you." (Lev. 19:19)

> "You shall not sow your vineyard with a mixture; lest the fullness of the seed which you have sown be forfeited with the production of the vineyard. You shall not plow with an ox and a donkey together. You shall not wear shaatnez, wool and flax together." (Deut. 22:9-11)

In the first verse, the rabbis have seen three distinct prohibitions: a prohibition on crossbreeding animals of diverse species, a prohibition on sowing together a mixture of seeds of diverse kinds, and a prohibition on wearing a garment containing a certain forbidden mixture of materials known as *shaatnez*. In the second set of verses, there is a specific prohibition of producing Kilayim in a vineyard, a prohibition on working with a team of yoked animals of distinct species, and a repetition and clarification of the *shaatnez*/clothing prohibition. Thus there are five separate prohibitions outlined in the text: Kilayim of seeds; Kilayim of the vineyard; Kilayim of clothes; and Kilayim of animals that includes two separate prohibitions, one of crossbreeding and one of plowing.

Since the Halakhah is a legal system, it has given precise definitions, parameters, and boundaries to these activities. In some cases their applicability is restricted and sometimes expanded. For example, the prohibition of Kilayim of seeds is seen to apply only to seeds of various grains and not to seeds of trees. On the other hand, the prohibition of Kilayim is extended to include grafting, and grafting would apply to trees as well. The grafting prohibition is derived by the sages in the Talmud (Sanhedrin 60a) from the juxtaposition of the two verses "you shall not mate your animal with a diverse species" and "you shall not sow your field with diverse species of seed." This indicates that there is a prohibited method of growing which resembles the coupling of animals, for example, grafting. Also, the prohibition is specifically on the physical act of planting two or more diverse seeds or the subsequent maintenance of such growth within a defined space limit of each other. However, regarding the products of such a forbidden mixture or a forbidden grafting, there is universal agreement that, post facto, such produce is permitted for consumption (see Talmud tractate Kiddushin 39a, for a discussion of the derivation of this principle). Similarly, directly bringing about the copulation of a horse and a donkey would be forbidden as Kilayim of animals, but the product of such a mating, the mule, is permissible for use.

It is also relevant to note that while many biblical laws such as the prohibitions of murder, robbery, and immorality are seen as being universally applicable, other laws such as Sabbath observance and dietary restrictions are only obligatory on the Jewish people. The laws of Kilayim appear to be of the latter variety; that is, restrictions that apply to Jews but not to the rest of humanity. The question would then be whether the production and growth of GM seeds violates any of the above restrictions, which would limit the participation of Jews in the production or planting of transgenic products.

Since the isolation, extraction, amplification, and transplanting of the desired gene are done in the laboratory under conditions that are totally dissimilar to the actual acts of planting diverse seeds in the field or the grafting of an existing branch or stem onto a preexisting plant, it would appear that the answer to this question is that there is no technical violation of the production of Kilayim when deriving and developing GM seeds. Therefore there would be no reason for a Jewish scientist to refrain from participating in such research, and there would be no reason for Jewish farmers to hesitate in using GM seeds.

An interesting question is, granted that there are no technical violations of the Halakhah in GM production, is there not somehow a violation of the spirit of the law? Jewish thought has always distinguished between different categories of biblical law. Some laws, defined by the Bible as Mishpatim, are laws whose basis in rational thought are such that had they not been divinely mandated, they would clearly have been humanly legislated, for example, murder, robbery, and the like. For a second group of laws called Chukim, the underlying reason or reasons are not immediately apparent, and these laws are followed simply because they reflect the divine will as revealed in Scripture. All of the laws of Kilayim fall in this latter category. While various sages throughout the centuries have attempted to suggest possible explanations for different Chukim, in such cases, the proposed explanations never become definitive. For example, regarding Kilayim, there is a fundamental disagreement between two of the greatest medieval Jewish savants, Nachmanides and Maimonides. Nachmanides proposed as an explanation for the laws of Kilayim, that they represent an illegitimate blurring of the boundaries between species that God created (see Nachmanides commentary on the Torah, Deut. 22:11). Maimonides vehemently disagrees with this explanation, and instead claims that the production of such mixed breeds of plants and animals was once a standard part of pagan religious practice and that this is the sole reason that it is forbidden by the Torah for the Jewish people (*Guide for the Perplexed*, Part III:37). Neither one of these explanations has been universally accepted as the reason by the normative Halakhah. Given this uncertainty of the underlying reason for Kilayim, there is no need to assume that it in any way violates the letter or the spirit of biblical law.

Dietary Laws—Kashrus

A second question of potential concern would be in the area of the various dietary restrictions that the Torah imposes upon Jews. In the animal kingdom, Halakhah limits food consumption to a relatively small group among the fish, birds, and mammals. For example, from those creatures that live in the water, consumption is limited to fish having fins and scales (only certain osteichthes), thus eliminating all shellfish, lobster, and squid from the kosher diet. Among terrestrial organisms, only some of the even-toed ungulates, specifically those that have both cloven hooves and chew their cud, are permissible as food and even then only if they are slaughtered in an appropriate fashion. As a result, cows are kosher, but pigs even though they have split hooves are not kosher, because they do not chew their cud. An additional dietary restriction forbids the cooking or consumption of meat and dairy products together, even when both come from kosher sources.

A question that has been raised is whether the transplanting of a gene such as the gene for growth hormone from a forbidden animal such as a pig, into another organism, would render the recipient transgenic plant or animal forbidden to the kosher consumer.

There are several precedents that are relevant to this discussion. As mentioned, the consumption of all insects or creeping organisms whether aquatic or terrestrial is forbidden under Torah law. By the late nineteenth century, the existence and ubiquity of microorganisms was becoming known, and Halakhic decisors had to decide whether these organisms fell under the definition of forbidden insects. In a representative and definitive ruling given by Rabbi Yehiel Michel Epstein in his monumental Aruch Ha-Shulchan (*Yoreh Deah, Laws of Forbidden Insects*, chap. 84:36) he writes: "It has been written in the name of some scientists who have looked through a magnifying device called a speculum/microscope and have seen that vinegar is filled with small creatures . . . and I have heard that all liquids, including [apparently pure] rain water are filled with small creatures invisible to the naked eye . . . and in my youth I met someone who in his travels had actually looked through such a microscope and told me that he had seen uncountable numbers of such organisms. . . . If so how is it possible to even drink water? But the correct answer is that the laws of the Torah regarding forbidden things do not apply to that which is invisible to the naked eye. The laws of the Torah were given to human beings, not to angels! Were this not the case how could we even take a single breath for the very air around us is filled with tiny, tiny organisms that we swallow every time we open our mouths. Certainly if the [unaided] eye can not perceive it, then it can be ignored."

The reasoning behind this insightful ruling did not originate with Rabbi Epstein, but rather is an extension of rabbinic thought that dates back to talmudic discussions on the question of mixtures of permitted and forbidden foods. The specific question is: How much of a forbidden product will render permissible food forbidden? The answer given is that if there is enough of the forbidden material such that it imparts some of its taste (*nosain taam*) to the final product, then the entire amount becomes forbidden. If it is less than this percentage, then the forbidden product becomes nullified in the whole. The rabbis estimated that in general this amounts to about one part in sixty. That is, if the forbidden material constitutes less than 1.7 percent of the whole, it is disregarded. The actual amounts vary depending on the nature of the product and under what category of food it is forbidden, but the general concept of nullification is a well-established Halakhic principle.

In the removal and even in the amplification of a gene (or even several genes) and its subsequent transfection in the target, both of the above principles are in effect. First of all, the gene is a molecule that is invisible to the naked eye and therefore is not subject to the rules of forbidden animals. Even were this not the case, the uptake of a single gene by the recipient cell would leave us well below the limits of Halakhic nullification. For these reasons, an animal or vegetable, say a potato, with a gene from a pig would be perfectly kosher.

There is yet another line of reasoning that I believe would also lead to a permissive ruling on the question of GM foods with a gene derived from a non-kosher source. The laws of nullification all apply to mixtures of materials that are essentially inert. I believe that a different and stronger argument than nullification can be made for a living organic system.

Jewish law forbids those who are Kohanim (direct descendents of Aaron, the bother of Moses) from voluntarily coming in contact with the dead. This would include at times even being in the same room with a dead body or a significant part of a body. In the twentieth century, with the advent of lifesaving organ transplant surgery, the question arose as to whether a Kohen would have any restriction on being in contact with the recipient of an organ transplant, if the organ had been taken from a cadaver. That is, does the transplanted organ still have the status of an organ from a corpse? A related question was asked about to whom does this organ ultimately belong, the donor or the recipient, when it comes time for appropriate burial arrangements to be made. The answer that has been given and widely accepted is that the transplanted organ, once it "takes" in the living recipient, becomes, for all intents and purposes, a permanent and integral part of the recipient's body. Therefore it does not impart the ritual impurity of the dead to a Kohen and remains with the recipient for burial. The underlying rationale for this ruling is that, as opposed to inert mixtures where a small amount of forbidden material can be considered to be nullified, when something becomes incorporated into a living organic organism, it is actually transformed and becomes a part of the recipient. It in effect assumes a new identity, as a living and vital part of its new host. This would lead to the conclusion that the transfected cell or organism is not to be viewed as a potato with a pig gene that we ignore, but rather as a potato with an additional "new" potato gene. At some point, a decision will have to be made regarding how much of a genome must be transferred to alter the essential identity of the new organism. Perhaps we will follow a simple majority, but Halakhic authorities have not yet addressed this question.

Conclusion

The perils of genetic engineering are great, but so are the commensurate rewards. Classic Jewish thought sees as the essence of being human to be in a creative dynamic flux, alternating between the poles of majesty and humility. The Psalmist (chapter 8) can sing, that on the one hand, "You have created him slightly lower than the angels" while at the same time ponder, "What is mankind that you (even) consider him?" There is no contemporary area of greater concern, requiring a delicate balance between bold enterprise and humble awareness of the finitude of human vision than genetic engineering. This dialectic is at the heart of Jewish ethical thought, and it is the role of Halakhah (literally "the pathway") to guide individuals and society in a fashion that best allows for the full flowering of human creative genius within a spiritually enriching environment.

Part 4

Labeling

Introduction

Genetically modified foods have already been labeled *Frankenfoods*. But how would you label genetically modified foods in a way that would be informative to consumers? Should GM foods be labeled as such? At least the answer to the second part of the question seems easy. Of course you should label such foods. But why? Well, obviously because people have the right to know! They should have the freedom to choose! But, continues the persistent questioner, what do people have the right to know and why do they need the freedom to choose or reject GM foods? This is no easy pair of questions to answer.

As Peter Spencer points out in his contribution to this section, even without genetically modified foods complicating the picture, there are serious and difficult issues to be tackled. No foodstuff is going to be absolutely pure—one hundred percent free of rat droppings or animal hair or microorganisms. When did you ever buy a pint of strawberries without a fly or a grub? Why should it be different for the package of frozen peas that you get from the supermarket? Or the can of baked beans? Of course, we try to keep these things down within limits (there are accepted limits for rat hair and insects), but there is only so far that you can go. And the same, Spencer feels, applies to GM foods. You must give the consumer some warning about what is being sold, but there comes a point when too much information is simply too much information and not helpful to anyone.

Alan McHughen continues with this theme. Suppose you have a bean modified with a tomato gene and a tomato modified with a bean gene. Separately, you should label them. Should you label a soup made from the two? In McHughen's eyes, what you are being served is no different from unmodified foods. The genes have been moved around, to be sure, but they are still making bean protein and tomato protein—just in different plants. The tricky question is this: If we end up with the same *products* of the genes, does it matter what *process* we use to get them? Conventional labeling practices register only the food's products, not the processes by which they are made. Food regulators must therefore decide on a controversial issue. Either genetically modified foods are the same as their conventional counterparts, and therefore do not require special labeling that goes beyond telling us what *products* are in the food, or they are categorically different—the *process* by which the food is created matters enough that we should break our old practices and include it on the label.

Franken-

A Monstrous Prefix Is Stalking Europe

William Safire

The hottest combining form in populist suspicion of science was coined in a letter to the *New York Times* on June 2, 1992, from Paul Lewis, professor of English at Boston College.

Commenting on an op-ed column criticizing the Food and Drug Administration's decision to exempt genetically engineered crops from case-by-case review, Professor Lewis held, "Ever since Mary Shelley's baron rolled his improved human flesh out of the lab, scientists have been bringing such good things to life." After this reference to Dr. Victor Frankenstein, who created the monster in Ms. Shelley's 1818 novel, the academic letter writer shot a bolt of juice into the lifeless coinage dodge with "If they want to sell us Frankenfood, perhaps it's time to gather the villagers, light some torches, and head to the castle."

And that's what they did! "Genetic engineering" was

not then a scary enough phrase. The science of making foods more productive or resistant to disease had noncontroversial roots in hybrid corn pioneered in 1923, but ethical concerns about cloning merged with worries about mad-cow disease and suited the promotion of "organic" foods. A frightening metaphor was needed, and the Franken- prefix did the trick.

A *Boston Globe* reporter wrote in 1992 that Frankenfood "summed up nicely the monstrous unnaturalness of such controversial new products as genetically enhanced tomatoes and chromosome-tinkered cows," and quoted the delighted Lewis, today the chairman of the English department, saying: "It has a phonetic rhythm, it's pithy and you can use the Franken- prefix on anything:

Frankenfruit. . . . Frankenair. . . . Frankenwater. It's a Frankenworld."

Since then, biotechnophobes and other members of the antigenetic move-ment have denounced Frankenseeds, Frankenveggies, Frankenfish, Frankenpigs, and Frankenchicken, lumping them together as fearsome Frankenscience. In the *Chicago Tribune*, David Greising wrote in 1999 of Frankenfarmers supported by Frankenfans arguing with Frankenprotesters about unfounded Frankenfears.

Frankly, there's no telling when or how it will end. It has enhanced the sales of the metaphysical novel that Ms. Shelley's husband, the poet Percy Bysshe Shelley, encouraged her to write, and has not banned sales at "Frank'n'Stein," the fast-food chain whose hot dogs and beer I find delectably inorganic.

At the American Dialect Society, Laurence Horn says: "I was hoping some-body might have coined Frankensense by now. This would be sort of the opposite of common sense, maybe as a description of politicians' motivations for a creatively stupid piece of legislation." But this play on frankincense, an aromatic gum resin used in religious ceremonies, has not caught on. "Alas," says the dialectologist Horn, "all the Web site usages I can find for Frankensense seem to be unintended misspellings of the traditional Christmas gift. You can tell because there's an equally orthographically challenged rendering of myrrh."

Biotech Foods

Right to Know What?

Peter Spencer

There's an argument tossed about these days that consumers have "a fundamental right to know what they eat" and so foods should be labeled accordingly. Sounds good. But what do we really need to know, or care, about?

Imagine, for a moment, picking up a bag of frozen broccoli at the store and reading "may contain up to 276 aphids"—the official limit allowed for this product. Or maybe pulling off the shelf a favorite box of macaroni and reading "may contain up to nine rodent hair fragments." What would you think if you read "may contain up to 10 mg of animal excreta" on the back of a can of beans?

This kind of information represents the safe and acceptable levels of insect parts, molds, rodent filth, grit, even maggots set for a range of foodstuffs the Food and

"Biotech Foods: Right to Know What?" by Peter Spencer. Reprinted with permission from *Consumers' Research*, October 1999. Copyright © 1999 by Consumers' Research, Inc.

Drug Administration (FDA) monitors for such "extra" ingredients—unavoidable in harvesting and processing foods. And it certainly indicates what can be in the food we eat. Yet the clamor to inform consumers does not include this reality of the food supply. Clearly, some facts about what we eat have no useful place at the point of purchase, even if they warrant regulatory oversight. Listing the potential of pest remnants would do little but confuse and alarm.

Of course, the push for "right to know" labeling nowadays concerns the food itself, not what gets mixed in with it. Even here, however, calls for mandatory labeling threaten to confuse more than inform.

The case in point involves genetically modified (GM) foods—a broad term, but now used popularly to signify foods made with the techniques of modern biotechnology, such as by gene-splicing (or recombinant DNA) techniques. Over the past couple of years, a controversy has been brewing over the marketplace introduction of these foods, stemming mainly from the widespread and growing use of biotech-derived corn and soy ingredients in food products. Most of the outcry has occurred in Europe, but is spilling over into the United States—and with it, an astonishing amount of misinformation about the appropriate approach to food risks and food labeling.

For consumers, especially those worried about the advent of biotech foods, it is essential to understand a few key points about how the Food and Drug Administration—the relevant authority in these matters—monitors the risk of new (and old) food varieties, and how it determines what warrants mandatory information on a food label. From this standpoint, as discussed below, the labeling of biotech foods just because they are "biotech" offers no meaningful information in terms of nutritional value or health risks, or even as a matter of identifying the food, all key factors behind labeling mandates.

No Longer a Remote "Concern"

The controversy has mounted as the fruits of modern biotechnology have practically exploded onto the agricultural scene and into the marketplace. GM versions of major crops, such as soybean and corn, first introduced just four years ago, will be planted on more than 70 million acres worldwide this year—a figure expected to triple in five years.

And there's no doubt that people have already been eating GM foods, or will do so soon. It's practically unavoidable. Fully half of the 72 million acres of soybeans grown annually in the United States will be sown with seeds genetically engineered to tolerate the biodegradable and environmentally friendly herbicide glyphosate, better known as Roundup, which is sold by Monsanto. About 40 percent of the 80 million acres of U.S. corn will be sowed from seeds engineered to produce plants resistant to major pests, such as the European corn borer. Some corn also will be "Roundup Ready," and other GM enhancements are just around the corner. And there are a number of smaller crops, such as potatoes, tomatoes, and oil seeds, being grown, with more waiting in agriculture research fields.

Even at this early stage, the agricultural benefits of GM crops are becoming clear. For example, a recent study by the National Center for Food and Agriculture Policy, in Washington, D.C., found that, in 1998 the GM corn crop resulted in two million fewer acres being sprayed with insecticides, and a 60-million-bushel increase in corn production. Overall, other industry estimates have found millions-of-gallons reduction in pesticide and herbicide use on cotton, corn, and soy—saving money, reducing chemical treatments, increasing yields, and producing other environmental-health benefits.

Obviously, all of this appears somewhat remote to the average food shopper. Consumer-oriented products like nutritionally enhanced foods, and with them voluntary labels plugging these attributes, are still in development, industry says. In the meantime, concerns about safety and environmental side effects dominate typical discussion.

Whatever the benefits, the argument goes, biotech foods pose unique environmental and safety risks and, therefore, consumers "have a right to know" if what they buy contains GM ingredients. Moreover, biotech critics add, government authorities should be extra vigilant in measuring the risks of the "new" crops, making sure they are "safe" prior to release to farmers.

No Meaningful Distinction

The threshold issue here is just what "genetically modified" signifies about a food. Biotech critics say GM labels would indicate foods that have been, as popular usage has it, derived from gene-splicing techniques, which enable crop breeders to copy specific genetic information from different species and genera to come up with desirable new crop and food varieties—with such "transgenic" products deserving extra regulatory scrutiny and consumer notification. But this turns out to be an arbitrary distinction with regard to how new food varieties are (and have been) developed and offers no meaningful information about the risks or attributes of particular new foods.

Essentially, this distinction turns on an assumption that the biotech process poses risks not created by the more traditional, or classical, methods of breeding. But this simply is not supportable by current scientific understanding.

New crop varieties, experts point out, should be viewed as part of a continuum of molecular and classical methods of breeding, which are widely accepted by the public. And, given that over the centuries all major crops have been refined through ever-more-precise breeding techniques, the term "genetically modified," despite its popular meaning, can apply more accurately to most all food plants—even "organically" grown plants. "It would also appear, now that we have about twenty years' experience," notes R. James Cook, scientist and professor of plant pathology at Washington State University, "that the new foods and other products of biotechnology raise no safety, ethical, or social issues that could not have been raised for food produced by the more traditional tools of breeding and genetics."

Indeed, whether the discussion involves environmental side effects such as "superweeds" or unanticipated toxins in new foods, on about any risk attributed to biotech products, the same could be raised for conventional techniques used to develop and bring hundreds of new varieties of food plants into the market each year. The Lenape potato, for example, was promoted as an excellent new variety for producing better chips. But the cross-breeding that produced it also produced potentially deadly levels of the glycoalkaloid solanine; it was removed from the market in 1970, and a policy was established to test for this toxin. As the Food and Drug Administration points out, "virtually all breeding techniques have potential to create unexpected effects," including multiple effects from a single genetic change. From this perspective, to single out the biotech process on a label, as genetically modified or anything else, is fundamentally misleading in terms of risk or food quality.

Ironically, genetic modification through conventional breeding processes such as hybridization, chemical mutagenesis, and tissue culture techniques are far less precise than gene-splicing in terms of the genetic (including "transgenic") information manipulated and the number of likely changes in the final product. So, researchers actually can identify and answer risk questions about biotech risks with far more precision than they can with conventional techniques, as the Congressional Office of Technology Assessment (OTA) and the National Research Council (NRC) have pointed out.

Extensive Experience with New Plant Varieties

The disconnect in popular discussion about biotech foods and the risks of traditional agricultural techniques also misleads about the vast amount of experience agricultural researchers have with experimental plant varieties, and, by extension, their understanding and ability to deal with biotech risks. Familiarity with introduction of new plant varieties forms the basis for how researchers and regulators approach risks.

The NRC offered some helpful perspective in its 1989 report on GM crops: "Extensive experience has been gained from routine field introductions of plants modified by classical genetic methods. For example, an individual corn, soybean, wheat, or potato breeder may introduce into the field 50,000 genotypes [new experimental varieties] per year on average or 2,000,000 in a career. Hundreds of millions of field introductions of new plant genotypes have been made by American plant breeders in this century. There have been no unmanageable problems from these field introductions through the use of established practices."

"The majority of genetic modifications that are being proposed for domesticated crops by molecular [biotech] methods are similar to those already achieved by classical means," the NRC added. "These include resistance to herbicides, pests, drought, and salt as well as compositional changes in the seed or other plant parts. The genes being used to obtain these traits may differ from those used in the past, but so far these genes have introduced traits with which we have considerable experience."

The NRC concluded, in part, that "crops modified by molecular and cellular methods should pose risks no different from those modified by classical genetic methods for similar traits. As the molecular methods are specific in terms of what genes are being added, users of these methods will be more certain about the traits they introduce into plants."

It is clear from reviews of agricultural practices by the OTA and the NRC that researchers, no matter the process, take considerable care with new crop introductions to ensure they get what the want. New plant varieties, for instance, typically take five to ten years of development at field test sites, where researchers gather information about the characteristics of plants and seeds, destroying those that exhibit undesirable traits. (Not to say there hasn't been controversy over how regulators oversee field trials; see Henry L. Miller, "Where Are the Promised Wonders of Biotech?" *Consumer's Research*, March 1998.)

Speaking on behalf of the Council for Agricultural Science and Technology before Congress recently, Roger N. Beachy, president of the Donald Danforth Plant Science Center in St. Louis, summed up the risk of biotech foods: "There appears to be little or no concern for the safety of genetically modified foods compared to those that are not thus modified. It is my opinion as an active practicing scientist that each of the safety concerns voiced by critics in the United States, Europe, and other parts of the world is not based on valid scientific concerns."

Risk-based Regulation

Which brings us to how the Food and Drug Administration monitors new food varieties, including biotech foods. In keeping with the scientific consensus and its statutory authority to ensure the safety of the food supply, the FDA focuses "safety evaluations on the objective characteristics of the food," not the process by which the new food variety was created. Thus, the agency approaches new biotech food varieties—fruits, vegetables, grains, and grain by-products such as oils—just as it would new varieties produced through classical breeding methods.

To do otherwise, as demanded by some activists, would be a waste of time and effort. As the agency explained when it first elaborated its biotech food policy in 1992: "Because of the limited nature of most modifications likely to be introduced, the FDA would waste its resources and would not advance public health if it were routinely to conduct formal premarket reviews of all new plant varieties. We will require such reviews before marketing, however, when the nature of the intended change in the food raises a safety question that the FDA must resolve to protect public health."

As with all food products, legal responsibility is on the developers and sellers of the new food to assure the safety of what they sell to consumers; the FDA has the authority to enforce this duty and remove food from commerce if there is a "reasonable possibility" that a substance added by human intervention might be unsafe, the agency explains. The FDA determines if a new food warrants "extra" scrutiny based on familiarity with food and with any new traits produced within it, as well as the fundamental approach to foods that the degree of oversight is commensurate with the risk.

Therefore, the agency can require prior consultation and tests if the history or characteristics or intended use of a new substance in a food warrants additional review. Greater scrutiny is applied if characteristics of a new food include the presence of a substance that is completely new to the food supply (and therefore lacks a history of safe use), the presence of an allergen in an unusual or unexpected way (such as an allergenic nut protein transferred to corn or soy), increased levels of natural toxins (such as with potato hybrids), or changes in the nutritional profile of the foods (such as fatty acid or carbohydrate makeup).

In short, any significant health risk posed by biotech-derived or any other new food falls under FDA authority and scrutiny.

In some instances, changes in food profiles brought by genetic modification will require new labels. As James Maryanski of the FDA's Center for Food Safety and Applied Nutrition explains: "If a gene from a food that commonly causes allergic reactions, like fish or peanuts, is inserted into tomatoes or corn, where people would not expect to find allergens, then the vegetables would have to be labeled to alert sensitive consumers. If companies can demonstrate scientifically that the allergenic component was not transferred to the vegetable, no special label will be required."

But the FDA will not simply require foods to be labeled biotech, just because they are biotech. "The law says labeling for foods must disclose information that's material, as well as avoid false or misleading statements," notes Maryanski. "It's our view that the method by which a plant is developed by a plant breeder is not material information in the sense of the law. For example, we do not require sweet corn to be labeled 'hybrid sweet corn' because it was developed through cross hybridization. . . . If genetic engineering or any other technique changes the composition of a tomato in a way that it's really not the same tomato any more, then it would have to be called something different."

"Right to Know" Broadly Understood

Most popular reports about GM foods fail to provide these important details about how the FDA approaches food risks or labeling, or how GM foods stack up against traditional food risks—all necessary for concerned consumers to be fully informed about these matters. Opinion surveys suggest that most people have little awareness of the issue, let alone any knowledge of the way authorities approach new food risks, even if they express strong "beliefs." (One recent Gallup survey reported that only 10 percent of Americans have heard "a great deal" about biotech food safety, but 27 percent still believe biotech foods pose a serious health hazard.) Rather than inform about the FDA's approach to GM foods—a "right to know" taken as a general educational approach rather than on the label—many reports actually reinforce misconceptions, including the notion that biotech poses unique food risks.

Meanwhile, Gallup's and other surveys suggest that the more people know about biotechnology, the more comfortable they are with it. In one series by North Carolina State University researchers, as consumers learned about the current regulatory practices, they wanted only "useful" information on the label. Moreover, informed consumers mainly wanted to know about taste, nutrition, safety, convenience, and price, whether the food was "produced by biotechnology or any other means."

Put another way: Unless the information is truly useful to consumers, there's no sense requiring it on a label.

Uninformation and the Choice Paradox

Alan McHughen

One of the few unifying features of the genetic food fight in industrial nations is that a majority of consumers appear to support mandatory labeling for products of genetic modification (GM) technology. Proponents of mandatory and indiscriminate GM food labeling congregate under the banner of "informed choice." They argue that if DNA is introduced into foods using recombinant DNA technologies, people ought to be able to know about it. Unfortunately, the problem with mandatory GM labels is that they inform no one and they diminish consumer choice.

Let us examine how this paradox arises. Following the European model, a food label might proclaim: "This product contains genetically modified organisms (GMOs)," or even, less assuredly, "This product may contain ingredients from genetically modified organisms."

What we need to look at is the manner in which discriminating consumers might be informed by such statements. For instance, some people want labels because they

Reprinted with permission from *Nature Biotechnology*, by Alan McHughen, "Uninformation and the Choice Paradox," vol. 18 (1 October 2000): 1018–19. Copyright © 2000 Nature Publishing Group.

consider DNA itself a food contaminant. In a recent survey, only 40 percent of respondents in the United Kingdom recognized that non-GM tomatoes contain genes. The suggested label formulations do not say much about DNA, and they certainly do not unpick the basic biological misconceptions of this group.

What about vegetarians? They surely are entitled to ask whether a particular food contains an animal gene or protein. Leaving aside the scientific niceties of species interrelationships or the technicalities of gene homology across kingdoms, we can still conclude that the labels would not be useful to vegetarians.

Other label-perusers might include the technologically curious who seek out and selectively purchase the ultramodern GM materials. The labels might help them, but only if they are truthful, which—as we will see—may not be the case.

The most vociferous groups want labels to avoid the dreaded "Frankenfoods" entirely. The "anti-GM" group can be divided into many factions. Process-based opponents are averse to GM technology, period. These consumers often base their opposition on ethical or religious beliefs, saying, like HRH the Prince of Wales, that we humans have no business tinkering in God's domain, regardless of how useful the end result.

Another group, probably encompassing the majority of consumers, is concerned about specific products of the technology. They might accept GM insulin or nutritionally enhanced GM rice, but don't want pesticide-resistant soybeans. Indiscriminate GM labels will be of no use to this group, because there's no distinction between the "acceptable" rice and the "unacceptable" soybean. Without any explicit distinction on a label, there can be no "informed choice" to satisfy these discriminating shoppers. Similarly, some differentiating consumers might wish to avoid eating GM DNA or protein, but might accept a refined GM food product devoid of DNA or novel protein. Lecithin, for example, a common food additive made from soybeans, contains no DNA or protein. Lecithin from GM soybeans is identical to lecithin from non-GM beans. Some consumers might look to their food labels to steer them away from foods containing lecithin from GM soybeans. The mandatory indiscriminate labels proposed would not help them at all.

The basic difficulty here is that the proposals for GM labels cannot take into account the range of purposes for which people might use them. Contrast that with, for instance, labeling information that indicates the presence of ingredients derived from peanuts or other nuts. The reason for the label there is to enable people who may suffer nut allergies to avoid foods that could cause them severe and possibly fatal anaphylactic reactions. Both the label's audience and the reason for their concern are abundantly clear.

But let us continue to consider the issue of GM labeling. To what kinds of products could they apply? Let's argue (for the hell of it and in direct contradiction of all the scientific evidence) that we can find a sensible reason to put a "contains GM" label on cross-species products. We might then label tomatoes modified with a bean gene, or beans modified with a tomato gene. But would we label vegetable soup containing the two types of GMOs? Before processing, both the tomatoes and the beans are detectably GM products. In the soup, however, the genes and proteins of both tomato and beans are blended together. If we labeled the soup, we would be doing so based solely on the process or method of crop breeding (an event that may have occurred many years before) and not on the composition of the final bouillon. GM labeling in this case would be entirely misleading and would contradict safety priorities, which should be based on the composition of the product.

There are, therefore, lots of reasons why GM labels are uninformative, and some reasons why they are downright misleading. But where is the paradox? How could it be, with strict regulations and penalties, that an unlabeled food product could carry GM ingredients? And how, with mandatory and indiscriminate labeling of all GM foods, can we avoid GM foods by choosing products labeled "Contains GM ingredients"?

"GM-free" Might Not Be

The paradox rests on the fact that negative labeling can never be justified. Those consumers who wish to avoid GM food—for whatever reason—constitute a legitimate market for non-GM products. Negative labels such as "GM-free!" or "This product contains no GM ingredients" could help such people identify desired foods. However, the retailers or manufacturers printing such labels could never prove their claims. In theory as well as in practice, negatives can't be proven.

In many jurisdictions, the onus is on the "labeler" to prove the claim before the label is allowed. Even without the legal aspects, we need to recognize the difficulty in assigning such labels. Vegetable oils, for example, may contain no detectable DNA or protein—GM or otherwise. The claim, "contains no GM ingredients," tells the purchaser nothing: if no DNA or protein, GM or otherwise, is detectable, how does anyone know for sure it didn't come from GM plants? Without a reliable means of detection, there's no means of enforcement. Shady oil processors could quickly debase the system by printing "contains no GM ingredients" on bottles of refined GM maize oil, content in the knowledge that regulators have no way to detect the origin of the product.

That hasn't stopped several groups from demanding mandatory labels for such undetectable products. For the labels to serve any purpose, officials would need to find a way of enforcing such a regulation. In most democracies, even an accused corporation is innocent until proven guilty, and legal proof of guilt would require evidence. Prosecuting authorities would have to develop ways of detecting an undetectable GMO. Legal technicalities dating back to the Magna Carta will be likely to interfere with effective prosecution. In any case, as long as there are GM-containing products, occasional mixing errors, intentional or otherwise, will ensure that no product can legitimately claim to be devoid of GM material.

The authorities cannot seek refuge in negative labels. But maybe that does not matter. After all, if a food is not labeled as "containing GM ingredients," then surely it must be GM-free? Unfortunately, things are not that simple. Current GM labeling regulations stipulate that foods containing GMOs, or detectable DNA and protein from GMOs, must be labeled. Products without labels are not necessarily GM free.

In the real world—the one we all share—there are few absolutes. In addition to the various exemptions from the labeling requirements (in the United Kingdom, for example, certain glass jars and single-serve packets), regulations dictate tolerance limits for every commodity. Every food carries contaminants, usually innocuous things like microscopic amounts of dirt, insect parts, bacterial spores, or fungal toxins. But regulations allow food processors a tolerance. As long as the contaminants are less than the regulated limit, there's no need to declare them. Organic producers, for instance, rather generously are allowed to include up to 5 percent nonorganic "contaminants" without risking their organic label. The same principle applies to GM ingredients. The European Union (EU) has approved a tolerance of 1 percent for GM ingredients. Thus a food carrying just under 1 percent of GM ingredients does not need a "GM" label.

GM ingredients might slip into other foods under conventional tolerances for impurities. Say, for example, regulations allow 2 percent impurities in a grade of conventional vegetable oil. Suppose now we have a batch of conventional rapeseed oil approved at 98 percent pure. It would avoid triggering the mandatory GM label. Can the remaining 2 percent be GM rapeseed oil without requiring a label? If yes, doesn't that defeat the purpose of the indiscriminate GM labeling? If no, doesn't that defeat the purpose of the 2 percent impurity allowance for this grade of oil? What if we don't know exactly what the impurity is? Does the approved batch get a GM label on the chance that it might be GM material? As long as "non-GM" products sell at a premium, unscrupulous processors will be tempted to carefully blend oil from GM varieties to just below the threshold level in a "conventional," non-GM batch, and sell it as "non-GM" oil.

The real world also must deal with costs. The cost of labeling is far more than just the ink and sticker. The expense comes in administering the segregated and monitored growing, transport, processing, testing, shipping, and marketing all of a particular product, say, tomatoes, and not just the GM ones. Each step adds an additional expense burden borne ultimately by the consumer. The administrative expense will be that much greater for multiple-ingredient processed foods. "GM-free" will be especially expensive, in that the burden to ensure purity within the tolerance limit at every step from growing, transport, processing, shipping, and marketing is particularly onerous. The premiums commanded by organic commodities may give a good guide to the magnitude of the cost.

Thus consumers wishing to avoid GM entirely will pay more for the privilege. But they're being misled and misinformed: they are still consuming at least small amounts of GM ingredients. Neither explicit nor implicit claims ("GM-free" labels, or the absence of "GM" labels) provide these people with the information they require.

Food Labeled "Contains GM" Might Not

We have already seen that the information value of GM labels is threatened by unscrupulous producers. Oddly, it is threatened as much if not more by impeccably scrupulous manufacturers or retailers, too. Regulations come with penalties that apply when they are broken. In the United Kingdom, for instance, food producers and retailers who fail to label GM foods face fines of up to £5,000 (U.S. $7,500) or incarceration. The scale of these official rebukes is, of course, next to nothing compared to the retribution that the market could exact on wrongdoers exposed in the media. In contrast, there is no penalty for putting labels saying "may contain GM" on non-GM foods. Many foods, especially processed or multi-ingredient foods, cannot readily be guaranteed GM-free. Food processors and vendors are not risk takers. If they are uncertain, they'll label the product to avoid risking prosecution or public reprisals, even at the risk of losing some sales. Thus, "may contain GM" will become the default status. Who is going to check that the food actually contains GM ingredients? No one. If there's no penalty for this kind of false labeling, there will be no enforcement.

Under EU regulations, in order for a tomato sauce to avoid the GM label, it must have documentation—a paper trail—to show it contains no GM products. This additional burden affects everybody, but whom does the paper trail hit the hardest? Certainly not the big corporations making branded products; they are used to the intricacies of regulatory compliance. Possibly not the large upmarket retailers either, because their customers have proved their willingness to pay inflated prices. It will, however, affect the businesses of "generic" low-cost manufacturers. It will also hit small producers supplying produce to local markets.

Where mandatory indiscriminant GM labeling is required, low-margin businesses will institute labels proclaiming "may contain GM." In all probability—though without certainty—the produce that arrives at local markets does not contain GM ingredients. Yet, as an insurance policy, it is among the most likely to be labeled. And consumers still will have no idea whether the product contains GMOs.

Mandatory GM Labels Satisfy No One

Mandatory GM labeling regulations satisfy no one. Those wishing to avoid all GM products, including "derived-from" foods, are annoyed at the exemptions for products lacking DNA or protein. They are also cheated by the tolerances allowing some GM content without triggering the required GM label. Those wishing to avoid only certain GM products (such as vegetarians) must eschew all such products, because the mandatory label does not specify the source of the novel DNA. The trendsetters actively seeking out GM foods might not, in spite of the GM label, be eating any. Ordinary consumers who don't have strong feelings one way or another are paying higher prices for the labeling bureaucracy, whether they buy foods labeled GM or not. And the poor, who must buy at the bottom of the market regardless of their personal opinions, pay a disproportionately higher share of the increased costs to the benefit of no one, especially themselves. No matter what your position, GM labels fail to provide their intended raison d'etre— informed choice.

Part 5

Law

Introduction

Conventional techniques for making high-yielding hybrid crops cost far more than the use of genetic engineering. Lest you think that agricultural biotechnology goes for fire-sale prices (or for free), stop to think about the difference between a product's development cost—say, a herbicide-resistant soybean—and its ability to make a return on investment. Annual soybean plantings number in the millions of hectares, so aside from any ethical argument that would defend innovators' property interest in their new soybean, there is a straightforward financial claim at stake.

Patents, which are supposed to protect the interests of innovators, are frequently criticized by opponents of GM food who argue that they are now more prevalent to secure greedy corporate control of agriculture. According to Jack Wilson, many patent systems have been around for a long time, so patenting agricultural biotech is an evolution of law in agriculture, not a revolutionary new way of doing things. Wilson refers to specific technologies as he explains how patents do their job, but notes that the rigorous pursuit and enforcement of patents can have unacceptable ethical consequences if they are enforced unscrupulously, skew research, or cause harmful inequities in biotech capacity between developed and developing countries.

Richard Gold discusses the most contentious of all aspects of patents—the ownership and control of parts of the food system by large corporations. He points out that while patent law and the technology it protects are often quite complex, the social benefits of patents (the protection of property and the promotion of innovation) are not complicated. The problem is that public interests are infrequently served by actual patent use, particularly when it comes to agricultural biotechnology. Past practices of structuring business plans around patent law should be replaced, Gold argues, by a branding model in which social and corporate interests are both met.

Wilson's and Gold's articles amply illustrate the depth of the debate on patent law in agriculture. The last reading in this section points to a different problem in law and agriculture that has been underappreciated. The 1992 Convention on Biodiversity and its 2000 Cartagena Protocol are the two treaties that protect biodiversity, particularly with respect to biotechnology's capacity to change ecosystems. Treaties are binding—like laws are—but they are enforced through economic and political suasion. Keith Culver examines international law's capacity to enforce treaties on biodiversity. He uses a hypothetical test case—damage caused to one nation by another's escaped transgenic salmon—to determine if international law has the capacity to redress harms caused by the release of transgenic organisms. Culver's view is that international law has not kept apace with the potential for environmental damage caused by genetically modified organisms—a warning about biotechnology running ahead of international law, and perhaps leaving undesirable moral remainders even if some kind of corrective justice is achieved.

18

Intellectual Property Rights in Genetically Modified Agriculture

The Shock of the Not-So-New

Jack Wilson

My grandfather once grew a crop of popcorn planted from a package of Orville Redenbacher corn purchased at the grocery store. His unusual appropriation of Redenbacher's snack exemplifies the kernel of the debate about intellectual property in agricultural organisms. Had he experimented with similar products today, he could violate federal laws governing intellectual property. Intellectual property rights in agricultural organisms come with special problems ranging from the fact that for many plants the food product contains the seed, which can be replanted and replicate the protected plant, to unprecedented restrictions on the use of seeds and plants by farmers. Though some form of intellectual property protection for agricultural plants has been available since 1930, the more controversial patent protection for plants came about at approximately the

Department of Philosophy, Washington and Lee University

same time as the first recombinant DNA plants were produced, and the two often face the court of public opinion together.

Food biotechnology has become a locus of financial, legal, ethical, and aesthetic controversy. It is not clear to what extent intellectual property plays into this broad area of dispute. Specific concerns about intellectual property rights in agricultural biotechnology are difficult to isolate from more general misgivings about agricultural biotechnology and industrial agriculture. The public tends to hybridize risk. People do not compartmentalize categories of risks and benefits of agricultural biotechnology; they consider them mixed together and then decide whether or not they find the resulting chimera acceptable.[1] Consider as an example of this phenomenon a quotation from Michael Pollan describing the experience of planting a row of potatoes genetically modified to contain genes from the bacteria *Bacillus thuringiensis* (Bt).

> By "opening and using this product," the card informed me, I was now "licensed" to grow these potatoes, but only for a single generation; the crop I would water and tend and harvest was mine, yet not also mine. That is, the potatoes I would dig come September would be mine to eat or sell, but their genes would remain the intellectual property of Monsanto, protected under several U.S. patents, including 5,196,525; 5, 164, 316; 5,322,938; and 5, 352,605. Were I to save even one of these spuds to plant the next year—something I have routinely done with my potatoes in the past—I would be breaking federal law.[2]

Pollan goes on to discuss his misgivings about the safety and aesthetics of eating the modified potatoes. He ultimately decides to not eat the potatoes he grows, but for reasons unrelated to their odd status as living patented inventions. Still, the alien plant is linked to an unfamiliar and intrusive legal regime. For the purposes of this paper, though, I will address only those concerns about genetically modified agriculture that deal specifically with intellectual property.

Intellectual Property Basics for Plants

Intellectual property rights grant inventors, discoverers, and creators of various kinds control over the use of their ideas, inventions, and discoveries. Narrowly conceived, there are only two ethical questions about intellectual property rights in agricultural organisms—who is entitled to intellectual property protection, and for what kinds of innovations and discoveries? These concerns are common to all systems of intellectual property. Almost everyone agrees that inventors deserve to have some degree of control over their inventions. Determining how much and what kind of control the inventor should have and determining what kinds of things should count as inventions are more problematic questions.

Similar patent systems have been built up from completely different philosophical foundations. The U.S. patent system, for example, is justified in the Constitution on strictly instrumental grounds. Patents are granted to encourage scientific and technological progress that would be impeded if inventors could not be given a chance to profit from their inventions. Some European patent systems, however, are partially justified through a basic right to property, intellectual or tangible, rather than through instrumental goods. It seems likely that both systems have been modified to include aspects of the other and in practical application balance these two desirable outcomes. In fact, there is often broad overlap in practice even where philosophical underpinnings are different. Patent policies are clearly intended both to recognize the special relation between the inventor and the idea that he or she has developed, and also to regulate the release and use of that information to allow others to benefit from the invention as well.

The general problems with enforcement and justification of intellectual property law are exacerbated when the invention is itself a living thing capable of self-replication.

There are at least six different ways to protect intellectual property rights in plants—plant patents, Plant Variety Protection Certificates, trade secrets, utility patents, technology use agreements, and genetic use restriction technologies. The first five are legal structures, and the sixth (genetic use restriction technology) is a biological mechanism that prevents unauthorized copying of the protected plant. Intellectual property protection for agriculturally useful organisms has a historical trajectory—it is not merely a legal innovation to deal with the special products of genetic engineering. Patents and patentlike forms of intellectual property protection for organisms have a long history in the United States. Revisiting this history enables us to anticipate the likely effects of intellectual property protection for genetically modified plants used in agriculture.

It surprises many people to discover that intellectual property protection for plants has a long history. Since the start of the twentieth century, plant breeders and nurserymen campaigned Congress for legal regulation of what they saw as the unfair appropriation of their work. Stark Brothers Nursery, for example, tried to protect the original Red Delicious Apple tree with a fence and sent buyers contracts forbidding unauthorized reproduction along with their saplings. But no legal protection for plant inventions or discoveries existed at that time.

Limited patent protection for some plants became available in the United States in 1930 when Congress passed the Townsend-Purnell Plant Patent Act and granted the first patent rights for agriculture in the world. The act allowed novel varieties of plants to be patented—the patent holder was protected from unlicensed asexual reproduction of the patented plant. A plant patent protects only against unauthorized cloning of a plant. In this sense, there is little "intellectual" property being protected. A plant patent does not allow the patent holder to prevent others from creating a plant with similar characteristics, as would a conventional patent. It only protects the actual plant and its clones, making it similar to the protection of tangible property. A plant patent protects a single plant and its asexually produced clones for twenty years from the date of application. Anyone who violates a plant patent can be made to pay compensation and be prohibited from further asexual reproduction of the plant. Plant patents were of limited utility because they did not protect most agricultural plants and were difficult to enforce effectively. Though the first intellectual property protection for plants was established in the United States, Europe took the next step.

In 1961 the International Convention for the Protection of New Varieties of Plants established the International Union for the Protection of New Varieties of Plants (UPOV) to provide a uniform set of standards for defining and classifying plant varieties and to reward their creators with intellectual property rights in new varieties that would be honored by all member nations.[3] The UPOV has approximately fifty member states, each of which commit to writing their plant breeding laws in accordance with this multinational accord. The UPOV gives the plant breeder the right to exclude others from commercially producing or selling a protected variety or *repeatedly* using the protected variety for the commercial propagation of another variety. Member states must offer these rights though they can also offer more expansive rights (as the United States now does through utility patents on novel plant varieties). The breeder's rights extend to "the protected variety, varieties *which are not clearly distinguishable from the protected variety*, varieties which are essentially derived from the protected variety, and varieties whose production requires the repeated use of the protected variety [italics added]."[4] There are three compulsory exemptions to the protection. Private growing with non-commercial purposes does not infringe, nor do acts done for experimental purposes or for the breeding or exploitation of other varieties. Further, there is an optional exemption that allows member countries to decide for themselves whether farm-saved seed violates breeder's rights. Minimum protection is set at twenty-five years for grapevines and trees and twenty years for other plants.

In 1970, to square U.S. protection for plant varieties with that offered by the UPOV, Congress passed the Plant Variety Protection Act (PVPA), which provides intellectual property protection for new, distinct, uniform, and stable varieties of plants that reproduce sexually.[5] The act allows a plant breeder holding a Plant Variety Protection Certificate to exclude others from exploiting the protected variety without the owner's permission. The PVPA includes research and farm exemptions absent from the PPA.[6]

The PVPA did not mark the fullest extent of protection. Plant inventions can now be protected with utility patents. In 1972, Ananda Chakrabarty applied for a U.S. patent on a bacteria he had developed that could digest the components of crude oil. When Chakrabarty applied for a patent, he claimed "a bacterium from the genus *Pseudomonas* containing therein at least two stable energy-generating plasmids, each of said plasmids providing a separate hydrocarbon degradative pathway" as well as a process using the bacteria. The patent examiner accepted his claim for a *process* that included the use of straw inoculated with the new bacteria to contain oil spills. Patents for *processes* involving microorganisms were well supported by precedent—several septic system and beer-making patents had claimed processes involving bacteria as subject matter. The examiner, however, rejected Chakrabarty's claim for a patent on the bacteria itself. He maintained that the microorganisms were "products of nature" and that the patent's claims were drawn to "live organisms." After a number of appeals, the case appeared before the Supreme Court.

In a 1980 five-to-four decision, the U.S. Supreme Court ruled that living organisms could be patented under the utility patent statutes as "compositions of matter" if they met the other requirements for patentability. The majority argued that the patent statute should be interpreted broadly so that manufacture and composition of matter include "anything under the sun that is made by man," including living organisms. Although the Court's decision was directed to Chakrabarty's patent for a variety of bacteria, it's expansive language was quickly interpreted as extending patent protection to multicellular plants and animals.

In 1985 the Patent and Trademark Office Board of Patent Appeals and Interferences ruled that nonnaturally occurring man-made multicellular plants are patentable under the conventional utility patent statutes as well as under the Plant Patent Act and the PVPA.[7] This decision marks the first time that multicellular plants were subject to general intellectual property protection rather than an alternate form of protection tailored to the peculiarities of plant breeding and the special considerations of protection for agriculturally useful organisms. Utility patents protect the general characteristics of the protected variety from copying for twenty years. These patents, unlike a PVP certificate, do not include exemptions for farmers to legally save and reuse the seed from a crop they have grown from patented seed, nor do they exempt researchers from licensing requirements. Although intellectual property protection for plants is in no way new, the extension of the utility patent statutes, originally designed to accommodate mechanical inventions, to new plant varieties has led to an unprecedented broad range of property rights for plant breeders.

A trade secret can protect some kinds of proprietary information, though their use in agriculture is circumscribed by the fact that most plant-based technologies deliver the relevant information along with the product sold. A trade secret can be maintained only as long as the secret can be held. Trade secrecy laws protect an inventor against persons who misappropriate a trade secret. Trade secrecy does not protect against reverse engineering or independent invention. Grain is not merely food or seed for one generation; it is also a means of reproducing the plant. The patent strains used to create a variety of hybrid corn could be protected through trade secrecy.[8] Another method to maintain trade secrecy in plant biotechnology is to release to consumers only products that do not contain whole fertile seeds, for example, Hopi blue corn only as cornmeal, GM canola as pressed oil. Trade secrecy can be used to protect both genetically modified and unmodified agricultural plants.

There are two other relatively new methods of protecting intellectual property in agricultural plants. They represent two radically different means of pre-venting the unauthorized use of an agricultural product, though they serve the same purpose—to limit the use of a plant to a single generation. The first is a *technology use agreement* (TUA). Technology use agreements are a form of license to use a patented invention. For agricultural plants, a TUA is "a contract specifying that farmers do not own the seed, that they just grow it under a contract and have to deliver all the grain in accordance with [the seed company's] wishes."[9] The farmer signing a technology agreement with Monsanto, for example, agrees to use the seed containing Monsanto gene technologies for planting a commercial crop in only one season; to not supply any of the protected seed to anyone else for planting, or to save any crop produced from the seed for replanting; to not use it or provide it to anyone for crop-breeding or research; and to spray only with Roundup if using a gylphosate-based herbicide.[10] The last caveat even allows Monsanto to continue to protect the market for Roundup after its patent on the product has expired. Current technological use agreements may well mark the high point of control of the farmer.

The second new method of protecting intellectual property allows plants to be altered to protect intellectual property investments. Recombinant DNA methods have been proposed to alter plants so as to protect an intellectual invest-ment through biology.

> Plant Genetic Use Restriction Technologies (GURTs) are a group of complex genetic transformations that insert a genetic "on–off switch" in plants to prevent the unauthorized use of genetic traits contained within, that is, GURTs ensure that farmers cannot replant genetically modified crops by simply saving seed.[11]

More than thirty patents for GURTs have been granted in the United States and Europe, but no seeds have yet been commercially released containing the tech-nology. GURTs come in two varieties, the variety-level GURT (V-GURT), which results in plant sterility, and the trait-level GURT (T-GURT), which limits the expression of a genetic enhancement but does not cause seed sterility. In 1998 the U.S. Department of Agriculture and Delta & Pine Land company (a cotton breeder) received a patent (U.S. Patent 5,723, 765—Control of Plant Gene Expression) for a V-GURT. Two months later, Monsanto bought Delta & Pine Land Co. and acquired the patent. Quickly dubbed "the Terminator" technology by its detractors, the GURT was subject to wide popular protest.[12] As a result, Monsanto agreed to at least temporarily suspend work on such technologies. The idea has been at least temporarily shelved more for public relation reasons than for any scientific or ethical problem with the technology, particularly when GURTs are compared with the legal control of farmers.

This technology presents some interesting advantages that seem to have been lost in the public controversy. It would stop the unwanted spread of GM crops either as volunteer plants or as a potential source of genes for flow into weedy relative species. And, it offers a way to protect intellectual property without impinging too far on the civil liberties of growers. It allows for a means of controlling intellectual property through the characteristics of the plant even when it is grown outside of the legal control of the patent holder.[13] GURTs provide an effective way to enforce patents or protect plants from unauthorized reproduction even where patent rights are unavailable or unenforceable. GURTs do not affect the civil liberties of farmers as technology use agreements do.

Intellectual property protection involves the manipulation of either the plant or the farmer, and the less one manipulates the first, the more one must manipulate the second, and vice versa. If a plant cannot be controlled—if the plant has a valuable trait that is reproduced in fertile seed, then farmers and consumers must be controlled if value is to be extracted from intellectual property rights in that plant. Plants that breed true with valuable traits require more legal protection and human vigilance if the technology is to be protected.

Viewed from the prospect of a new century, the changes in biology and the intellectual property system that has grown to embrace them, in the twentieth century are amazing. At the start of that century, farmers grew and exchanged most of their own seeds, aided by free seed developed by government plant breeders and free foreign seed distributed by the U.S. Patent Office. The close of that century saw the development of sophisticated new techniques of plant breeding and genetics, and the increasing domination of the seed industry by just a few multinational corporations protected by elaborate intellectual property laws.[14]

Why Is Agriculture Particularly Vexing Even among Patents on Living Organisms?

Genetically modified organisms—bacteria, plants, and animals—are all currently subject to intellectual property protection in fields outside of agriculture. Medically useful compounds such as insulin are produced by genetically modified bacteria and expressed in the milk of "pharm" animals; genetically modified mice are commonplace in biomedical research. What distinguishes these uses from agricultural biotechnology is the degree to which the patented organism can be controlled. Experiments take place in laboratories; test greenhouses are protected from outside vandalism, industrial espionage, and escape of the organisms not cleared for environmental release; fermentation vats are secured within industrial compounds. Agriculture is different. Millions of modified organisms are released into the relatively insecure environment of farmers' fields. Modified grain is transported with the carelessness common to low-value bulk commodities—it blows out of trucks and into fields. The seed is sold directly to consumers and farmers who are as able to illegally reproduce a patented invention as my grandfather was able to plant popcorn.

In general, if plant breeders are to make a profit, either the biological characteristics of agricultural plants or the social system of control that surrounds them must prevent the unauthorized copying of the plants. The advent of genetically modified agriculture only raises the stakes in this balance between biology and social practice. Technical means can alter crops so that they better fit within an intellectual property regime. Social and legal means can alter the actions of people. The easier a plant is to control through its method of reproduction, the more a breeder can count on the biology of the plant to protect the investment. It is not a coincidence that most of the plants that were profitable to market before intellectual property protection for plants were those plants whose biology—either through slow growth or means of reproduction—did not allow for an unlimited supply to be quickly grown and disseminated.[15]

Three Potential Problems for Intellectual Property in Genetically Modified Agriculture

Intellectual property rights in agriculture are not new, which means that we can learn from the past. Recombinant DNA technology changes the landscape in only a few ways. Patent rights for genetically modified organisms require considerable effort to enforce against unauthorized use and reproduction, but policing the spread of plants has never been easy. Research priorities can be altered to protect investments rather than to produce the best product for farmers and consumers. Also, the use of genetic resources from the developing world can expand as genes have agricultural value even if they cannot be crossbred with crop plants. We must be careful not to exploit this relationship. All of these problems arose before the introduction of genetically modified agriculture, but there is good reason to think that the new technology and the relatively recent introduction of utility patent protection will cause another round of trouble.

Enforcement of Patent Rights

Self-replicating inventions are difficult to control, but if patent rights are to be enforced, the plants must be controlled somehow. If plants do not contain GURTs, and the public reaction to the "Terminator gene" suggests that the public is not ready for this technology, then legal means will have to fill the void left by the plants themselves. For example, Percy Schmeiser, a Bruno, Saskatchewan, canola farmer, was recently fined for growing Monsanto's patented genetically modified canola without a license.[16] In March 2001, Schmeiser was found guilty of infringing on Monsanto's patent by "planting in 1998, without leave or license by the plaintiffs, canola fields with seed saved from the 1997 crop which seed was known, or ought to have been known, by the defendants to be Roundup tolerant and when tested was found to contain the gene and cells claimed under the plaintiff's patent."[17] Schmeiser had been a canola farmer since the 1950s and had been saving and reusing his own seed for years. He had not purchased canola seed since

1993. In 1997, Schmeiser had noticed some volunteer canola plants that seemed to be resistant to Roundup. As an experiment, he sprayed approximately three acres of one of his fields with the herbicide, and approximately 60 percent of the plants survived the application of the herbicide. One of his farmhands later swathed and combined that part of the field and used it as a part of the seed for the 1998 crop. Acting on an anonymous tip, agents working for Monsanto discovered that Schmeiser was growing the patented canola and successfully sued him.

Schmeiser's case was an unusual one. Schmeiser was unusual in his farming practices in that he saved his seeds and replanted them rather than buying new seed each year. The initial source of the Roundup-ready canola in his fields was never resolved, but Schmeiser did what would most concentrate the number of surviving plants (spray with Roundup and then use that seed). Yet another strange feature of this case is that Schmeiser did not use Roundup as an herbicide for his 1998 crop, which was a mix of both modified canola resistant to Roundup and conventional canola that would have been killed by the application. After Roundup-ready canola of unknown origin contaminated an earlier crop, Schmeiser did the one thing that would ensure the selection of that crop (spray with Roundup), then used that seed to plant his next year's crop (as opposed to seed from one of his other fields), and then took no advantage of the benefits offered by the crop. Schmeiser was ordered to discontinue using saved-seeds that might have been contaminated with the patented technology. This case raises some serious concerns about the means necessary to protect self-replicating inventions, but the extent of this kind of problem remains to be seen.

Industry-informed Research Biases

Some people say information wants to be free.[18] Though there is an element of truth to this aphorism, at a deeper level, the situation is not so straightforwardly tilted toward free dispersion. Some information is created because it can be sold rather than given away. It appears as this dynamic affects agricultural innovation. Private resources tend to concentrate where the resulting product can be protected; private companies are committed to making profits and invest their research capital accordingly. Where intellectual property rights are important as they are in agricultural biotechnology, they will play a role in these decisions. Developments that would better serve farmers or consumers might not be explored because they cannot be adequately controlled or licenses cannot be obtained for the relevant technology. This concern must be weighed against the idea that incentives for invention lead to invention. Some agricultural researchers now have difficulty getting access to patent-protected materials and techniques for plant breeding. There is also evidence that the availability of plant variety protection has led to the frivolous proliferation of plant varieties.[19]

Taking Advantage of the Developing World

Almost uniquely among valuable things, plant genetic resources have been considered as common human heritage and been relatively freely exchanged. Whatever the merits of this system in the past, it is open to some troubling abuses once some of those resources are privatized through intellectual property law. Private agricultural companies in the developed world have benefited from the genetic diversity of the developing world. Many valuable crop plants originated in areas of the developing world and have long been a source of genetic diversity for disease resistance and other commercially valuable properties. Where these were once introduced through conventional crossbreeding techniques, individual genes from sexually incompatible species can now be used. This vastly increases the potential range of plants and other kinds of organisms that can be considered valuable for agricultural breeding programs. The loss of biodiversity worldwide combined with the increased number of agricultural uses for that biodiversity have combined to create a situation in which the supply is shrinking and demand is growing. Most developed countries are relatively gene-poor and many of the most gene-rich areas are found in the tropical developing world.[20]

In addition to the normal problems associated with interactions between private companies in the developed world and the developing world, new problems can arise with the advent of intellectual protection for GM plants.[21] Richard Lewontin has noticed perhaps the most important one. "Much of the agricultural economy of these countries depends on growing specialty commodities like lauric acid oils, used in soaps and detergents, once found only in tropical species. Now with recombinant DNA, these are produced by canola.[22] Countries that do not seek patents on their genetic resources or prevent others from doing so may find themselves at a disadvantage in a global agricultural market. Those that do seek patents find themselves pulled into the kinds of intellectual property systems set up in the developed world. Members of the World Trade Organization, for example, are required to protect agricultural innovation at least up to the UPOV standards.

Conclusions

Concerns about intellectual property protection for genetically modified organisms used in agriculture pull together many issues that seem separable in theory. But, if one must judge a hybrid entity—an organism that depends for existence not only on the sun and soil, but laboratory scientists, and a financial and legal regime to call it into being—it may be foolish to divide what is seamlessly conjoined. The creation and marketing of GM agricultural organisms and the resulting food products have developed as they have not merely because of the properties of those organisms but also because intellectual property rights allow companies to make profits on them. Although every method of protecting intellectual property in GM agriculture I describe has also been applied to non-GM organisms, it is their application to the former which seems to have drawn the most public attention. I have tried to separate out those issues specific to intellectual property in plants rather than deal with the larger concerns about GM foods. But, uneasiness about GM foods easily extends to every part of the system that delivers it to our tables. The future's judgment of the current state of intellectual property law regarding agricultural plants will hinge on its satisfaction with the total food-production system those legal structures have helped to build.

Notes

1. Paul Thompson, *Food Biotechnology in Ethical Perspective* (London: Blackie Academic and Professional, 1997), 232–33.

2. Michael Pollan, *The Botany of Desire* (New York: Random House, 2001), 190.

3. The perceptive reader may wonder how the name of the organization relates to its acronym. The acronym is derived from the French name of the organization, *Union internationale pour la protection des obtentions végétales.*

4. http://www.upov.int/eng/about/protect.htm.

5. This act was passed in response to the 1961 International Union for Protection of New Varieties of Plants. For exceptions to the PVPA (carrots, celery, cucumbers, okra, peppers, tomatoes—all repealed in 1980), see Lawrence Busch et al., *Plants, Power, and Profit: Social, Economic, and Ethical Consequences of the New Biotechnologies* (Cambridge, Mass.: B. Blackwell, 1991), 27. The initial legislation had prompted little public comment, but the revision did, perhaps because issues surrounding the patentability of living organisms had become more prominent as Chakrabarty's case made its way to the Supreme Court.

6. For more information on this history of intellectual property protection for plants in the United States, see Glenn E. Bugos and Daniel J. Kevles, "Plants as Intellectual Property: American Practice, Law, and Policy in World Context," *Orisis* 7 (1992): 75–104, or Cary Fowler, "The Plant Patent Act of 1930: A Sociological History of its Creation," *Journal of the Patent and Trademark Office Society* 82 (2000): 621–44, and Jack R. Kloppenburg, *First the Seed: The Political Economy of Plant Biotechnology, 1492–2000* (Cambridge: Cambridge University Press, 1998).

7. The first utility patent on a plant was for a "Tryptophan Overproduced Mutant of Cereal Crops," U.S. Patent 4,581,847, a maize plant produced through artificial selection.

8. Example is from Donald D. Evenson, "Patents and Other Private Legal Rights for Biotechnology Inventions (Intellectual Property Rights—IPR)," in *Agriculture and Intellectual Property Rights: Economic, Institutional, and Implementation Issues in Biotechnology,* Vittorio Santaniello et al. eds (Wallingford, U.K.: CABI Publishing, 2000).

9. Alan McHughen, *Pandora's Picnic Basket: The Potential and Hazards of Genetically Modified Food* (Oxford: Oxford University Press, 2000), 249.

10. Summarized from a Monsanto Technology Agreement.

11. Alejandro E. Segarra and Jean Rawson, "The 'Terminator Gene' and Other Generic Use Restriction Technologies (GURTs) in Crops" [online], http://www.cnie.org/nles/ag-83.html, 1.

12. The Rural Advancement Foundation International nicknamed the technique the

"Terminator Technology." See RAFI, "Terminator Seeds Rejected by Global Network of Agriculture Experts, news release, 2 November 1998 [online], http://www.rafi.org/pr/release23.html.

13. U.S. farmers have consistently complained that they bear more than their fair share of the research and development of new agricultural varieties because the United States has a relatively strong system of intellectual property rights and enforcement when compared with many other countries. See, for example, Michael Holmberg, "Cheaper Roundup-Ready Seed Gives Argentina an Advantage," *Successful Farming* (2000): 9–10.

14. This story is told in great detail in Kopplenburg's *First the Seed: The Political Economy of Plant Biotechnology, 1492–2000.*

15. The tulip is a good example. Tulips are bulb-propagated. Seeds from a tulip do not produce plants identical to the parent, only bulbs and their offsets do, and they are slow to reproduce. "Broken" tulips, the most valuable ones, produce smaller, weaker, and fewer offsets. Thus the biological groundwork for the tulip bubble in Holland.

16. The patent violated was Canadian patent number 1,313,830, issued in 1993.

17. Judge Andrew McKay, "Reasons for Judgment," *Monsanto Canada Inc. and Monsanto Company v Percy Schmeiser and Schmeiser Enterprises Ltd,* Federal Court of Canada 256, 2.

18. This aphorism has been attributed to Stewart Brand in a 1984 speech, but has spread widely.

19. See Jean-Pierre Berlan and Richard Lewontim, "Breeders' Rights and Patenting Life-Forms," *Nature* 322 (1986): 785–88.

20. Jack R. Kloppenburg and Daniel L. Kleinman, "Seeds of Struggle: The Geopolitics of Plant Genetic Resources," *Technology Review* 90 (1987): 48–56.

21. The normal problems include determining a fair value for raw germplasm, a rough interface between countries with intellectual property laws and those without them, and the role of indigenous people who may lack adequate government representation.

22. Richard Lewontim, "The Genetic Food Fight," *New York Review of Books* 48 (2001) 81–84.

19

Merging Business and Ethics

New Models for Using Biotechnological Intellectual Property

E. Richard Gold

Introduction

Too often, we focus on the inner workings and logic of patent law and forget to examine how industry actually uses patents. We thus debate how much an inventor must know about the function of a biotech invention before we grant a patent,[1] whether patents ought to be withheld over immoral inventions,[2] or whether a gene patent provides its holder with the ability to prevent others from growing plants containing that gene.[3] On the other hand, we have very little knowledge about such general questions as whether and under which circumstances biotech patents actually increase innovative activity[4] and more particular questions relating to how industry uses the biotech patents it has.

The reason for this myopia is that we too-easily relegate discussion about biotech patents to the "experts" since patent law appears to be a highly technical area of law dealing with a highly technical field of scientific knowledge. While patent law is technical, as is biotechnology itself, its implications are not. In fact, patent law is, at its base, supposed to be an instrument for achieving the social good. Defining this social good is not, therefore, something to leave to the experts; it is something over which the public ought to take control.

This chapter represents a first step in enlarging our examination of patents by placing them in the context in which we actually use them. Whatever the experts may claim patents to be about, the use of patents is determined by those who hold them. Simply assuming, as we usually do, that patent holders will use their patents rationally tells us little about actual practice. Not only are there many different ways of being rational, but we are also often irrational. Consider, for example, the position of several biotechnology industry representatives at a consultation meeting in 2000. These individuals argued that it was difficult to attract research and development money to Canada because international industry viewed that country's patent laws as being anti-biotech.[5] While at first glance this may seem an economically rational response, it is not. This is because patent law has nothing to do with where research and development is done. Through international conventions, an inventor is entitled to get a patent in all major markets almost no matter where the research is conducted. This means that the decision about where to conduct research ought to be evaluated on its own merits—for example, based on scientific expertise, access to research subjects, and so forth—rather than on how strong that country's patent system happens to be.

Once we move away from assumptions about the technical nature of patents and the rationality of those who use them, we can begin to examine industry practice with respect to biotech patents. I begin this analysis by describing two principal business models that industry uses to exploit biotech patents. While, as I argue, industry has used each of these models at times, neither model has been described as such in the literature. I hope to remedy this oversight.

Each of the two models draws on its own set of internal assumptions and each has significantly different outcomes for both industry and society. As I will argue, the more conventional model not only leads to greater ethical concern but, in consequence, often to greater government regulation and thus greater anxiety for industry. While the second model avoids (or at least lessens) these difficulties, it has not been fully adopted because it requires a change in attitude by industry.

Fortress Model

Explanation

While the origins of the modern patent system reach back hundreds of years, interest in intellectual property—the legal regimes we use to allocate rights over knowledge—as a business asset is a fairly recent phenomenon. Prior to the early

1990s, business looked upon intellectual property as a formality handled by lawyers. Patents were not something that affected the bottom line; they were simply devices that assisted the business enterprise in carrying out its function.

Once it became evident, however, that intellectual capital (inventions, written work, know-how and so on) was more valuable to business than physical capital in the so-called new economy, business leaders suddenly woke up to the need to develop business models around this newly discovered asset. Intellectual property, they realized, need not be relegated to a supporting role; intellectual property could be exploited directly. With this understanding that patents affected their business's bottom line, businesspeople started reading *Wired* magazine and talking of distributed networks and disintermediation. Business journals, such as the *Harvard Business Review*, began featuring articles on developing and capturing the value of intellectual property. In short, the business world woke up to the potential of intellectual capital.

Up until this awakened interest in intellectual capital, there was an absence of overt business modeling surrounding intellectual property. Even large producers of intellectual property, for example IBM, frequently collected intellectual property without knowing what to do with it. Often, unless the enterprise could use the intellectual property within its own organization, it left the intellectual property on the shelf, wasting away.

With interest in intellectual capital mounting in the 1990s, business started to view its intellectual property as an asset from which it could extract profits. The first thing the progressive business leader would do, therefore, was to quantify the intellectual property that existed within the organization. Thus, businesses created internal inventories of intellectual property. This consisted of knowing not only which patents the enterprise had obtained, but asking those employed in the organization, particularly in the research departments, whether they had developed anything that could be subject to a patent, trade secret protection, or copyright. After this first step, the enterprise had two choices. It could either use the intellectual capital itself or it could sell it to others. In the first case, the intellectual capital was viewed as a way of doing things more efficiently, thus reducing costs and making the enterprise more profitable in comparison with its competitors. In the second case, in which the intellectual capital had little utility within the organization, the enterprise would license or sell it to those who could use it to reduce their costs of production.

An even more lucrative use of intellectual property was to use it to catch the unwary. Because intellectual property permits the holder of that property to prevent all others from using, making, selling, or importing an invention, the holder could use its patents to threaten anyone who made use of the invention with legal action. The beauty of this tactic was that the suit would be successful whether or not the person using the invention knew that it was subject to patent protection. Through this technique, intellectual property holders could ambush those who had used the technology. This ability to collect on the mistakes of others soon turned certain internal law departments into money-making centers. One of the best known companies to use this approach was Texas Instruments.[6]

The common element in this use of intellectual property is to extract value directly from it so as to contribute to the enterprise's bottom line. The enterprise does so by quickly establishing control over its intellectual capital by seeking patent protection and then aggressively attempting to extract revenue from those patents. That is, the organization constructs high and strong walls around its inventions and then charges all those who wish to access those inventions. Thus, industry uses intellectual property as if it were a fortress that is expensive for those on the outside to penetrate.

The problem with this "fortress" model is that, while a patent-fortress mindset may have worked in the past for less ethically charged inventions, it fits only awkwardly with the particular ethical and social concerns that surround modern biotechnology. In particular, there are three major difficulties with the fortress model with respect to biotechnology.

First, simply because an enterprise holds a patent doesn't mean that the patent is valid. Statistics in the United States show that only about two out of three patents in all fields survive court review when challenged.[7] While one could argue that these statistics simply reflect some inherent weakness in those patents that are challenged in court, this is probably not the case. Patents are challenged because they block competitors from accessing technology. Thus, those challenging a patent do so on the basis of the importance of the invention, not on the weakness of the patent.

If patents in general face a one-third invalidity rate, the situation is likely worse for biotechnological inventions. This is because the standards that patent offices and courts use to evaluate these inventions are still in flux. For example, the United States Patent and Trademark Office issued new guidelines on the application of one of the patent standards—utility—in January 2001.[8] Most observers believe that these standards make it harder to obtain patents over genes than previously. If the courts accept this change in standard, then many previously issued gene patents will be held invalid if challenged. This will lead to a higher rate of invalidation.

Second, as Robert Merges and Richard Nelson have argued,[9] those who succeed in gaining a biotechnology patent may find that the value of that patent is lower than they had anticipated. This is due to the fact that biotechnology is not only relatively new as a commercial endeavor, but is also rapidly advancing. The combined effect of these two factors is that more people receive conceptually more limited patents while competing with a greater number of individuals. This statement needs to be unpacked a bit.

Because applied biotechnology is a relatively new field, there are initially few disclosed inventions in existence. Since there is little existing disclosed art in the area, a first-generation inventor can claim the widest plausible breadth to his or her patent claim. This phenomenon repeats itself with respect to every new technology: the first entrants into the field receive broad patent protection. Second-generation inventors are not so lucky, especially in a quickly developing field such as biotechnology. When these inventors seek patent protection, they are faced with all the broad, first-generation patents. This means that their patents will be more restricted in scope. Since second-generation inventors compete not only with each other, but also with first-generation inventors, they are at a disadvantage because of their (relatively) smaller patent claims.

Even first-generation inventors do not extract as much value from their patents as they may have hoped. Because biotechnology is advancing so quickly, these inventors must share existing license fees with a large pool of inventors, decreasing returns to everyone.[10] In other fields, this problem is significantly reduced by the fact that much of the basic technology is off-patent and by the slower growth rate in patents granted.

Third, biotechnology patents may be rendered unusable because of public fears and concern. Agricultural biotechnology has not taken off in Europe because of public opinion, not technology.[11] European consumers see no benefit and at least theoretical risk to genetically engineered crops. Because ignoring customer concern may result not only in public criticism, but also in boycotts, European stores have taken genetically engineered foods off of their shelves, and several large food production companies have vowed not to include genetically engineered ingredients in their products.[12] It is thus apparent that even if you have a patent, there is no guarantee that you can do anything with it.

The effects of public fear and concern over biotechnology reach beyond lowering the value of a particular patent. Since the patent holder of an invention is easily identified, the public and governments can hold that patent holder responsible for remedying the concern. This can take the form of consumer boycotts of the patent holder's goods and services, court action in tort or nuisance to prevent the sale of the product or to recover the costs of removing the biological product produced through the patent, or the imposition of severe regulation on the sale and use of the products in question. As the examples described below illustrate, this is far from being simply a question of academic interest.

Examples

In order to understand the limitations of the fortress model, it is worth examining a few examples of the use of that model and the problems that have resulted from that use. These examples are not meant to be exhaustive or even representative. Rather, they provide evidence that the fortress model has important limitations.

Monsanto

Monsanto is probably the most obvious example of a biotechnology company that has suffered by following the fortress model. In the late 1980s, Monsanto developed several biotechnology products. One of the products was seed that had been genetically modified to be resistant to one of its own herbicides, Roundup.

Farmers who used these seeds could spray them with Roundup, killing every plant but the one desired.

At first, Monsanto moved cautiously, believing that it needed to educate consumers about the benefits of genetically engineered plants and bring opponents of biotechnology on board as consultants before aggressively introducing its seeds onto the market.[13] This changed in the early 1990s when a new management team took over at the company. This team decided to introduce the seeds quickly in order to maximize returns from its patents. That is, management chose to follow the fortress model in respect of its patents. This turned out to be very harmful to the company.

In marketing its seeds in Europe in this way and in selecting herbicide-resistant crops as the subject for its first major marketing foray, Monsanto alienated both its direct customers (farmers) and its ultimate customers (the public). By ignoring the interests of both of these groups, Monsanto had few allies when public pressure led the European Community to implement a moratorium on new genetically engineered plants and led several countries to prohibit them altogether.[14]

In a classic fortress-model approach, Monsanto skillfully arranged its affairs so as to maximize its returns not only from its genetically engineered seeds, but from its Roundup herbicide. Roundup was going off patent, leading Monsanto to take measures to ensure that it would continue to be able to protect its market share in the herbicide business. Monsanto therefore formulated a plan to license, rather than sell, seed to farmers. This enabled Monsanto to impose restrictions on the way that farmers used the seeds. In return for using the seeds, farmers had to agree not to grow any seeds collected from the initial harvest, to accept regular inspections from Monsanto, and, most important, to only use Monsanto's Roundup herbicide rather than a generic equivalent. Monsanto was thus able to extend its protection over not only its seeds, but also its Roundup product. From a fortress-model point of view, this was a brilliant strategy. Unfortunately, it also led to farmer dissatisfaction toward Monsanto, a factor that cost it dearly later.

In pursuing the fortress model, Monsanto also ignored the European public's concern over its food supply. By introducing genes with unknown effects into the food supply, especially without any reciprocal benefit to the public and without labeling, Monsanto scared the public.[15] While Monsanto had examined the safety of its products, it did a poor job of communicating this to the public. Concern mounted to such a level that both national governments and the European Community were forced to take action in the form of a moratorium. This action was out of proportion with the real dangers posed by the technology, but because Monsanto had alienated those who could normally be counted upon to support it—farmers—there was little call for a more proportionate response to public concerns and fear.

In the end, Monsanto's share price fell dramatically while the technology markets in general rose to ever giddier heights.[16] The situation became so serious that the Deutsche Bank advised investors to sell their investments not only in Monsanto, but in all companies involved with genetically engineered products.[17]

Finally, Pharmacia bought out Monsanto, took over its pharmaceutical division, and spun a new Monsanto back out with a new management team. Meanwhile, despite continued discussions, the European moratorium continues.

Myriad Genetics Inc.

Myriad Genetics Inc. is another company that has implemented the fortress model, leading to much criticism, again mostly in Europe. Myriad developed a genetic screening test for women concerned about whether they have a genetic predisposition to breast and ovarian cancer.[18] Women in general have a one in eight chance of contracting breast cancer in their lifetimes; for women carrying a mutation in one of two identified genes, this risk jumps to as high as eight in ten.[19] However, the risk of a woman carrying a mutation is very low: approximately one in four hundred.[20] This is because only 7 percent of breast cancer cases are linked to one of these mutations.[21]

Myriad and the Cancer Research Campaign, through the Haddow Institute in England, both claimed to have discovered the second of the two breast cancer genes, BRCA2. Myriad is, and continues to be, a for-profit biotechnology company that is in the business of providing breast cancer screening tests. The Haddow Institute is a not-for-profit research center while the Cancer Research Campaign is a charity that aims at eliminating cancer.

What is interesting is what each group hoped to do with its discovery of the BRCA2 gene. The two groups certainly had much in common: both wanted to make a breast and ovarian cancer test available to help women plan their lives. The test presents women who are found to carry a mutation with a chance to take preventive steps against breast and ovarian cancer. It also allays the fears of those women who do not have the mutation, but thought that they were at risk because a close family member became ill with breast or ovarian cancer.

Apart from this overall agreement, however, the approach taken by the two organizations was different. The scientists at the Haddow Institute who discovered the gene wanted to ensure that the test would only be used to further women's health. They were afraid that if the test became routine, it would actually do more harm than good.[22] For the vast majority of women, the test results would not only be useless but might, in fact, be misleading. This is because 399 out of 400 women would be told that they did not have the mutation. Unless careful genetic counseling is given (and it rarely is), women may wrongfully believe that a negative test for a mutation means that they will not get breast cancer.[23] In fact, their chance of getting breast cancer remains at one in eight. Therefore, these women should still take steps (diet, exercise, examinations, and so on) to reduce their risk of developing breast cancer. Because of this concern, the Cancer Research Campaign attempted to require its licensee to ensure that genetic counseling would be conducted, that only women with a strong family history of breast cancer would be given the test, and that there would be no advertising of the test.[24]

Myriad, on the other hand, does not impose any restrictions on those taking the tests. To be fair, Myriad recommends the test for women with a family history of breast or ovarian cancer and recommends that women seek genetic counseling in conjunction with taking the test. But neither of these recommendations is required. Despite these recommendations, Myriad spends a significant amount on advertising the test and encourages investors to believe that the breast-screening test will become "routine," opening up a U.S. market of $150 to 200 million per year.[25] Such revenues could only be achieved if the test were given to a far greater number of women than have a strong family history of the disease.[26]

The criticism of Myriad focuses on three factors. The first of these concerns access to the genetic tests. Myriad charges approximately U.S. $2,580 for a full test.[27] This puts the test beyond the means of many people who must pay for it themselves and represents a significant cost to health care systems where such tests are insured. Second, Myriad has told certain independent laboratories that had developed their own tests for mutations in the BRCA1/2 genes that they must either purchase the right to use the genes or cease their research work. These labs have complained that the terms proposed by Myriad are unacceptable.[28] Third, Myriad has been criticized for the way it markets its tests. As mentioned earlier, Myriad's marketing strategy envisions that many more women than may need the test will have to take it in order for Myriad to meet its sales targets. Myriad also introduced the BRCA1/2 test onto the market before good data existed as to how many women possessing a mutation in one of the genes would actually contract cancer. When Myriad first marketed the test, it pointed to research that suggested that approximately 84 percent of women with a mutation would contract the illness.[29] While this number is accurate for women with a strong family history of breast or ovarian cancer, more recent studies indicate that a rate of 40 to 45 percent is more realistic for most women with a mutation in the gene.[30]

Myriad's strategy has not endeared itself to governments administering public health-insurance plans. These governments have not shown themselves to be keen on covering the tests as part of the public health system. The government of Ontario, for example, only covered the costs of the tests after being forced to do so by the courts,[31] while several hospital administrators in Britain expressed concern over the availability of the tests in that country.[32]

Gene Patenting

Several companies involved with sequencing the human genome in competition with the publicly funded Human Genome Project also followed the fortress model. These companies developed the strategy of patenting potentially promising human genes as they were sequenced, often with very little knowledge of their use. For example, Human Genome Sciences Inc. patented the CCR5 gene. When it did so, it thought, based on computer modeling, that the gene was likely a receptor gene but had no information as to why this receptor might be important. At about the same time as Human Genome Sciences Inc. won its patent, researchers unconnected with the company announced that the gene was linked to the infection of cells by HIV.[33] This gave rise to considerable concern because Human Genome Sciences Inc. would effectively receive the benefit of this discovery even though it had no part in it.

In another example linked to the identification of a disease-linked gene, patients suffering from Canavan disease gave blood, urine, skin, and (after death) brain tissue to Miami Children's Hospital researchers. The researchers used these materials to identify the Canavan-linked gene, which the hospital patented. The hospital then adopted a restrictive licensing policy that some say inappropriately restricted access to a diagnostic gene test, hurting the very community that had donated the material.[34] In response, patient groups commenced a lawsuit against the hospital for failing to provide them with access to the test.

These and other similar examples provoked concern among researchers and patients afraid that their access to both basic scientific research tools and genetic tests would be limited. The concern became so serious that President Bill Clinton and Prime Minister Tony Blair issued a joint statement in March 2000 that there should be widespread access to the human genome. It also led the United States Patent and Trademark Office to announce new and tougher patenting standards for genes, in January 2001.[35]

AIDS Drugs in the Third World

Despite the tragic effect of AIDS in Africa, the pharmaceutical industry had for years refused to sell AIDS drugs into Africa at cost out of fear that those drugs would be imported into developed countries, undermining pharmaceutical companies' profits in the latter countries. This strategy is a prime example of the fortress model: the industry's focus on maintaining direct returns from its patents led it to effectively ignore the plight of developing nations. The industry ran into difficulty when Brazil and certain African countries threatened to allow generic versions of the AIDS drugs to be sold despite the existence of patents on those drugs. These countries based this threat on language that exists within the World Trade Organization agreements allowing countries to permit the use of patented inventions by others in case of emergency.

The industry's initial reaction to this move was twofold. First, it argued that access to the AIDS drugs would not address the AIDS crisis in the developing world because developing nations did not have the internal infrastructure necessary to distribute the drugs and ensure compliance with the complex instructions on taking the drugs. This argument was dispelled by Brazil's success in effectively distributing the drugs and achieving compliance rates comparable to those of the United States.[36] Second, the industry brought legal action against some of the countries in question. As public disapproval mounted (including in the developed world), the industry found itself in the embarrassing position of challenging the only effective method known to address the AIDS crisis in the developing world. With bad public press and pressure from across the world, the pharmaceutical industry finally started entering into agreements with developing nations to permit them to sell the drugs at cost.[37]

Industry's current reevaluation may have come too late, however, for it to protect its patent strategy. The example set by Brazil and Africa may have already established a precedent permitting countries to override patent rights in certain circumstances.

Branding Model

Explanation

The mistake of the fortress model is that it assumes that biotechnology is fundamentally the same, ethically and socially, as previous technologies. As the examples discussed above illustrate, this is simply not the case. Biotechnology trenches on some of our most cherished beliefs and social practices. It affects the food we eat, the way we view ourselves and our bodies, health care, and environmental safety. By ignoring this effect, the fortress model gives rise to marketing and regulatory approaches that are not, in the long term, advantageous either to industry or to society.

A different model is therefore called for. Monsanto, in fact, initially followed this model in the late 1980s until its fatal decision to proceed with the fortress model. According to then Monsanto President Earle Harbison Jr., management developed a strategy to reach out to farmers and environmental groups and include them in policy making surrounding the use of genetically engineered crops.[38] Years later, Monsanto would return to this policy once it recognized the failure of the fortress model. In fall 1999, then Monsanto President Robert Shapiro appeared before a meeting of Greenpeace to reach out to the environmental group. In October of that year, Monsanto announced that it was dropping the so-called terminator technology that had caused a furor among environmentalists. Then, in summer 2000, Monsanto announced that it would make its vitamin A "golden rice" freely available to all. Unfortunately for Monsanto, this change of strategy occurred too late to offset the growing resistance to genetically engineered crops.[39]

The essential characteristic of the new model is that, instead of attempting to extract value directly from the patent, companies try to convert their patents into something of longer term value: customer and supplier loyalty and brand name. Thus, instead of trying to set up walls around an invention, the patent holder invites others to share in the benefits arising from the invention so as to create public acceptance of the technology and consumer loyalty. By working with governments and environmental groups to implement reasonable rules and regulations, industry is able to avoid the more damaging effects of moratoria and draconian regulation.

A second advantage of this "branding" model is that it turns a time-limited asset—the patent—into something of long-term value: brand name and loyalty. Patents, by their nature, last only for twenty years from the date a patent application is filed. Given the time it takes to develop an invention so that it is ready for market, the effective commercial lifespan of an invention is less than this. Brand name and customer and supplier loyalty have no inherent time limitation. If, by showing a willingness to share the direct benefits of a patent with others, a company is able to establish greater loyalty among its consumers, the company is likely to be better off.

The advantages discussed so far relate to industry. Perhaps the branding model's greatest benefits accrue to society. This is because the model calls for industry to proactively consult with consumers and governments in order to identify social and ethical concerns arising from biotechnology. By creating consensus or at least facilitating discourse prior to the use of new technologies, society will be in a better position to receive the benefits of biotechnology while mitigating its potential harms.

Examples

Because the branding model is only recently emerging as an alternative to the fortress model, there are fewer examples of its use. Further, since it has not yet been recognized as a model per se, it is sometimes difficult to identify those companies that use it. Therefore, the examples I discuss below represent some of the known uses of the model, but may not fully describe the scope of its use by industry.

Merck and INBio

Merck, a United States–based pharmaceutical company, has used its patents to build links with supplier countries and with consumers by actively promoting environmental protection. Merck did so by entering into an agreement with INBio, a Costa Rican nonprofit governmental organization charged with exploiting and protecting the country's natural biological resources.

The Costa Rican government established INBio in 1989 to enter into agreements with those companies wishing to access Costa Rican biological resources.[40] Because these resources are particularly rich in Costa Rica, industry was interested in gaining access to them.

As part of Merck's continuing drug development programs, it sought out new plants and animals that have some pharmacological effect. It was therefore interested in gaining access to Costa Rica's biological resources to build up a bank of material that it could then screen for such an effect. In 1991, Merck entered into an agreement with INBio for access to these resources.[41] Under the agreement, Merck reportedly agreed to pay INBio an initial fee of $1 million (U.S.), for research equipment, and a royalty based on the sale of any product that Merck developed as a result of its access to Costa Rican resources, and education and knowledge transfer to Costa Rica with respect to bioprospecting.[42] The agreement benefits Costa Rica in three ways. First, under the agreement, INBio collects the samples used by Merck. It thus employs individuals who would otherwise likely work for industries involved with cutting down Costa Rica's forests. Second, INBio must, by law, provide financial support for the protection of Costa Rica's natural environment. Third, INBio is able to create an inventory of Costa Rica's biological heritage.

While the Merck-INBio agreement has not been universally praised, it is surprising how little criticism it has received. After all, Merck, a large North American pharmaceutical company, is "bioprospecting" in a developing nation. Previous attempts by others to "bioprospect" led to vigorous public outcry.[43] Merck successfully avoided any public backlash over its activities by publicly and actively working with the authorities in Costa Rica. By acknowledging concerns over benefit sharing at the beginning of the agreement, Merck was able to turn a potentially negative situation into a positive one. Now that Merck has a track record of being willing to share the benefits of its patents, it is in a good position to enter into similar agreements with others. In fact, Merck has done so with the New York Botanical Gardens.[44]

It is too early to tell what the financial repercussions of the Merck-INBio agreement will be for either Merck or INBio. While Merck has collected all the samples to which it is entitled under the agreement, it will take many years for it to screen those samples for pharmacological effect. We do not, therefore, know how profitable the agreement will turn out to be for either Merck or INBio. Nevertheless, the agreement does illustrate that the branding model preserves opportunities for companies such as Merck to continue with their activities despite public concern over bioprospecting.

deCode

Iceland represents a treasure trove of information for genetics researchers and biotech companies. Not only is its population genetically homogeneous, but the country's well-maintained genealogical records provide researchers with the opportunity to conduct population studies in order to link genes and diseases.

An Icelander living in the United States, Kari Stefansson, developed the idea of creating a population health and genetic database to discover genetically linked diseases. He formed deCode in Iceland to follow through on this idea. His plan was for deCode to enter into an arrangement with the government of Iceland under which the company would access health information regarding all Icelanders. Under a separate arrangement, deCode was to obtain access to genealogical records. With these two banks of information, deCode would solicit Icelanders willing to undergo genetic tests to participate in the study. By putting these three sources of information together, deCode would be able to find genes that influenced disease formation.[45]

Researchers from deCode approached the Icelandic government in order to obtain access to Icelanders' health information. In order to provide deCode with this access, the Icelandic government needed to pass legislation. This legislation became the focus of vigorous discussion through various means in Iceland. In response to this discussion, the government adopted amendments to the legislation to better protect the confidentiality of health information.[46] Finally, on the eve of the passage of the legislation by the Icelandic Parliament in fall 1998, opinion polls suggested that 75 percent of Icelanders were in favor of it.[47] Despite criticism of the project,[48] public support for the law increased to 90 percent by April 2000.[49] In fact, according to Stefansson, the company was voted the most popular Icelandic company in 2000 by a large margin.[50]

Once the legislation was passed, deCode entered into an agreement with the Icelandic government. In return for its access to health information—on an opt-out system—deCode promised to provide the government with a centralized health-information database and free access to any pharmaceuticals developed as a result of the use of the health information.[51] deCode then entered into an agreement with Hoffman-La Roche under which the latter would commercialize deCode's findings.[52] The partnership has already produced important findings.[53]

As had Merck, both deCode and Hoffman-La Roche avoided any serious criticism and gained significant public support for their efforts. In general, far from being seen as plunderers of the woes of others, deCode's and Hoffman-La Roche's willingness to provide free access to technology and a health database have led to significant governmental and popular support for their efforts. By giving away some of the value of their patents—through the grant of free access to medications—the companies have built up loyalty and regulatory approval.

Tonga

Several other countries have decided to follow Iceland's lead by providing biotechnology companies with access to their populations' health information. One such example is Tonga, a small island in the Pacific.

Tonga has provided Autogen, an Australian company, with access to genetic information that has been gathered on a voluntary basis. With its small population and close family links, Tonga provides Autogen with the possibility of creating a population database to study the link between genes and disease. Although the island is run by a monarchy, DNA samples are only to be collected with fully informed consent on a voluntary basis.[54] Under the agreement, Tonga will maintain ownership of the genetic database, but Autogen gains the exclusive right to commercialize and sell collected information to medical researchers.[55] In return for these rights, Autogen will provide annual research funding to Tonga's Ministry of Health and royalties on revenues generated from commercialized discoveries. If any pharmaceutical drugs result from the deal, they will be provided to Tonga at no charge. Autogen will also provide Tonga with training and work for its scientists and modern equipment for its hospital.[56]

PXE

In contrast with the patient groups involved in the identification of the Canavan disease gene discussed earlier, a group representing patients suffering from pseudoxanthoma elasticum (PXE) entered into an agreement with the University of Hawaii under which patients donated blood and tissue samples for research purposes in return for joint control over the eventual patent of the PXE-related gene. The patients did so in order to ensure that genetic tests would be inexpensive and widely available.[57] Now that the research has resulted in the identification of the gene, patient groups and researchers are able to work together on the distribution of genetic tests.

Using Legal Tools to Encourage Use of the Branding Model

This chapter has provided evidence to support a business case in favor of what I have called the branding model for the use of biotechnological intellectual property. The evidence suggests that those companies that attempt to hoard their inventions and protect them through strong enforcement of intellectual property rights suffer by alienating their customers and supporters. Because of this alienation, governments often step in with regulations that undermine the long-term interests of these companies. A branding approach offers the companies an alternative way of viewing their intellectual property. By opening up access to their patents and by proactively addressing social concerns surrounding the use of their technology, these companies are not only able to capitalize on their intellectual property, but build support among consumers and governments.

Because of the significant social benefits that the branding model offers to society, we should not simply leave it to the private sector to recognize the benefits of the model. In order to encourage conversion to it, we ought to ensure that our laws provide industry with incentives to abandon the fortress model in favor of the branding model. It is therefore appropriate to either limit patent law itself or to regulate the exercise of rights by patent holders in order to counter these incentives.[58] I will discuss two possible approaches in this section: tort liability for patent holders who fail to abide by ethical standards and amending patent legislation to include ethical considerations.

Tort Liability for Patent Holders

Because patents play a critical role in the commercialization of biotechnological inventions, we ought to explore ways of harnessing the powerful economic incentives attached to patents as a means of mitigating the ethical concerns associated with the use of biotechnology.[59] For example, we could attach liability to patent holders for the unethical and negligent use of biotechnological inventions. This is particularly useful with respect to the use of genetic tests where market factors may lead to restricted access or premature use of the tests.

As Timothy Caulfield, Mildred Cho, and I have argued elsewhere, this approach is not beyond the stretch of current tort law theory.[60] Since the patent holder has the ability to prevent all others from using an invention, the patent holder has the power to ensure compliance with ethical norms. The patent holder can, through licensing arrangements, ensure that licensees of a technology take steps to ensure compliance with standards of access and reliability of information given to patients. This option not only encourages companies to take a branding approach to their intellectual property, but provides an alternative to government regulation that, in the long term, may cause greater harm to industry.

Patent Law Changes

As this chapter has illustrated, intellectual property regimes and their use by industry are not ethically neutral. It is therefore appropriate to wonder whether we can make changes to our patent laws that further ethical conduct without undermining any commercial incentive provided by patents.[61]

The changes to patent law can take many forms. These include making it more difficult to gain patent protection, making it easier to access patents for research purposes, legislating compulsory licensing of important technology, and introducing ethical considerations into the patent-granting process.[62] In selecting options, we must be mindful not only of ethical and social concerns, but of commercial reality. The aim of introducing ethical and social concerns into the patent system is to encourage compliance with the branding model and not to provide a forum for those completely opposed to biotechnology to block or delay research, development, and distribution.

Thus, we must take care in our approach. Nevertheless, carefully selected changes to patent law can provide an added incentive to industry to adopt the branding model of intellectual property.

Conclusion

Once we look past the internal workings of intellectual property law and examine how industry actually uses patent protection, we can find opportunities to address social and ethical concerns in a constructive manner. The experiences of companies in the biotechnology field provide us with a basis for constructing models for exploiting intellectual property. As the discussion in this chapter has illustrated, these models address social and ethical concerns in fundamentally different ways. By uncovering these differences, we are better able to develop policies at both the company and societal levels with respect to biotechnology.

Although the predominant model that industry has adopted with respect to the exploitation of intellectual property has been the fortress model—under which companies seek to maximize their direct returns from their intellectual property—there is reason to believe that, with respect to biotechnological inventions, it is better for both industry and society for industry to adopt a branding model. Under this model, companies not only permit others to access their technology, but proactively engage in discussion with respect to social and ethical concerns to which the technology gives rise. This gives rise to greater long-term relationships between industry and consumers, allowing industry to proceed with the introduction of biotechnology in a stable and publicly acceptable manner.

The advantages of the branding model over the fortress model reach beyond the company itself. Society benefits from the branding model since it encourages resolution of ethical and social concerns before the introduction of technology. This permits society to take advantage of the benefits of biotechnology while minimizing the harms. It also prevents the need for governments to take strong regulatory measures to address public fears and concerns over biotechnology.

Given the social benefits accruing from the branding model, it is appropriate for governments to take measures to encourage industry to adopt the branding model. In considering these measures, governments ought to be careful to ensure that they do not unduly delay or increase the cost of introducing new biotechnologies.

Notes

1. See, for example, United States Patent and Trademark Office Utility Examination Guidelines, 66 Fed. Reg. 1092 (2001).

2. Directive 98/44 of the European Parliament and of the Council of 6 July 1998 on the Legal Protection of Biotechnological Inventions, 1998 O.J. (L213) 13.

3. See, for example, *Monsanto Canada Inc.* v *Schmeiser* [20011 F.CJ. No 436 (Can.)].

4. E. Richard Gold, "Finding Common Cause in the Patent Debate," *Nature Biotechnology* 18, no. 11 (2000): 1217.

5. Canadian Biotechnology Advisory Committee, "Summary Report of President/CEO Industry Hearing to CBAC," Canadian Biotechnology Advisory Committee, Ottawa, 2000 [online], http: // www.cbac-cccb.gc.ca/documents/Report-En.lisbl.pdf [7 May 2001].

6. See, for example, Bill Roberts, " A Tale of Two Patent Strategies," *Electronic Business* 25 (1999): 79.

7. Donald R. Dunner, Michael J. Jakes, and Jeffrey D. Karceski. "A Statistical Look at the Federal Circuit's Patent Decisions: 1982–1994," *Fed. Circuit B.J.* 5 (1995): 151.

8. United States Parent and Trademark Office Utility Examination Guidelines.

9. Robert P. Merges and Richard R. Nelson. "Market Structure and Technical Advance: The Role of Patent Scope Decisions," in *Antitrust, Innovation, and Competitiveness*, eds. Thomas M. Jorde and David J. Teece (New York: Oxford University Press, 1992), 185.

10. See Robert Hunt, "Patent Reform: A Mixed Blessing for the U.S. Economy?" *Business Review—Federal Reserve Bank of Philadelphia* (November/December 1999): 1–9.

11. See, for example, Kempf, "OGM: l'UE adopte sans le dire un moratoroire sur les organismes génétiquement modifies," *Le Monde,* 25 June 1999.

12. See, for example, Veronique Lorelle, "L'arrogance de Monsanto a mis à mal son rêve de nourrir la planète," *Le Monde,* 8 October 1999, 20; David Barboza, "Monsanto Faces Growing Skepticism on Two Fronts," *New York Times,* 5 August 1999, C1; Craig Whitney, "Europe Loses Its Appetite for High-Tech Food," *New York Times,* 27 June 1999, D3; Warren Hoge, "Britons Skirmish over Genetically Modified Crops," *New York Times,* 19 August 1999, A3; François Dufour, "Les savants fous de l'agroalimentaire," *Le Monde,* July 1999, 1.

13. Kurt Eichenwald, Gina Kolata, and Melody Petersen, "Biotechnology Food: From the Lab to a Debacle," *New York Times,* 25 January 2001, C6 [online], http://www.biotech info.net/lab_debacle.html.

14. Ibid.; See also, Barboza, "Monsanto Faces Growing Skepticism on Two Fronts"; Paul Brown and John Videl, "GM Investors Told to Sell Their Shares," *Guardian* (London), 25 August 1999; Whitney, "Europe Loses Its Taste for High-Tech Food."

15. Eichenwald, Kolata, and Petersen, "Biotechnology Food: From the Lab to a Debacle."

16. During a six-month period ending on August 24, 1999, Monsanto's shares lost 11 percent of their value. Brown and Videl, "GM Investors Told to Sell Their Shares."

17. Ibid.

18. "Terms of Payment and Reimbursement," Myriad Genetics, Inc., 22 August 2000 [online], http://www.myriad.com/gtpatb20.html.

19. "Questions and Answers About BRAC Analysis," Myriad Genetics, Inc., 11 August 1999 [online], http://www.myriad.com/gtpatb21.html.

20. Dr. Michael Stratton of the Haddow Institute, U.K., personal interview, 12 December 1998.

21. "Genetic Analysis for Risk of Breast and Ovarian Cancer," Myriad Genetic Laboratories, Inc., 1997 [online], www.myriad.com.

22. Stratton, personal interview.

23. See, for example, Francis Giardello et al., "The Use and Interpretation of Commercial APC Gene Testing for Familial *Adenomatous polyposis,*" *New England Journal of Medicine* 336 (1997): 823–27.

24. Stratton, personal interview. Although this clearly was the intention of the scientists involved, the actual contract with the Cancer Research Campaign's licensee was less stringent. Guy Heathers of the Cancer Research Campaign Technology, U.K., personal interview, 9 August 1999.

25. Myriad Genetics, Inc., SEC Form 10-K, 25 (24 September 1998); Myriad Genetics Inc., SEC S-3 Registration Statement no. 333-16143 (14 November 1996); Stone and Schmidt, "Myriad Genetics ($15)," *Biotechnology Quarterly* 7 (1998): 112; "AMA Introduces New Physician Guide on Myriad Genetics Breast Cancer Test-Increased Testing Likely as Patient Inquiries Swell," Myriad Genetics, Inc., 2 June 1999 [online], http:// biz.yahoo.com prnews/ 990602/ ut__myriad_1 .html.

26. Michael Stratton estimates that the U.S. market for the screening test, if it were only administered to those truly at risk, would be approximately $5 million (Stratton, personal interview); See also, Newman et al., "Frequency of Breast-Cancer Attributable to BRCA1 in a Population-based Series of American Women," *JAMA* 279 (1998): 915 ("These data suggest that in general U.S. population, widespread screening of BRCAl is not

warranted."); K. E. Malone et al., "BRCA1 Mutations and Breast Cancer in the General Population—Analyses in Women before Age 35 Years with First-degree Family History," *JAMA* 279 (1998): 922 ("These findings on women drawn from the general population suggest that it may be difficult to develop BRCAI mutation screening criteria among women with modest family history profiles.")

27. This information was valid in August 2000. Myriad actually offers several different tests based on the BRCA I and BRCA2 genes. The Comprehensive BRAC Analysis test is the most comprehensive, costs $2,580, Myriad Genetics, Inc. "Terms of Payment and Reimbursement." Myriad justifies the high cost of the test because of the complexity involved, Myriad Genetics, Inc., SEC S-3 Registration Statement no. 333-16143.

28. Steve Bunk "Researchers Feel Threatened by Disease Gene Patents," *The Scientist* 13 (1999): 7.

29. See the study by Ford et al., "Genetic Heterogeneity and Penetrance Analysis of the BRCA1 and BRCA2 Genes in Breast Cancer Families," *American Journal of Human Genetics* 62 (1998): 676.

30. J. L. Hopper et al., "Population-based Estimate of the Average Age-Specific Cumulative Risk of Breast Cancer for a Defined Set of Protein-Truncating Mutations in BRCA1 and BRCA2," *Cancer Epidemiology Biomarkers & Prevention* 8 (1999): 741; Antoniou et al., "Risk Models for Familial Ovarian and Breast Cancer," *Genetic Epidemiology* 18 (2000): 173.

31. Timothy Caulfield, "Gene Testing in the Biotech Century: Are Physicians Ready?" *CMAJ* 161 (1999): 1122.

32. Christine McGourty, "Will the Legal Minefield of Gene Patenting Harm Patients?" *Electronic Telegraph* 8 (June 2000) [online], http://www.telegraph.co.uk:80/et?ae=004935905919217&rtmo=aC9l34CuJ&atmo=rrrrrrtq&pg~/et00/6/8/ecfpato8.html.

33. Eliot Marshall, "Gene Patents: Patent on HIV Receptor Provokes an Outcry," *Science* 287, no. 5457 (2000): 1375.

34. Eliot Marshall, "Families Sue Hospital, Scientist for Control of Canavan Gene," *Science* 290, no. 5494 (2001): 1062.

35. United States Patent and Trademark Office Utility Examination Guidelines.

36. Tina Rosenberg, "Look at Brazil," *New York Times Magazine,* 28 January 2000, 26.

37. See Reuters, "Kenya says Will Import Cheap Generic AIDS Drugs," 6 March 2001; Melody Peterson and Donald McNeil Jr., "Maker Yielding Patent in Africa for AIDS Drug," *New York Times,* 15 March 2001.

38. Eichenwald, Kolata, and Petersen, "Biotechnology Food: From the Lab to a Debacle."

39. Ibid.

40. Barbara L. Kagedan, *The Biodiversity Convention, Intellectual Property Rights, and the Ownership of Genetic Resources, International Developments* (Ottawa: Intellectual Property Policy Directorate of Industry Canada, 1996), 70.

41. Ibid.

42. Ibid.

43. See, for example, "Bioprospecting/Biopiracy and Indigenous Peoples, Rural Advancement Foundation International," 30 November 1994 [online], http://64.4.69.14/web/alipubone.shtml?df1=allpub.db&tfl=allpub-one-frag.ptml&operation=display&ro1=recNo&rf1=44&rtl=44&usebrs=truce.

44. Kathryn Munoz, Merck & Co. Public Affairs, personal interview, 2 February 2000.

45. Jeffrey R. Gulcher and Kari Stefausson, "The Icelandic Healthcare Database and Informed Consent," *New England Journal of Medicine* 342, no. 4 (2000): 827.

46. See, for example, Bartha M. Knoppers, "Sovereignty and Sharing," in *The Commercialization of Genetic Research: Ethical, Legal, and Policy Issues,* eds. Timothy A. Caulfield and B. William-Jones (New York: Kluwer Academic/Plenum Publishers, 1999), 1; Martin Enserink, "Iceland OKs Private Health Databank," *Science* 283, no. 5398 (1999): 13.

47. Gulcher and Stefansson, "The Icelandic Healthcare Database and Informed Consent."

48. George J. Annas, "Rules for Research on Human Genetic Variation—Lessons from Iceland," *New England Journal of Medicine* 342, no. 24 (2000): 1830.

49. Gulcher and Stefansson, "The Icelandic Healthcare Database and Informed Consent."

50. Karl Stefansson, Conference presentation at the HGM2000 meeting in Vancouver, Canada, 10 April 2000.

51. See Knoppers, "Sovereignty and Sharing."

52. DeCode, press release, 2 February 1998 [online], http://www.decode.com/news/releases/older/item.ehtm?id=1561.

53. See, for example, "deCode and Roche Announce Major Progress in Turning Their Genomic Discoveries in Schizophrenia and PAOD into Novel Drugs and Diagnostics," deCode, 14 February 2001 [online], http://xvxvw.decode.com/news/releases/itcm.ehtm?id=7462.

54. Uri Juhn, "Tonga Agrees to DNA Commercialization," *Pacific Islands Report* 29 (November 2000): 5.

55. Kim Griggs, "Tonga Sells Its Old, New Genes," *Wired News,* 27 November 2000 [online], http://www.wirednews.com/news/technology/0,1282,40354,00.html.

56. Ibid.

57. Paul Smaglik, "Tissue Donors Use Their Influence in Deal over Gene Patent Terms," *Nature* 407, no. 6806 (2000): 821.

58. See, for example, E. Richard Gold, "Moving the Gene Patent Debate Forward," *Nature Biotechnology* 18, no. 12 (2000): 1319.

59. Ibid.; E Richard Gold, "Making Room: Reintegrating Basic Research, Health Policy, and Ethics into Patent Law," in *Commercialization of Genetic Research: Ethical, Legal, and Political Issues,* eds. Timothy A. Caulfield and B. Williams-Jones (New York: Plenum Publishers, 1999); Timothy A. Caulfield and E. Richard Gold, "Genetic Testing, Ethical Concerns, and the Role of Patent Law," *Clinical Genetics* 57, no. 5 (2000): 370–75; Timothy A. Caulfield, E. Richard Gold, and Mildred Cho, "Patenting Human Genetic Material Refocusing the Debate," *Nature Reviews Biotechnology* 1 (2000): 227.

60. Caulfield and Gold, "Genetic Testing, Ethical Concerns, and the Role of Patent Law"; Caulfield, Gold, and Cho, "Patenting Human Genetic Material Refocusing the Debate."

61. Gold, "Moving the Gene Patent Debate Forward."

62. See, for example, E. Richard Gold, *Patents in Genes* (Ottawa: Canadian Biotechnology Advisory Committee, 2000); E. Richard Gold and Alain Gallochat, "The European Biotech Directive: Past as Prologue," *European Law Journal* 7, no. 3 (2001): 331–66; B. M. Knoppers, M. Hirtle, and K. C. Glass, "Commercialization of Generic Research and Public Policy," *Science* 286, no. 5448 (1999): 2277.

Returning to Normal

Can Corrective Justice Be Achieved When Genetically Modified Salmon Escape and Do Damage?

Keith Culver

The Food and Agriculture Organization of the United Nations reports that 70 percent of the world's fish stocks are now fully or seriously depleted. There is no sign of a substantial recovery.[1] New marine sources of food will necessarily come from aquaculture. The leading edge of aquaculture development is found in intensive farming of salmon in states of the developed world, especially Canada, the United States, Britain, Norway, Chile, and Argentina. Yet aquaculture's bright prospects rely in large part on an anticipated shift toward a tremendously controversial practice: genetic modification of fish to produce individuals with commercially

---

desirable characteristics.[2] One worry about this practice is driven by concerns about potential damage done by genetically modified (hereafter GM) fish which escape. If we accept that aquaculture is here and farming of GM fish is next, we must accept also that accidents will happen and GM fish will escape.[3] Damage of various types will be done, and someone will want compensation. One tricky part about understanding this damage in the context of GM salmon comes with a characteristic of all salmon—they are "anadromous" creatures who begin in rivers, go to sea, and return to rivers. Naturally, they do this without passports,[4] and often interact at sea with other fish whose original river homes are located in several different coastal states. The hypothetical question I want to consider arises from the possibility that escaped GM fish might cross international boundaries and cause a variety of damages. Can corrective justice be achieved under current international mechanisms for redress of damages suffered in transboundary contexts? Can there be a return to normal?

My argument points to three sources of doubt regarding international law's ability to deliver corrective justice. I shall point to a shortfall in the biological knowledge needed to understand a return to normal, a problem with international law's capacity to generate enforceable obligations, and practical and moral remainders left by the available legal points of view to damage done by escaped GM fish. Taken together, these three doubts sound an important alarm. If we want to build corrective justice in our shared future, we must begin to set down adequate foundations now.

Risk and Threat of Escape of Genetically Modified Fish

The importance of answering the question I have set rests in large part on the plausibility of the scenario I have sketched, where escape of GM fish leads to damage.[5] Threats posed by escaped GM fish can be modeled on the threats posed by escaped farmed salmon of ordinary provenance.[6] Escaped fish may pose a *competitive* threat, as they occupy scarce spawning space or compete for food or evade predators especially successfully. *Interbreeding* between escapees and wild fish may reduce biodiversity, and reduce adaptive advantage in at least the first generation of interbred offspring. A slightly different danger arises in the possibility that escapees may successfully displace wild fish from spawning areas and establish *feral populations*. In addition to these competitive and genetic risks, the relatively crowded conditions of fish farming leave farmed fish prone to *diseases* that escapees can carry to wild or other farmed-fish populations.

Unhappily, the risk of damage from escaped GM salmon is substantial if current production methods continue.[7] Escapes into the wild are common, in numbers ranging from a few thousand to escapes in excess of a quarter of a million fish.[8] In British Columbia, scientists have observed escaped Atlantic salmon destroying nests of wild stocks by "overcutting"—establishing egg nests in places already occupied by wild salmon eggs.[9] Undetected interbreeding may exist on a very large scale in areas of the Atlantic coast (primarily the Bay of Fundy) where local fish stocks used as farm fish escape and are not recovered. Feral populations of escaped salmon are also thought to exist in the Atlantic region, and stocks of brown trout native to Europe are known to have established feral populations in Newfoundland. While data on transmission of disease from farmed to wild salmon is slim, the mere fact of escapes of fish that carry disease opens the possibility of farmed fish serving as delivery mechanisms for diseases. In fact, with respect to nearly all of these measurements of the magnitude of the risk posed by escaped salmon, data is slim and consequently predictive modeling of the impact of threats is difficult. For example, some salmon from Atlantic Canada spend their sea-winters off the coast of Greenland in the company of salmon from Europe, and a range of competitive pressures apply equally to Canadian and European salmon, with effects throughout the food chain. It is certainly possible that escaped GM Canadian salmon might carry out the same migration as their wild Canadian cousins, and arrive off Greenland with quite different adaptive fitness. With what effects on the populations of wild salmon? Or on the food chain? No one knows with certainty, because the data is not yet in.

Current Regulatory Approaches

What can be done if GM fish *do* escape and *do* cause damage in a transboundary context? This question can be answered in two ways, sometimes simultaneously. The first answer is located thoroughly within the sovereign power of the state where the damage is thought to have occurred. The twentieth century saw an increasing willingness of domestic courts to claim jurisdiction over situations in which their own nationals are harmed outside state boundaries, with a variety of not insuperable problems regarding enforcement of judgement.[10] The second answer is found outside sovereign jurisdiction, in the less certain arena of international law. If the state conceives of the problem as something larger than damage to one of its citizens, the state may choose to take up the matter itself. This question of state action will be examined here, in part because it is already well-established that domestic courts sometimes claim jurisdiction in these cases, and in part because the magnitude of potential damage makes state-level intervention likely.

It is worth putting this discussion in context by identifying some crucial features of international dispute resolution, beginning with the presumption that states are the main actors in international law, sovereign over their internal affairs and limited only by agreements into which they have entered voluntarily. The fundamental norm of state sovereignty does not collapse the authority of general principles of international law that govern interactions between states, but it does provide a kind of guarantee of state autonomy against interference from other states. The effect of this fundamental norm on adjudication of disputes between states is quite striking. General principles of international law contain no provision for general courts of compulsory jurisdiction—courts which apply rules that compel states to come before those particular courts. States who do choose to resolve disputes before the International Court of Justice do so on an immediately voluntary basis, or on a less immediate basis generated by an originally voluntary agreement to a treaty that requires certain disputes to be referred to the International Court of Justice. Voluntary submission of disputes to the International Court of Justice is certainly not unheard of in the context of fisheries, as, for example, Canada and Spain took their dispute over capture fisheries off the Grand Banks of Newfoundland to the International Court of Justice. But international law that applies to fisheries is not simply a matter of general principles, such as the established international duty to withhold from interference with the sovereign rights of other states.[11] Fisheries are a complex topic, and a range of international declarations, protocols, and conventions may apply. Here the problem becomes more complex, in at least two ways: the problem of the practical force of international law, and the problem of determining which laws apply.

The Problem of Normativity

The first trouble with recovery for damages done by GM fish comes out of the *nature* of international law. There is substantial disagreement as to whether the patchwork of charters, treaties, conventions, and so forth actually add up to something properly called "international law" and generating nonoptional obligations. The debate is driven in part by the wide variation in normative force between the instruments claimed to be pieces of international law. The traditional sources of international law are just four: treaties, customary international practice, general principles of international law, and judicial opinions of international dispute-resolving bodies such as the International Court of Justice.

In the context of our problem of how to recover for damages done by escaped GM fish, treaty-based international law provides the best and only hope for concrete, clear, and so *justiciable* legal norms to govern the complexities of the problem. Treaties are usually the most practically forceful of international norms, largely because they have received explicit consent from signatory states that have participated in the drafting of terms—unlike, for example, relatively vague customary international law. Resolution of disputes can sometimes involve little more than a state-to-state reminder of the terms of a treaty agreement. But things are not always so simple. Treaties come in a rainbow of varieties, and their interpretation and application is not governed by a universally accepted set of interpretive practices. As Paul Reuter neatly puts it,

> There is no precise nomenclature for international treaties: "treaty," "convention," "agreement," or "protocol" are all interchangeable. Furthermore the meaning of most of the terms used in the law of treaties is extremely variable, changing from country to country and from Constitution to Constitution; in international law it could even be said to vary from treaty to treaty: each treaty is, as it were, a microcosm laying down in its final clauses the law of its own existence in its own terms. The uncertainty in wording is a result of the relativity of treaties.[12]

Additional problems beyond interpretive uncertainty arrive with the absence of any standard form or force for international treaties. Some treaties are aspirational statements setting benchmark standards to be striven for in the domestic legislation of signatory states. These treaties often lack a clear conception of binding norms together with a dispute-resolution mechanism to promote enforcement of those norms. Multilateral treaties of the sort needed to govern the multiple states involved in salmon aquaculture can be further hampered in their practical effectiveness by individual states' use of specific opt-out provisions called "reservations," which function as a kind of notwithstanding clause, allowing signatory states to withhold from accepting specific parts of treaties. In a similar vein, it is important to understand a practical dimension of treaty-formation: the drafting and approval of a treaty by a set of authorized representatives is usually only the first step toward a treaty having force and effect. Treaties typically require further ratification by governing bodies of states party to the treaty, and this can take a considerable amount of time.

It is sometimes asked why states do not work more quickly toward framing and ratifying treaties that are plainly in everyone's best interests. The answer here, as in so many other areas of international law, is bound up with state sovereignty. States traditionally attempt to maintain as large a measure of sovereignty as possible, and for that reason are sometimes keen to avoid losing autonomy to restrictions imposed by treaties. States are much more willing to participate in consultative processes that generate so-called soft law—declarations, resolutions, statements of principle, and so forth. These instruments tend heavily toward promotion of ideal standards that form a blueprint for desirable change in municipal legislation of states, or for more-concrete treaties. But these documents are rarely enforceable, in large part because the promotional language of their provisions sets political commitments to developing guiding norms rather than providing those guiding norms, and because these documents rarely if ever contain practically workable dispute-resolution mechanisms.[13]

Some writers suppose that the "relative normativity" of international legal standards with varying force should not be taken too seriously, especially in view of the successes enjoyed by "soft" international legal standards in recent years.[14] This is not to be shrugged off—states seem to be more willing to sign on to general declarations whose intent can be incorporated into municipal law at leisure, and this may be leading to faster achievement of consistency of norms between states than might have been reached using the unwieldy development of international law through multilateral treaties. Yet none of this blunts the force of the observation that soft-law approaches leave so much room for state maneuver that it is reasonable to worry that soft approaches may not result in justiciable and enforceable norms to govern disputes.

Framing the Problem

Even if we assume that there are or can be sufficiently forceful treaties to carry out the requirements of corrective justice, a prior conceptual question still needs an answer to drive the choice of legal category with which to account for the damage done. What, exactly, is the "damage" done by escaped GM fish? Harm to property? Harm to a coastal ecosystem? Harm to biodiversity? A common view sees international legal norms as a web of options. Damage can be conceived through nearly unlimited combinations of international treaties on fisheries, borders, biodiversity, environment and pollution, less formal protocols and declarations, and national legislation on animal care, animal health, the environment, human health, and so forth. In some areas of international life, practices have solidified out of this muddle to the point where particular groupings of norms are treated as the appropriate point of view to a particular problem. Yet this appearance of solidity and absence of vagueness comes at a certain cost. Practical solutions based on this kind of conception of international law have more than a whiff of the ad hoc, and may make for bad law—momentarily workable solutions that do not export readily as general principles to new fact patterns in similar situations, and may in fact generate conflict of laws. What is worse, accepting international law as an apparently infinite recombination of norms can tend to obscure deeper conceptual problems. Without a well-justified conception of *what* is damaged, redress that achieves the immediate purpose of stating a solution to a problem has the potential to become merely a formally stated approximation whose precision is without accuracy. Practical and moral remainders of ad hoc solutions may include ecosystems only partially restored, or legally uncompensated destruction of future generations' access to viable capture fisheries. I shall suggest in what follows that existing international legal instruments are usefully viewed as displaying three individually and perhaps collectively incomplete approximations of the kind of damage done by GM fish. I shall call these overlapping legal points of view to damage done by fish the "sovereignty model," the "regional model," and the "global-biodiversity model." Corrective justice without practical or moral remainder will require at very least that we eliminate the inconsistencies between these points of view, and better yet, that we integrate these points of view by rebuilding a unified international legal point of view to GM escapees.

The Sovereignty Model

On the sovereignty model, fisheries are natural resources of sovereign states. Within constraints provided by specific treaties and customary and general principles of international law, sovereign states retain full control over their internal waters and territorial seas to the extent of the Exclusive Economic Zone (EEZ) they can claim through a mechanism of the United Nations Convention on the Law of the Sea (1982) (hereinafter UNCLOS).[15] Wild salmon are within the scope of a further specific conservation and harvest regulatory purview of the coastal state of origin,[16] subject to further specific regulatory measures contained in the UN Convention on Straddling Fish Stocks and Highly Migratory Fish Stocks (1983) and the 1995 agreement for implementation of that convention.[17] Domestic policy (distinct from law) is often driven by at least lip service to the "precautionary principle" most recently endorsed in the context of aquaculture by the Bangkok Declaration's promotion of "precautionary, safe, and practical" aquaculture.[18] (It is worth noting that while the precautionary principle is very popular, its proper application is a matter of fierce dispute).[19]

On this view, damages done to wild or aquaculture fish or their environment is primarily a matter of damage to sovereign property rights to fish stocks. "Stocks" are rather ambiguously understood sometimes biologically as referring to the river of origin and sometimes economically as a stock of goods. The sovereign state's vital interest, from this point of view, is preservation of sovereign resources *and* preservation of territorial sovereignty from a kind of invasion. This is, of course, something of an oversimplification insofar as specific provision is made under UNCLOS for collaborative management of overlapping fish stocks;[20] but even this provision assumes that sovereign states collaborate to negotiate preservation of sovereign rights to resources. As in other international disputes, applicable international legal obligations are interpreted against the backdrop of the UN charter's requirement that states resolve disputes peacefully. Specific provision is made under both the UN charter and the UNCLOS for establishment of dispute-resolution tribunals. The precise workings of these tribunals, standards of evidence, and other mechanical details are often left to the specific tribunals. And, once again, these tribunals lack compulsory jurisdiction and ready enforcement power. Foot-dragging on the part of one of the disputing states could be interminable, and damage done by escaped GM fish could continue unabated.

This point of view has several strengths. The sovereignty model is consistent with the presumption that states are the central actors in disputes under international law. Within this assumption, the sovereignty model contains well-used practices of dispute resolution via the International Court of Justice and other arbitral tribunals. Yet this consistency with orthodoxy comes at the cost of an accurate fit between facts and obligations. Most important, the sovereignty model's current treaty sources fail to recognize the dynamic nature of anadromous fish that exist in a regional and global environment whose boundaries are immune to legislative direction—it is not as though we can direct salmon to swim only within specified Exclusive Economic Zones. The sovereignty model also assumes the practical viability of solving the problem as a matter between nations whose interests are readily quantifiable in largely economic terms of fish stocks. This assumption omits consideration of the place of GM fish in a chain of competitive relations in the much larger global environment and its functional requirement of biodiversity of some unknown range. Solutions acceptable on the largely economically and border-focused sovereignty model may fail completely to engage the larger problems of damage to regional-ecosystem health and global biodiversity. And perhaps most painfully for the individual fisher, there is no restoration of his or her position prior to the damage done by escaped GM fish, except, perhaps, by a trickle-down effect as the state compensates the individual fisher. It is probably not an exaggeration to say that this is justice deferred, if not denied.

The Regional Model

On the regional model, it is much more difficult to express the nature of the damage done by escaped GM fish. Fish are still conceived as stocks of property to which sovereign states retain rights of exploitation, but the essentially biological notion of an ecosystem is recognized as the productive source of natural wealth, and the extra-legal quality of an ecosystem as a distinctive biological region is given priority. Damage is accordingly understood as done to both fish stocks and the regional ecosystem in which they participate as consumers and prey, without regard for state borders. Not surprisingly, since this model requires decreased concern with sovereignty and increased concern with regionality, the applicable hard norms of international law are quite slim. Regional norms are plainly guided by the peremptory norm of customary international law that a sovereign state's conduct must not harm other states. Regional fishers are more specifically guided by a mixture of UNCLOS provisions requiring joint governance by neighboring coast states over regional resources, and UNCLOS provisions promoting development of regional regulatory bodies.

In the context of the North Atlantic salmon–aquaculture industry and capture fishery, the principal source of international law is the North Atlantic Salmon Conservation Organization (NASCO)-sponsored Convention for the Conservation of Salmon in the North Atlantic (1982). This convention is directed primarily toward the capture fisheries, distinguished as geographically defined areas under the charge of Commissions for North America, West Greenland, and the North-East Atlantic.[21] Aquaculture-specific normative approaches are found in a subsequent series of resolutions, protocols, and guidelines of uncertain normative force. The Oslo Resolution[22] of 1994 enjoins states parties to the NASCO Convention to "take measures, to the full extent practicable to: minimize escapes of farmed salmon. Minimize the straying of farmed salmon. Minimize adverse genetic and other biological interactions from enhancement activities."[23] Further determination of the meaning of "full extent practicable" and "minimize adverse . . . interactions" caused by GM is not demonstrably improved by the subsequent Agreement on Implementation of the Oslo Resolution.[24] Much of the Agreement on Implementation serves as lamentable evidence of the normative softening that accompanies each fresh plea for implementation, descending from the relatively hard normativity of the ratified convention to the softness of an open-textured set of guidelines. Article 3, for example, states that "In order to have confidence that wild stocks are protected from irreversible genetic change, from ecological impacts, and from impacts of diseases and parasites, the measures in the Oslo Resolution *should be fully implemented.*" And at Article 8: "[C]ontainment measures are currently not adequate to deal with the problem [of escapes]. *Renewed efforts* should, therefore, be made to minimize escapes and a *more effective enforcement policy should be adopted* by the Parties." The accompanying Guidelines for Action on Transgenic Salmon are similarly normatively uncertain, requiring that states parties "take *all possible action* to ensure that the use of transgenic salmon, in any part of the NASCO Convention Area, is confined to secure, self-contained, land-based facilities" [emphasis added throughout].[25]

The general trend of these regional approaches is clearly aspirational rather than regulatory, but this normative softness should not blind us to the virtues of the regional approach. It improves on the sovereignty model by understanding anadromous fish as occupying ecosystems that stretch across state boundaries, and relies on a biological understanding of fish stocks rather than the limited economic understanding of fish stocks as sovereign resources of a particular state. There regional view's reliance on benchmark norms may also serve as a practically viable way of gaining support of sovereign states who may be more willing to comply voluntarily with guidelines than to enter into autonomy-reducing multilateral treaties.

This cautious optimism should be counterbalanced by recognition of the seriousness of the NASCO norms' shortcomings. States are required only to take measures "to the extent practicable" to "minimize" ill-defined "adverse . . . interactions" in a regional area whose isolation from global biodiversity is left unaddressed. Perhaps more important, the perceived need for a subsequent Agreement on Implementation serves as important evidence of practical noncompliance. It is difficult to judge whether noncompliance is due to states' reluctance to surrender autonomy, or whether instead the NASCO norms are so open-textured as to be virtually impossible to render into national legislation likely to be consistent with other NASCO states' use of those norms. This latter worry is given some weight by the absence of a regular arbitral tribunal in this regional scheme. In the absence of a dispute-resolving body, it seems plausible to suggest that regional mechanisms of this sort are only capable of serving as benchmarks, leaving adjudication of disputes to the usual options of domestic courts claiming jurisdiction or to the International Court of Justice. Corrective justice is unlikely to be served by either of these options. Courts of sovereign states adjudicate internal matters, and matters made internal to the state by the court's claiming jurisdiction. Weighing and balancing of regional (international) concerns of ecosystem health so carefully addressed by the benchmarks set by NASCO are not easily justiciable by national courts. The possibility of appealing to the International Court of Justice to rule using customary norms of international law is equally unlikely to lead to corrective justice. Again, only states have standing with the International Court of Justice, so regional ecosystems are left untreated, and the position of individual fishers is again subsumed under national interests. Moreover, the International Court of Justice is itself restricted to trying disputes governed by justiciable norms. The imprecisely iterated norms of NASCO's Agreement for Implementation are unlikely to provide a clear basis for the International Court of Justice to resolve specific disputes.

The Global-Biodiversity View

The global view is a product of relatively recent emergence of the idea that biodiversity is universally important. Within the global–biodiversity view, fish are conceived as part of biodiversity, which is a precondition of sustainable development. Biodiversity is conceived largely as existing in ecosystems that are best protected under gradually emerging international environmental law and biodiversity law. Damage done by escaped GM fish is accordingly harm to an ecosystem whose value and preservation are of universal interest. It is important to observe the tension between the global nature of biodiversity, and the existing state-based international legal system's need to understand damage to biodiversity in terms of state interests, perhaps mapped to individually isolable ecosystems containing biodiversity. This last "container" view involves a good deal of wishful thinking, since the continuous and dynamic nature of interlocking global ecosystems may preclude assigning "shares" in biodiversity to states who might then seek redress for damage to their share alone. Perhaps because of these difficulties, adjudicative and compensatory mechanisms regarding the global impact of GM fish have been slow to develop. Arguably the "hardest" sources of international law in this area are the Convention on Biological Diversity (1992), and its Cartagena Protocol on Biosafety (2000),[26] which commits the states parties to:

> adopt a process with respect to the appropriate elaboration of international rules and procedures in the field of liability and redress for damage resulting from transboundary movements of living modified organisms, analyzing and taking due account of the ongoing processes in international law on these matters, and shall endeavor to complete this process within four years.[27]

At the first conference of the states' parties (which resulted in the Montpellier Declaration), no specific normative framework emerged. Indeed, familiar tensions resurfaced, as some representatives "supported the imposition of sanctions" while others "stressed the facilitative and non-confrontational nature of Article 34 of the Protocol."[28] All parties were mindful of difficulties in obtaining compliance in the special situations of developing states. In a word, there is no *authoritative* agreement on the norms and dispute-resolution mechanisms needed to guide corrective justice.

It is clearly a strength of the convention and protocol that it recognizes explicitly the global nature of biodiversity, the regional and global nature of ecosystems, and the universal vital interest of states in biodiversity as a precondition of sustainable development. It is a further strength of this strategy that it at least aims to provide mechanisms for liability and redress. Yet this strategy encounters the same kinds of difficulties as we have seen in local-sovereignty and regional contexts. It is far from clear that the conception of fish as part of biodiversity and GM fish as a threat to biodiversity is adequate to provide corrective justice when questions of liability and redress arise. Are fish individual-wealth stocks, part of an ecosystem, or units of biodiversity? How are findings of liability and orders for redress at the level of biodiversity to mesh with the the legitimate purposes aimed at by the sovereignty and regional models? And can we understand adequately the damage done to universally important biodiversity without ascribing guilt? Here we must return to the idea of corrective justice.

Corrective Justice and Room for Guilt

So far my use of the idea of corrective justice has relied on the view that corrective justice has been done when the wronged party is restored to its original position or some reasonable approximation of what its position might now be in the absence of the wrong actually suffered. Further discussion requires a slightly more robust understanding of what corrective justice requires in practice. Traditionally, application of corrective justice has three elements. First, the wrongdoer or wrongdoers must be properly identified together with the extent of their contribution to harm suffered, to ensure that blameworthiness and sanctions or a duty of reparation are attached to the party reasonably held responsible for the harm suffered. Second, "horizontal consistency" requires that the corrective measures undertaken must be equivalent to measures taken in like cases. Third, "vertical proportionality" requires that corrective measures taken must be proportionate to the harm done, leaving the wronged party neither behind nor ahead of its position prior to suffering harm. This sketch is compatible with the idea that in some situations of correction or restoration, there is a kind of moral remainder of guilt. This conception of corrective justice with space for guilt is somewhat morally thicker than a model of tortious wrongdoing which supposes that certain kinds of wrongs can be fully compensated for. But this may be precisely the shade of justice needed given the universal importance of biodiversity, and potentially universal suffering on the part of species, ecosystems, or biodiversity.

In applying this conception of corrective justice, let us assume counterfactually that disputing states are bound by or agree to the jurisdiction of a court, and that decisions of the court have theoretically and practically binding force. Even given these assumptions, corrective justice will be difficult to reach under existing legal points of view. First, identification of wrongdoers may be tremendously difficult in disputes involving GM fish, for simple empirical reasons to do with stating reliably the causal relation between harm done and the escaped GM fish identified as the causal agent. As discussed previously, escaped fish are thought to cause damage through competitive outperformance of wild stocks or competitive displacement of other local species, through interbreeding, or through transmission of disease. It is currently very difficult to state with precision whether and to what extent specific escaped GM fish have harmed some state's, or region's, interests or biodiversity as a whole through any one or all of these potential harms.

The second requirement of corrective justice, horizontal consistency with like cases, is also likely to be difficult to obtain under current approaches. It is a familiar feature of modern legal systems that law is sometimes stretched very far when it is applied to genuinely novel situations, and reasoning by analogy sometimes becomes quite strained where needed factual information is uncertain and conceptual distinctions are poorly cast. This strain is likely to be especially evident in questions of liability and redress for damage done in transboundary contexts, which are inherently dynamic contexts as geological time marches on, natural selection continues, and courts struggle to develop solutions out of the mesh of domestic and international norms capturing parts of the biological context within with damage is done. Corrective justice will require more and better biological data to undergird arguments for consistency across like cases, and greater consistency across domestic jurisdictions whose norms must intertwine with international tribunals' decisions.

"Vertical" measurement of the seriousness of the damage done similarly needs conceptual, biological, and social accounts of what the damage consists of. Yet again, relevant factual knowledge is incomplete, so we are factually ill prepared to enter ontological debates about what the results of corrective justice as restoration ought to resemble. Even if needed facts were available and the ontological debates were settled, the fact remains that existing legal conceptions of damage are individually incapable of ascribing liability and ordering redress in a manner consistent with comprehensive corrective justice. Local-sovereignty, regional, and global approaches are each individually nonexhaustive and leave both practical and moral remainders. The practical remainders consist in damage unaddressed because it exists outside the conceptual purview of the particular level. A settlement between immediately involved states, for example, may leave damage done to a regional ecosystem incompletely assessed and incompletely redressed insofar as redress stops at the boundaries of the particular states which have reached a settlement. It is very unlikely that state, regional, and global views of escaped GM fish are likely to converge in a way that will lead arbitral tribunals at each level to aim to match precisely their orders for redress with the orders for redress reached at other levels, for reasons including the failure (so far) of the global view to determine whether it will resolve disputes in an adversarial or a conciliatory manner. It is worth noting again the moral remainder that accompanies the inability of these different legal points of view to deliver corrective justice. Individual fishers may be affected profoundly by the results of their states' suffering damage from escaped GM fish, and compensation may eventually be offered. But in the interim, a great deal of pain may be felt by fishers who suffer disruption of their current livelihoods, their future prospects, and their self-conception as self-sustaining individuals. This kind of morally relevant suffering may not fit comfortably in a tort model of wrongdoing, but it does fit in the model of corrective justice I have employed. The threat posed by GM fish may be so significant that failure to manage risks in a reasonable way might reasonably attract attention from the emerging norms of international criminal law. Biodiversity is of huge importance to us all, and it is in this core idea of values of universal importance that we find the basis for criminalization.

The Way Forward

After all of this doom and gloom, I owe at least a general indication of how liability and redress for damage done by GM escapees ought to be handled in ideal theory and the nonideal world. This indication is likely best framed as a return visit to the three animating doubts of my preceding discussion: an observation about a shortage of needed factual and conceptual knowledge, a doubt about the force of international law, and a worry about the remainders left by existing legal avenues. On the empirical front, further research is needed to better understand the effects of escape of aquaculture organisms. The difficult empirical and conceptual question of the nature of redress appropriate to this context must also be taken up. I suspect something like a return to sustainability might emerge as the appropriate understanding of redress, and sustainability may be recognized as a property of ecosystems that are often regional and only sometimes national. In this context of a broader understanding of sustainability, we are overdue for a break from the international practice which resists the idea of punishment for infractions of environmental law. Biodiversity has special, universal value and may deserve special protection in a novel category of international law to capture crimes against biodiversity.

None of these insights will be of any use if left in abstract. Here I can only sketch the argument I intend to make in future work, suggesting that our best hope lies with a specific, comprehensive multilateral treaty on liability and redress for damage done by escaped aquaculture organisms. I suspect that a multilateral treaty is the only international instrument with sufficient normative force to escape the problems of enforceability I sketched earlier, and a specific treaty regarding aquaculture organisms is the only kind of agreement whose provisions are likely to be stated with sufficient precision to ensure justiciability. This treaty must capture concerns legitimately present in each of the local-sovereignty, regional, and global legal points of view. Interstate disputes over liability and redress ought to be resolved under a treaty framework whose provision for adjudication of disputes provides also for that dispute to be conceived in supranational terms, and for judgements which can require redress beyond that due the immediately damaged state, perhaps extending to ascription of guilt to individual wrongdoers. This sketch is only a small beginning, but there is some comfort in the fact that there is still time in which to do this work prior to widespread use of GM aquaculture organisms.

Notes

1. FAO Fisheries Circular no. 920 FIRM/C920, "Review of the State of World Fishery Resources: Marine Fisheries," Marine Resources Service, Fishery Resources Division, Fisheries Department, FAO, Rome, Italy, 1997.

2. Increase in growth rate is just one commercially valuable feature sought by genetic modification programs. At least one firm, Aquabounty Farms of Prince Edward Island, Canada, has developed salmon which grow four to six times faster than salmon of ordinary provenance. See http://webhost.avint.net/afprotein/ bounty.htm.

3. This assumes that some combination of industry lobbyists, new farming methods, and the global need for new sources of protein will overcome the currently popular strategy of requiring land-based farming of GM fish.

4. In the early days of aquaculture in North America, safety concerns motivated a practice of tagging farmed fish (kept captive in sea-cages) and ranched fish (released, and caught on return); but cost considerations have resulted in the suspension of this practice.

5. It is worth separating this scenario into dimensions of risk and threat. These two separable ideas are too often run together, except, notably in intelligence work, where spies of various sorts are careful to distinguish "risk" as a measurement of the actually likelihood of some identified "threats" of varying seriousness actually occurring.

6. By "ordinary provenance" I mean simply that farmed salmon are selected from existing stocks known to have commercially desirable traits. So, for example, most farmed salmon in Canada are Atlantic salmon that originate in the Saint John River.

7. In some jurisdictions, it is likely that GM fish will be raised (by legal requirement) in land-based containment facilities, to avoid the problem of escape. Whether this requirement withstands the efforts of aquaculture lobbyists to promote cheaper, possibly comparably safe containment in improved sea-cages remains to be seen.

8. *Report of the Auditor General of Canada*, December 2000, chap. 30, "Fisheries and Oceans—The Effects of Salmon Farming in British Columbia on the Management of Wild Salmon Stocks," p. 69: "The number of escaped Atlantic salmon in the northeast Pacific coastal waters may be increasing, although no clear trend has been established. Farm salmon routinely escape from their enclosures, either in small numbers or in large events as a result of net failures or tears. From 1991 to 1999, more than 345,000 Atlantic salmon, both juveniles and adults, reportedly escaped from B.C. salmon farms. While this number may appear relatively insignificant given the high level of production, even at this level of escapes

Atlantic salmon are entering and reproducing in B.C. streams. In fact, some very large escapes have occurred. For example, in the United States on 19 July 1997, 370,000 Atlantic salmon escaped from a farm site in Rich Passage, Washington. Atlantic salmon have been caught off the coast of Washington, B.C., and Alaska, and their presence has been documented in seventy-nine rivers and streams in B.C."

9. *Report of the Auditor General of Canada*, December 2000, chap. 30, p. 50, "There is recent evidence on the presence of Atlantic salmon in B.C. rivers and streams. At the time of the B.C. Salmon Aquaculture Review, Atlantic salmon ready to spawn had been observed in B.C. streams, but no juveniles from spawning Atlantic salmon had been captured. Since then, Atlantic salmon have successfully reproduced in at least two rivers on Vancouver Island, the Tsitika and the Amor de Cosmos. In view of the possible effects on life history processes of wild Pacific stocks, this information needs to be taken into account in assessing potential effects of an expanded industry. An increase in the number of salmon farms could result in an increase in the overall number of escaping Atlantic salmon. The larger the number of Atlantic salmon present in B.C. waters, the greater the chance of their disrupting the life history processes (migration, reproduction, competition, and predation) of the wild Pacific salmon."

10. The locus classicus for this is *Lotus*, PCIJ Series A, 1929.

11. The locus classicus for this principle is *Trail Smelter Arbitration (USA v Canada)* (1941) 3 RIAA 1905, 1965-66 (Arbitral Tribunal). This case dealt with sulphur dioxide emissions from a Canadian smelter damaging American territory. See also the earlier *Corfu Channel (Merits)* [1949] *ICJ Rep*, 4.

12. Paul Reuter, *Introduction to the Law of Treaties* (London: Pinter Publishers, 1989), 23.

13. Consider some operative phrases from a principle of the well-known *Rio Declaration on Environment and Development*:

Principle 11

States shall enact effective environmental legislation. Environmental standards, management objectives, and priorities should reflect the environmental and developmental context to which they apply. Standards applied by some countries may be inappropriate and of unwarranted economic and social cost to other countries, in particular developing countries. (*Rio Declaration on Environment and Development* UN Doc A/CONF 151/5/Rev 1, 13 June 1992.)

Even quick analysis reveals significant interpretive problems in this principle. On what basis might states be sanctioned for failing to provide "effective" legislation? How are contexts for environmental standards to be determined—economically, by geopolitical region, by ecosystem? And why might environmental concerns be legitimately sacrificed in favor of development in developing states? As laudable as the intention of the first sentence might be, the impetus it provides toward domestic legislation and effective multilateral treaties is severely weakened by linguistic imprecision and compromises whose specific application is deferred to further national and international decision-making bodies. In fact, some of these nontreaty instruments of international law may fail to be "law" as "law" is understood in municipal systems. These international instruments may violate a fundamental demand of the ideal of the rule of law: the requirement that legal obligations be framed in such a way that their subjects can be guided by them.

Consider further the kind of flexible commitment offered by Dr. Jacques Diouf, Director-General, Food and Agriculture Organization of the United Nations, in "Food for the New Millennium: Innovation in Nutrition, Safety, and Biotechnology" lecture of the Director-General on the occasion of the International Food and Nutrition Conference Tuskegee University, United States, 9 October 2000:

> With the rapid growth of aquaculture, the fisheries sector has also recognized that GMOs are a diverse class of organisms that share many common features with introduced or alien species. FAO's Regional Fisheries Bodies have adopted, in principle, codes of practice on the use of introduced species and GMOs. The general principles in such codes of practice have been incorporated into the FAO Code of Conduct for Responsible Fisheries. They deal with environmental assessment, contained use, advanced notification, and the application of the precautionary approach.
>
> I would like to take the opportunity of this conference to assure the international community that, through *holistic and multidisciplinary scientific* [emphasis added] approaches of evaluation, risk assessment, management, and communication, FAO will continue to address all issues of concern to its constituents, regarding biotechnology and its effects on human, plant, and animal health. In view of the importance of harmonizing regulations related to the testing and releasing of GMOs, FAO will continue, at the national, sub-regional, and regional levels, to strengthen its normative and advisory work, in coordination and cooperation with other international organizations.
>
> Harmonization of regulations would first address protocols for risk assessment for testing and releasing GMOs. Biosafety issues, pertaining to food safety, will continue to be addressed in the context of the Codex Alimentarius. As recent advances bring into agricultural production environments a diverse set of GMO-based technologies and transgenic animals, there will be the need for a more systematic consideration of the biosafety questions involved.

14. On the question of relative normativity, see especially Prosper Weil, "Towards Relative Normativity?" *American Journal of International Law* (1983): 413–42.

15. United Nations Convention on the Law of the Sea 1982, Art. 56(1)(a). This treaty is in fact the third revision of the long process of codification of the Law of the Sea, and for this reason is often referred to as UNCLOS III. The device of the Exclusive Economic Zone emerged in response to actions of some states such as Peru and Chile who sought to extend their "belt" of domestic jurisdiction into waters customarily regarded as "high seas" whose use was to be open to all states to be used equitably according to need. The existence of an EEZ is determined by claim. More than seventy states have made such a claim. For academic treatment of the emergence of the EEZ, see Shigeru Oda, *International Control of the Sea Resources* (Boston: Martin Nijhoff, 1989). For international case law development of the doctrines of internal waters, territorial waters, and high seas, see *UK v Iceland* [1974] ICJ Rep; *Gulf of Maine Case* [1984] ICJ Rep; *Libya v Malta (Continental Shelf)* [1985] ICJ Rep; *Guinea/Guinnea-Bissau Maritime Delimitation Case* [1985] ICJ Rep.

16. Arts. 66(1), 66(2), 66(3)(a) UNCLOS 1982.

17. *Agreement for Implementation of the Provisions of the United Nations Convention on the Law of the Sea of 10 December 1982 relating to the Conservation and Management of Straddling Fish Stocks and Highly Migratory Fish Stocks.* New York, 4 August 1995.

18. Bangkok Declaration, p. 18.

19. A superb example of the sovereignty model may be found in the *Council on Environmental Quality and Office of Science and Technology Policy Assessment: Case Studies of Environmental Regulations for Biotechnology, Case Study No.1 Growth Enhanced Salmon* (Washington, DC: Office of Science and Technology Policy, 2000). The objectives of the case study focus "on environmental oversight of the potential production of transgenic Atlantic salmon in net pens or other ostensibly contained conditions in or near the Atlantic or Pacific coastal waters of the United States, including tank rearing and hatchery operations associated with aquaculture production." Further, "This case study is aimed at illustrating the types of environmental safety considerations that would go into a U.S. government evaluation of a request for approval of a transgenic Atlantic salmon variety for use in aquaculture, and the government agencies and authorities involved. This case study is intended to give an overview of the federal oversight process, to point out any gaps, weaknesses, or ambiguities in that process, and to facilitate improvements in it." It is noteworthy that there is no reference to regional considerations or any international norms—the entire discussion is carried out within the assumption of sovereign isolation.

20. Arts. 63(1) and (2); 64(1) and (2) UNCLOS 1982.

21. Convention for the Conservation of Salmon in the North Atlantic (1982) Art. 3.

22. Resolution by the Parties to the Convention for the Conservation of Salmon in the North Atlantic Ocean to Minimize Impacts from Salmon Aquaculture on the Wild Salmon Stocks, Oslo (1994).

23. Ibid., Art. 2.

24. Agreement on Implementation of the Oslo Resolution, 1999.

25. NASCO Guidelines for Action on Transgenic Salmon, 1997.

26. The Protocol describes "emergency measures" to be taken in the event of "unintentional transboundary movement." Cartagena Protocol on Biosafety (2000), Art. 17.

27. Cartagena Protocol on Biosafety (2000), Art. 27. Article 34 provides further that state parties shall "consider and approve cooperative procedures and institutional mechanisms to promote compliance with the provisions of this Protocol and to address cases of non-compliance."

28. Report of the Intergovernmental Committee for the Cartagena Protocol on Biosafety on the Work of Its First Meeting, UNEP/CBD/ICCP/1/9, 2 February 2001, p. 32.

Part 6

Food Safety and Substantial Equivalence

Introduction

Are genetically modified foods categorically different from conventional foods, just a bit different, or are they the same? Those arguing that they are categorically different sometimes base their views on the fact that the products of transgenesis share genes from utterly different species. It may be true that so-called wide-crosses in conventionally bred hybrids can result in offspring from unrelated species, but the genetic differences are not as great as they are with, say, transgenic plants containing genes from bacteria.

Others argue that GM foods are either the same as their conventional counterparts, or perhaps just a bit different in virtue of the transgene that they carry. But, you must be thinking, the same in virtue of *what*? Surely the presence of a soil bacillus gene in a corn plant will make that plant materially different from a plant without the gene. Yet physical comparability between GM and conventional foods is not the criterion for measuring food safety. Rather, the criterion is how the food *functions* in our bodies once it is consumed. A GM food may be physically different from a conventional food, but if it looks, tastes, smells, feels, and digests the same, it is the same as the conventional food.

Or, more precisely, the GM food is thought to be *substantially equivalent* to the conventional counterpart—a misleading term, for *substance* is not the issue, *function* is. So long as the functions are *substantially* equivalent, which is to say, not discernably different in function, then the food is thought to be safe.

In the first reading, Nick Tomlinson documents how this criterion came to be the standard for food-safety assessments, and explains its current use. In the second, Henry Miller, an outspoken advocate of GM food, takes issue with critics who challenge substantial equivalence as a scientific concept. Miller, like Tomlinson, argues that it was never intended to be a scientific concept or protocol in its own right, but that it forms part of a larger food-safety assessment procedure. You will have to decide if the concept does indeed assume this more modest role, or if it is nevertheless expected to do the work of a rigorous scientific concept in actual practice. Ultimately, the litmus test for substantial equivalence as a food-safety criterion is described in the reading by Bob Buchanen that describes how the food-safety issue has come to focus on allergenicity. Is substantial equivalence a fine enough net to catch an allergenic—and potentially lethal—novel protein in a genetically modified food?

21

The Concept of Substantial Equivalence

Its Historical Development and Current Use

Nick Tomlinson

Introduction

This chapter describes how and why the concept of substantial equivalence was developed to facilitate the nutritional and safety assessment of foods derived from biotechnology. The chapter sets out a definition of substantial equivalence and illustrates how the concept is used by regulatory agencies worldwide.

Background

For many years the practical difficulties of obtaining meaningful information on the safety of whole foods from

Reproduced by permission of the Food and Agricultural Organization of the United Nations Joint FAO/WHO Expert Consultation on Foods Derived from Biotechnology, "Topic 1: The Concept of Substantial Equivalence, Its Historical Development and Current Use," 19 May–2 June 2000.

conventional toxicology studies have been well recognized. This became particularly apparent from the vast number of animal-feeding studies conducted to assess the safety of irradiated foods.

Animal studies are a major element in the safety assessment of many compounds such as pesticides, pharmaceuticals, industrial chemicals, and food additives. In most cases, however, the test substance is well characterized, of known purity, of no nutritional value, and human exposure is generally low. It is therefore relatively straightforward to feed such compounds to animals at a range of doses, some several orders of magnitude greater than the expected human-exposure levels, in order to identify any potential adverse effects of importance to humans. In this way it is possible, in most cases, to determine levels of exposure at which adverse effects are not present, and so set safe upper limits by the application of appropriate safety factors.

By contrast, foods are complex mixtures of compounds characterized by wide variation in composition and nutritional value. Because of their bulk and effect on satiety, they can usually only be fed to animals at low multiples of the amounts that might be present in the human diet. In addition, a key factor to consider in conducting animal studies on foods is the nutritional value and balance of the diets used, to try to avoid the induction of adverse effects which are not related directly to the material itself. Picking up any potential adverse effects and relating these conclusively to an individual characteristic of the food can therefore be extremely difficult. Another consideration in deciding the need for animal studies is whether

Table 1: Differences Between Chemical and Food Toxicity Evaluation, based on a paper by Dr. P. J. Rodgers[1]

Chemical	Food
Material usually simple, chemically precise substance	Complex mixture of many compounds
Highest dose level should produce an effect	Effects improbable at the maximum dose level that can be incorporated in the diet for the test species
Small dose (usually less than 1 percent of diet)	High intake (usually greater than 10 percent)
Easy to give excessive dose	Intakes above those normally present in the diet difficult
Acute effects obvious	Acute effects difficult to produce (usually absent)
Generally independent of nutrition	Nutrition dependent
Specific route of metabolism simple to follow	Complex metabolism
Cause/effect relatively clear	Cause/effect, if observed at all, may be confused

it is appropriate to subject experimental animals to such a study if it is unlikely to give rise to meaningful information.

The main differences between the toxicological evaluation of chemicals and whole foods are set out in table 1. In practice, very few foods consumed today have been subject to any toxicological studies yet are generally accepted as being safe to eat.

In developing a methodology for the safety assessment of new foods, it was essential to establish a benchmark definition of safe food.

Development of a Safety-Assessment Framework for GM Foods

Recognizing that the development of GM foods was progressing rapidly, the FAO and the WHO convened an expert consultation in 1990 on the "Assessment of Biotechnology in Food Production and Processing as Related to Food Safety."[2] The consultation recognized the limitations of traditional toxicological test methods when applied to whole foods and recommended that a more structured approach to safety assessment should be developed. The 1990 consultation identified the comparative principle whereby the food being assessed is compared with one that has an accepted level of safety, as being of considerable importance.

In 1993 the OECD published a report[3] on the safety evaluation of foods derived by modern biotechnology. This report, which was based on a number of intergovernmental consultations, included a definition of safe food.

> Food is considered safe if there is reasonable certainty that no harm will result from its consumption under anticipated conditions. Historically, food prepared and used in traditional ways is considered safe on the basis of long-term experience, even though it may naturally contain harmful substances. In principle, food is presumed to be safe unless a significant hazard has been identified.

This is not to say that many foods that are already widely consumed would not show adverse effects in animal studies if they could be fed at high enough doses. Equally, given the many adverse effects that can be observed with existing foods, it would be unreasonable to require a demonstration of absolute safety for novel foods.

The 1993 OECD report also expanded upon the comparative principle identified by the 1990 FAO/WHO consultation and formulated the concept of substantial equivalence.

The WHO and the FAO refined the concept further at an expert consultation meeting held in Rome in 1996.[4]

What is Substantial Equivalence?

Substantial equivalence is not a substitute for a safety assessment, but a part of the assessment process. As such, it provides a useful framework for regulatory scientists. Underlying the concept is the requirement that any safety assessment should show that a genetically modified variety is as safe as its traditional counterparts, through a consideration of both intended and unintended effects. This involves consideration of a wide range of information, including agronomic properties, phenotypic changes, and compositional data on critical nutrients and toxicants.

In the report of the 1996 expert consultation, substantial equivalence was identified as being "established by a demonstration that the characteristics assessed for the genetically modified organism, or the specific food product derived therefrom, are equivalent to the same characteristics of the conventional comparator. The levels and variation for characteristics in the genetically modified organism must be within the natural range of variation for those characteristics considered in the comparator and be based upon an appropriate analysis of data."

Critical nutrients and toxicants are components of a particular crop known to be relevant to human health, as determined through our knowledge of the unmodified crop and its related species. Comparative assessment of these components and their potential for change as a result of genetic modification, together with a wide range of other information on agronomic, phenotypic, and other properties, permits an assessment of the likelihood of unintended effects in a modified crop. For example, where the level of expression of a particular gene is altered, this is likely to be reflected in other changes in either the crop's composition or appearance.

In the application of substantial equivalence, these key components are considered on the basis of the long history of safe use of the traditional counterpart, and any differences are identified. The defined differences are then the subject of safety assessment, which can include nutritional, toxicological, and immunological testing as appropriate.

The concept of substantial equivalence has been used extensively as a tool in assessing the safety of GM foods. In comparing a GM food with a conventional counterpart, consideration is given to both intentional and unintentional effects. A wide range of information is used in these comparisons ranging from agronomic data such as crop height, yield, flowering pattern, disease resistance, and so on, through to compositional data on key nutrients and toxicants. In this context, key nutrients are those food components which may have a major impact on the total diet and include fats, proteins, and carbohydrates as well as minerals and vitamins. The comparison can result in one of three conclusions:

- The GMO or food product obtained from it is substantially equivalent to a conventional counterpart.

- The GMO or food product obtained from it is substantially equivalent to a conventional counterpart except for a few clearly defined differences.
- The GMO or food product obtained from it is not substantially equivalent to a conventional counterpart—either because the differences cannot be defined or because there is no existing counterpart to compare it with.

Where a food can be demonstrated to be substantially equivalent, it is considered to be as safe as its counterpart, and no further safety assessment is required. Where there are clearly defined differences between the GM food and its conventional counterpart, the safety implications of the differences need to be fully assessed. Where a food is not substantially equivalent, it does not mean that the food is unsafe. However, there would be a need for extensive data to be provided to demonstrate its safety.

How Substantial Equivalence Is Currently Used in Practice?

The concept of substantial equivalence has been integrated into safety-assessment procedures used by regulatory authorities worldwide. In many countries, for example, the EU, Australia, Canada, and the United States, the safety-assessment process has been formalized in a series of decision trees which guide regulators and potential applicants through the various stages of the comparative process.

At present, each regulatory authority could, in theory, determine a unique set of components they wish to see analyzed. In practice, there is already broad agreement on key components, although work is progressing to develop international consensus on a core set of components on a crop-by-crop basis. Although the concept has been interpreted in slightly different ways by various regulatory authorities, the overall approach to the safety assessment is very similar in all countries. The area where there is perhaps the greatest scope for divergence is which of the three categories identified above that a particular novel food is assigned to. This is largely due to differing interpretations as to what constitutes a difference from a conventional counterpart.

While the foods derived from biotechnology that are on sale have all been assessed for safety using the substantial equivalence concept and the results endorsed by the respective governments, there have been some published criticisms of substantial equivalence. Many of these criticisms are based on a misunderstanding of the concept. Nevertheless, such criticisms provide a useful stimulus to ensure that safety-assessment procedures are kept at the forefront of scientific knowledge.

Conclusions

The concept of substantial equivalence has been used by regulatory authorities worldwide for approximately ten years. In this time, some forty products have been assessed, and the complexity of the genetic modifications has increased. For some products, particularly commodity crops such as soya, many millions of tons have been consumed with no evidence of any adverse health effects. The concept has stood the test of time, although it is important that it is kept under review to ensure it remains the most appropriate mechanism for assessing the nutritional and food-safety implications of foods derived from biotechnology.

Notes

1. OECD, Food Safety Evaluation, OECD, Paris, 1996.

2. WHO, "Strategies for Assessing the Safety of Foods Produced by Biotechnology," report of a joint FAO/WHO consultation, WHO, Geneva, 1991.

3. OECD, "Safety Evaluation of Foods Produced by Modern Biotechnology—Concepts and Principles," OECD, Paris, 1993.

4. "Biotechnology and Food Safety," report of a joint FAO/WHO consultation Rome 30 September–4 October 1996, FAO Food and Nutrition Paper 61, 1996.

Substantial Equivalence

Its Uses and Abuses

Henry Miller

These are trying times for biotechnology applied to agriculture and food production. Most recently, the accepted paradigms for genetically modified food risk assessment and management have come under attack from the fringes of the scientific community. These attacks have been based on the overinterpretation of experiments, flawed data, and false assumptions about risk and risk assessment. They have included the overinterpretation of a laboratory experiment that showed a modest effect on monarch butterflies of larvae fed milkweed dusted with pollen from recombinant corn containing a *Bacillus thuringiensis*[1] toxin, and a gratuitous controversy about methodologically flawed experiments by Arpad Pusztai that purportedly showed toxicity in rats fed recombinant, lectin-enhanced potatoes.[2]

Reprinted with permission from *Nature*, by Henry Miller, "Substantial Equivalence: Its Uses and Abuses," vol. 17, no. 11 (November 1999): 1042–43. Copyright © 1999 Nature Publishing Group.

The latest attack comes in the form of a commentary by Erik Millstone et al. criticizing the application of a concept called "substantial equivalence" to the risk assessment of foods derived from recombinant DNA–manipulated organisms.[3] Deriding substantial equivalence as a "pseudo-scientific concept because it is a commercial and political judgment masquerading as if it were scientific," Millstone et al. appear to misunderstand the concept, its origins, and its purpose. Their arguments are symptomatic of ideological opponents of the new biotechnology torturing logic and science in order to manipulate government regulation to obstruct the use of a technology that they dislike for any number of reasons.

Millstone et al. also appear to be unaware of the prevailing regulatory standards for new foods, and of the experience with and the scientific consensus about assessing the safety of products derived through recombinant DNA techniques.

The history of the term "substantial equivalence," first applied to food by work at the Paris-based Organization for Economic Cooperation and Development (OECD), is important. In 1986 the OECD's Group of National Experts on Safety in Biotechnology reached a consensus that

> While rDNA techniques may result in the production of organisms expressing a combination of traits that are not observed in nature, genetic changes from rDNA techniques will often have inherently greater predictability compared to traditional techniques, because of the greater precision that the rDNA technique affords; [and] it is expected that any risks associated with applications of rDNA organisms may be assessed in generally the same way as those associated with non-rDNA organisms.[4]

Others echoed this consensus. In 1992, *Nature* editorialized that

> the same physical and biological laws govern the response of organisms modified by modern molecular and cellular methods and those produced by classical methods. . . . [Therefore] no conceptual distinction exists between genetic modification of plants and microorganisms by classical methods or by molecular techniques that modify DNA and transfer genes.[5]

This language was virtually identical to that in a landmark 1989 report of the U.S. National Research Council, which went even further, observing that

> Recombinant DNA methodology makes it possible to introduce pieces of DNA, consisting of either single or multiple genes, that can be defined in function and even in nucleotide sequence. With classical techniques of gene transfer, a variable number of genes can be transferred, the number depending on the mechanism of transfer; but predicting the precise number or the traits that have been transferred is difficult, and we cannot always predict the phenotypic expression that will result. With organisms modified by molecular methods, we are in a better, if not perfect, position to predict the phenotypic expression.[6]

This last quotation is remarkable: It expresses the widely held scientific consensus that our ability to predict "phenotypic expression," the very essence of risk-assessment related to environmental protection and public health, is superior for gene-spliced foods!

The OECD's experts group subsequently took up food safety specifically, concluding in a 1993 report that

> Modern biotechnology broadens the scope of the genetic changes that can be made in food organisms, and broadens the scope of possible sources of foods. This does not inherently lead to foods that are less safe than those developed by conventional techniques. Therefore, evaluation of foods and food components obtained from organisms developed by the application of the newer techniques does not necessitate a fundamental change in established principles, nor does it require a different standard of safety.[7]

In that same report, the group described the concept of substantial equivalence in new foods not, as asserted by Millstone et al. as a scientific principle, but merely as a kind of regulatory shorthand for defining those new foods that do not raise safety issues that require special, intensive, case-by-case scrutiny. (The appropriation of the concept and the name—both borrowed from the U.S. Food and Drug Administration's (FDA) definition of a class of new medical devices that do not differ materially from their predecessors, and thus, do not raise new regulatory concerns—was suggested by a member of the OECD experts group, senior U.S. White House adviser John J. Cohrssen.)

The OECD has continued to explore the concept of substantial equivalence, another expert group concluding in 1998 that

> While establishment of substantial equivalence is not a safety evaluation per se, when substantial equivalence is established between a new food and the conventional comparator [that is, an antecedent], it establishes the safety of the new food relative to an existing food and no further safety consideration is needed.[8]

It bears repeating that substantial equivalence is not intended to be a scientific formulation; it is a conceptual tool for food producers and government regulators, and it neither specifies nor limits the kind or amount of testing needed for new foods.

The FDA's 1992 policy on foods from "new plant varieties" is instructive in this regard, because although the agency does not formally use the term, it applies the concept in its risk-based, scientifically defensible approach.[9] This policy applies irrespective of whether the plant arose by gene-splicing or "conventional" genetic engineering methods. The FDA does not routinely subject foods from new plant varieties to premarket review or to extensive scientific safety tests. Instead, it considers that the usual safety and quality-control practices used by plant breeders—mostly chemical and visual analyses and taste testing—are generally adequate for ensuring food safety.

The FDA's policy defines certain safety-related characteristics of new foods that, if present, require greater scrutiny by the agency. These include the presence of a substance that is completely new to the food supply, an allergen presented in an unusual or unexpected way (for example, a peanut protein transferred to a potato), changes in the levels of major dietary nutrients, and increased levels of toxins normally found in foods. Additional tests are performed when suggested by the products composition, characteristics, or history of use. For example, potatoes are generally tested for the glycoalkaloid solanine, because this natural toxin has been detected at harmful levels in some new potato varieties that were developed with conventional genetic techniques.

The absence of such characteristics that are correlated with enhanced risk, in effect, defines foods that are substantially equivalent to antecedent products. Foods lacking characteristics that raise safety issues are not subjected to premarket FDA review.

By considering all recombinant DNA–mediated genetic changes—but only those—as novel, Millstone et al. are inconsistent.

They seem unimpressed by the fact that thousands of foods containing gene-spliced ingredients have for years been consumed routinely and safely by consumers in Europe and North America. Likewise, they ignore the fact that many products on the market are derived from "wide crosses," hybridizations in which genes are moved from one species or one genus to another to create a variety of plant that does not and cannot exist in nature. They demand extensive, difficult to perform, hugely expensive "biological, toxicological, and immunological"[10] testing of foods from recombinant plants, but not of other foods from the dozens of new plant varieties improved with far less precise traditional techniques of genetic modification, such as hybridization, that enter the marketplace each year without premarket review or special labeling.

But the testing of fundamentally benign whole foods—gene-spliced or not—in animal-feeding studies, for example, is limited by factors such as the animal's qualitative and quantitative feeding preferences, and by the levels of nutritional and nonnutritional substances that are present in the food under study. When Harry Kuiper of the State Institute for Quality Control of Agricultural Products in Wageningen, the Netherlands, attempted to determine the toxic threshold for a recombinant tomato by feeding rats freeze-dried tomato extract, the experiments were limited to the equivalent of thirteen tomatoes a day because of the negative effects of naturally occurring inorganic compounds such as potassium in the tomato powder. But, as noted by Kuiper, "toxicologists still said we hadn't fed them enough to get a meaningful result."[11]

Wholly ignoring such empirical experience, as well as scientific consensus, Millstone et al. suggest that gene-spliced food should be treated "in the same way as novel chemical compounds, such as pharmaceuticals, pesticides, and food additives, and [requiring] a range of toxicological tests, the evidence from which could be used to set acceptable daily intakes (ADIs)."[12] Then, of course, we would need "regulations . . . to ensure that ADIs are never, or rarely, exceeded."[13]

This sort of argument illustrates the fallacy that underlies many of the unscientific attacks on the new biotechnology—the assumption that somehow gene-splicing systematically introduces into organisms (and the foods derived from them) greater uncertainty or risk than other, older, less precise techniques. As described above, neither scientific consensus nor empirical evidence supports that view.

If new and Draconian regulatory regimens are appropriate for the new gene-splicing biotech, they are certainly also applicable to the old traditional biotech. In that regard, one must wonder how we would calculate the ADI for the mutant peach called a nectarine, or the tangerine-grapefruit hybrid called a tangelo. Such an exercise would clearly be absurd. And where it is not absurd—such as when estimating the acceptable intake of foods known to have high endogenous levels of solanine, such as potatoes—the exercise has nothing to do with the method of genetic manipulation used to construct the plant.

A central theme of Millstone et al.[14] is that the widespread endorsement and use of substantial equivalence is, in effect, the result of a conspiracy by industry and government to avoid the adequate testing of foods from recombinant organisms. Not only are they inaccurate in suggesting that the testing of the foods themselves is deficient, but they have completely ignored the quality assurance that is part of the certification of plant seeds sold to agricultural producers or growers, in order to prevent any compromise of seed quality or consistency.

In California, for example, oversight is performed by the nonprofit California Crop Improvement Association (CCIA), which provides a voluntary quality-assurance program for the maintenance and increase of crop seed. (CCIA is the designated California authority for the international seed-certification scheme administered by OECD in forty countries.) Each variety that enters this program is evaluated for its unique characteristics such as pest resistance, adaptability, uniformity, quality, and yield.

Seed production is closely monitored by CCIA to prevent outcrossing, weed, and other crop and disease contamination that may negatively affect seed quality. Seed movement is monitored from field harvest, through the conditioning plant, and into the bag. Samples can be rejected if "off-type" seeds are found at a percentage that is greater than standards permit, as is occasionally the case with beans, cereals, and sunflowers.

Healthy skepticism is necessary for the evolution of scientific thought and discussion, to be sure, but so are consistency and the application of accurate assumptions, and of late we have seen far too little of these latter elements in the debates about the new biotechnology used for agriculture and food production. Discussions of public policy, like those about science, cannot tolerate those who take lightly the moral obligation to report strictly what is true, and in the proper context. Everyone must be held to that standard.

Notes

1. John E. Losey, Linda S. Rayor, and Maureen E. Carter, "Transgenic Pollen Harms Monarch Larvae," *Nature* 399, no. 6733 (1999): 214.

2. Ehsan Masood, "The Search for Missing Mass Finds Funds for UK Researchers," *Nature* 398, no. 6723 (1999): 98.

3. Erik Millstone, Eric Brunner, and Sue Mayer, "Beyond 'Substantial Equivalence,'" *Nature* 401, no. 6753 (1999): 525–26.

4. *Reconbinant DNA Safety Considerations* (Paris: Organization for Economic Cooperation and Development, 1986).

5. "U.S. Biotechnology Policy," editorial, *Nature* 356, no. 6364 (1992): 1–2.

6. *Field Testing Genetically Modified Organisms: Framework for Decisions* (Washington, D.C.: National Academy Press, 1989).

7. *Safety Evaluation of Foods Derived by Modern Biotechnology* (Paris: Organization for Economic and Cooperation and Development, 1998).

8. "Report on the OECD Workshop on the Toxicological and Nutritional Testing of Novel Foods," SG/ICGB,OECD, Paris, 1998

9. U.S. Food and Drug Administration statement of policy, *Foods Derived from New Plant Varieties*, Federal Register 57, 22984-23005 (1992).

10. Millstone, Brunner, and Mayer, "Beyond 'Substantial Equivalence.'"

11. Debora MacKenzie, "Unpalatable Truth," *New Scientist* 162, no. 2182 (1999): 18.

12. Millstone, Brunner, and Mayer, "Beyond 'Substantial Equivalence.'"

13. Ibid.

14. Ibid.

23

Genetic Engineering and the Allergy Issue

Bob B. Buchanan

Although much has been learned since the field was put on a scientific basis at the turn of the last century, our knowledge of food allergies is far from complete. It is still unclear, for example, why only certain individuals are affected and why, even among them, the problem is often restricted to childhood.

It is also not clear why the allergies caused by various nuts and aquatic animals tend to persist and be lifelong. Milk, egg, soy, and wheat are the major food allergies in children, whereas peanut, tree nuts, shellfish, and fish are most prevalent in adults.

The field is complicated by the fact that many more people believe they suffer from food allergies than is actually the case. Thus, although up to 20 percent of Americans have a perceived food allergy, the problem can be medically diagnosed in only about 2 percent of the population.[1] The issue is further clouded by confusion with food intolerance and by evidence that allergies are increasing rapidly in developed countries for reasons that are only beginning to be understood. These factors collectively contribute to the lack of understanding that has long been a part of the food allergy field.

Aside from limited attention drawn to the increased prevalence, food allergy has historically attracted little notice. However, with the advent of genetic engineering and its application to the production of food, the situation has changed dramatically.

The development and commercialization of a variety of food crops with transgenes has thrust the allergy issue onto a public stage and given the field unprecedented exposure worldwide. Although not yet apparent, I believe the allergy and food-technology fields will benefit from this attention in the long term, akin to the progress made in understanding the cellular immune system as a result of publicity brought by the acquired immune deficiency syndrome epidemic.

Why the Sudden Interest?

The increased public awareness of food allergy has arisen from a combination of three factors: reasoned concern, fear through ignorance, and political motivation.

The first two factors are expected and limited in scope. The third, which was unanticipated and amplifies the second, stems from the goal of certain individuals and environmental organizations to delay the commercial development of genetic engineering, especially as applied to food. The allergy issue was selected because of its vulnerability: In addition to its enigmatic nature mentioned previously, opponents of genetic engineering recognized early on that it is difficult to determine with absolute certainty whether a protein introduced into a food by genetic engineering is a potential allergen. In retrospect, one wonders why the allergy issue was not raised earlier—for example, in the countless plant-breeding programs since World War II—that significantly have not converted nonallergenic into allergenic foods. A new allergen has been introduced independently of plant breeding. The introduction of kiwi, a relatively obscure fruit, led to the development of a new allergy in the general population of the developed world.

Interest in the allergy issue has been heightened by knowledge that a protein known to be an allergen in one species remains an allergen when transferred by genetic transformation to a second species. An example of such a protein, now widely known, is the Brazil nut allergen (2S protein) transferred to soybean.

The allergenicity associated with the original 2S protein in Brazil nut was found to be retained after it was overexpressed in soybean.[2] Although not surprising, this example is reassuring in documenting that the scientific community is capable of detecting and identifying a known allergen that has been transferred from one species to another by genetic engineering. As a result of the allergy tests, the transgenic soy product in question was not further developed as a commercial product.

In this commentary, I shall identify the issues surrounding the allergy issue and discuss their scientific validity, rather than the production of hypoallergenic foods by genetic engineering—a research focus of a number of laboratories, including ours. I then turn to a discussion of the tools available to address the concerns and where we are in their resolution. It will be seen that a solution to this problem appears to lie on the near horizon.

What Are the Issues?

Concern about the genetic modification of food appears to stem from three questions: Is the protein of interest an allergen? Has the protein of interest become an allergen as a result of the transformation and selection process? Has the transformation and selection process in some unknown way altered a normal cellular protein so that it has become an allergen?

Scientific Basis for the Concerns

The first question, whether a particular protein is an allergen, is valid and should be answered. The second question, based on the conversion of the protein of interest into an allergen (for example, by glycosylation), also relates to a change that is biochemically feasible. One would think that indications of such a change would have surfaced with significantly abundant proteins in earlier plant-breeding programs. Nonetheless, this point should be tested, at least until we have a greater understanding of the fate of transgenic proteins in plants. The last question, which raises the possibility that a given protein of the cell could become an allergen as a result of transformation and selection, is less tenable. However, this question, like the other two, will continue to be raised until additional experience has been gained and consumers have expressed confidence in genetically modified foods, especially those based on a protein to which the human population has not been previously exposed.

Current Tools for Solving the Problem

The question of whether a transgene product is an allergen or whether its presence unintentionally renders a food product more allergenic than its nonengineered counterpart is addressed in several ways, including (a) comparing the predicted amino acid sequence of the transgene product with that of known food allergens; (b) determining the abundance of the protein in food as significant food allergens typically represent one percent or more of the total protein; (c) examining the expressed protein for characteristics often associated with known food allergens, such as glycosylation, heat stability, and presence of disulfide bonds; and (d) monitoring the digestibility of the transgene product in simulated mammalian gastric and intestinal fluids.

Although numerous nonallergens show one or more of the properties often associated with allergens, each analysis provides indirect evidence that is of some predictive value. Moreover, the tests to determine these properties were included in a decision tree that was proposed by Metcalfe et al.[3] As far as I know, the protocol suggested in that tree has been closely followed in the industrial development of transgenic food products.

However, as a result of recent problems in introducing new transgenic foods, it has become clear that an additional test is needed, namely an animal model for testing genetically modified products.

An animal model is needed to provide a direct test of the allergenic properties for proteins showing potential evidence of allergenicity. Such tests cannot be done on humans directly, ethical considerations aside. Present populations have not been exposed to the engineered food in question and, as a result, would not show an adverse reaction, even if the food contained an allergen. In developing the decision tree, Metcalfe et al. pointed out the desirability of including an animal model, but did not do so because none "have been shown to predict the allergic potential of introduced proteins." Animal models were also a major topic of discussion at a recent conference dedicated to allergy issues, "Assessment of the Potential Allergenicity of Genetically Engineered Foods," held December 5 and 6, 2000, at the National Center for Food Safety and Technology (Summit-Argo, Illinois). The advantages and disadvantages of each model were considered at the meeting: Brown Norway rat, guinea pig, dog, pig, and various mouse models. To be beneficial, it was considered that an animal model should: (a) show an allergic response to allergens in humans, but not to nonallergens; (b) show an allergen profile similar to that of humans—for example, the response to a strong allergen (peanut) > moderate allergen (milk) > a nonallergen (spinach leaf); (c) have a gastrointestinal system similar to humans; and (d) ideally, show an epitope response similar to humans. This latter feature was considered a desirable but not a mandatory feature in view of the wide range of epitopes that humans can recognize.

The advantages, disadvantages, and current status of each model were discussed in Summit-Argo. It was agreed that, although decisive progress has been made, none of the current models meets these criteria because characterization and testing is still ongoing. Therefore, at this point it is not clear which of the models will prove to be of most value in detecting and assessing food allergens.

I am personally prone to the dog because, perhaps as a reflection of having a gastrointestinal system similar to humans,[4] it is unique among animal models in having natural allergies as far as is known. The dog shows clinical symptoms typical of food allergy in humans, that is, vomit and diarrhea.[5] Advances made using the dog will, therefore, benefit dogs as well as humans because of similarities in their allergic response. In recognition of these features, our laboratory started a project to determine the suitability of the dog as a predictor of allergens in humans in collaboration with Dr. Oscar L. Frick (University of California, San Francisco) and Drs. Laura Privalle and Greg delVal (Syngenta, Research Triangle Park, North

Carolina). Initiated three years ago, this study is now entering its final stage and is yielding encouraging results. The results, which will be published when the study is complete, suggest that the dog will be useful as an animal model. That point withstanding, the other models mentioned above warrant continued study, because, in the end, each of several could present a particular advantage in detecting and characterizing allergens in humans.

One precautionary note seems in order. While proceeding with allergy testing, we must be careful not to overregulate and impose undue restrictions to stifle innovation. Rather, we should seek to formulate a balanced policy that insures food safety without hindering product development.

Closing Comments

Great strides have been made in our understanding of food allergy since the problem was originally recognized by Hippocrates almost 2.5 millennia ago. Despite this rich history, large gaps remain in our knowledge, and they are of such nature as to lend an element of mystery to the field. These features have led certain individuals and environmental groups to target food allergy in an effort to slow the commercial development of genetically modified crops and foods and, at the same time, utilize the issue as a fund-raising mechanism. Their efforts have been successful not only by having the intended effect, but also by negatively influencing science funding, especially in Europe. The net result has been that the participating organizations have experienced financial gain and genetically modified crops derived from research in developed countries are now being grown disproportionately in the developing world. For example, between 1999 and 2000, the area used for growing transgenic crops increased by 2 percent in industrial countries, whereas the area in developing counterparts, although still relatively small in total hectares, grew by 51 percent.[6] The long-term economic effect of the shift in emphasis to developing countries could significantly impact research on transgenic crops in developed countries unless the situation changes. Such an impact on research would eventually adversely affect hunger and nutrition worldwide because, as recently pointed out in this series,[7] continued progress in the genetic engineering of crops is critical to feeding future world populations.

I believe, however, the problem to be transitory and that, once appropriate allergen-testing capability is in place, health concerns will abate and the development of transgenic foods will continue apace. As seen above, the needs to bring about this change are not extensive. What seems to be most lacking at this stage is an animal model to identify transgenic plant proteins that either are, or have become, allergens in humans. Such a model is especially important for proteins to which humans have not been exposed. Had a reliable model been available, it is likely that StarLink corn could have avoided current problems.[8] Animal test data would have been available to allay consumer concern once the product was on the market. I am confident that, with progress now being made, one or more animal models will soon be available to serve as a reliable indicator of allergens in humans and that a safe but reasonable testing policy will be formulated. Once such testing capability is in hand, the public will respond in a positive manner. In the long term, the food-allergy and technology fields will likely benefit, rather than suffer, from this pause in their development.

Notes

1. Daryl R. Altman and Lawrence T. Chiaramonte, "Public Perception of Food Allergy," *Journal of Allergy and Clinical Immunology* 97, no. 6 (1996): 1247–51.

2. Julie A. Nordlee et al., "Identification of a Brazil-Nut Allergen in Transgenic Soybeans," *New England Journal of Medicine* 334, no. 11(1996): 688–92.

3. Dean D. Metcalfe et al., "Assessment of the Allergenic Potential of Foods Derived from Genetically Engineered Crop Plants," *Critical Review of Food Science and Nutrition* supplement 36 (1996): S165–86.

4. Donald R. Strombeck and W. Grant Guilford, *Small Animal Gastroenterology,* 2d ed. (Davis, Calif.: Stonegate Publishing, 1990), 346–55.

5. Richard Ermel et al., "The Atopic Dog: A Model for Food Allergy," *Lab Animal Science* 47 (1997): 40–49; Gregorio del Val et al., "Thioredoxin Treatment Increases Digestibility and Lowers Allergenicity of Milk," *Journal of Allergy and Clinical Immunology* 103 (1999): 690–97.

6. Clive James, "Global Review of Commercialized Transgenic Crops, 2000," International Service for the Acquisition of Agri-biotech Applications, no. 21 [online], http://www.isaa.org/publications/briefs/Brief_17.htm.

7. Norman E. Borlaug, "Ending World Hunger: The Promise of Biotechnology and the Threat of Antiscience Zealotry," *Plant Physiology* 124, no. 2 (2000): 487–90.

8. David Barboza, "Negligence Suit Is Filed over Altered Corn," *New York Times,* 4 December 2000, C2.

Part 7

Risk Assessment and Public Perception

Introduction

Genetically modified foods are supposed to bring a wide variety of benefits to people and the environment. They also pose risks as we saw in the previous section. There are other kinds of risk in addition to food safety, such as uncertain environmental impacts and the potential harmful effects on food security. Gabrielle Persley and James Siedow think these risks are inherent in the technology itself. Other risks they call "technology transcendent" risks, and these arise in the context in which the technology is deployed. GM crops affect groups of people differently, depending on their sociopolitical contexts. Anticipating these effects requires that we go beyond the technology itself to consider the people who will be affected by it. If the public is going to be affected by GM foods, presumably they ought to be able to say for which benefits they are willing to take risks.

Sagar, Daemmrich, and Ashiya expand on this point by noting that the vast majority of people—the "commoners"—are not consulted about the adoption of biotechnology in agriculture. They will be affected by it, but they are often presumed to not really understand the nature of scientific risk analysis. Scientific risk assessments on the health and environmental risks are necessary, but for the vast majority of people in this world, the real threat is that widespread adoption of agricultural biotechnology will deepen existing wealth and knowledge inequities between developed and developing countries. The idea of the "biotech divide" is fully developed in Juma and Fang's paper later in the book. In Sagar et al., the problem is that commoners—regular folk who make up the majority—are tyrannized by pro- and antibiotech minorities. Are commoners' concerns ever properly addressed?

Wolfgang van den Daele describes a participatory technology assessment in Germany, the point of which was to find out what the unheard commoners wanted of biotechnology. Van den Daele reports that while no "conclusive reasons" to ban herbicide-tolerant GM crops within existing regulations were found, the process was derailed when critics of genetic engineering failed to attend the final, resolution-setting session. Was it a failure? Van den Daele thinks that consensus will not always be reached with participatory processes. So if the "commoners" are at the table with all other groups, consensus may not be reached, and political decisions might have to be "taken in dissent."

24

Applications of Biotechnology to Crops

Benefits and Risks

Gabrielle J. Persley and James N. Siedow

Introduction

The purpose of this chapter is to summarize the recent scientific developments that underpin modern biotechnology and to discuss the potential risks and benefits when these are applied to agricultural crops. This introduction is intended for a general audience who are not specialists in the area but who are interested in participating in the current debate about the future of genetically modified crops. This debate is particularly timely with the forthcoming discussion of a new round of international trade talks in Seattle in December 1999 where international trade in genetically modified organisms (GMOs) will be an issue. This chapter is restricted to genetically modified crops. It is the intention of the Council for Agricultural Science and Technology (CAST)

Council for Agricultural Science and Technology. Issue Paper No. 12. "Application of Biotechnology to Crops: Benefits and Risks," December 1999.

to produce a series of subsequent papers that will address some of these issues in more detail and in the broader context of genetic modification beyond crops.

Terminology

Biotechnology refers generally to the application of a wide range of scientific techniques to the modification and improvement of plants, animals, and microorganisms that are of economic importance. *Agricultural biotechnology* is that area of biotechnology involving applications to agriculture. In the broadest sense, traditional biotechnology has been used for thousands of years, since the advent of the first agricultural practices, for the improvement of plants, animals, and microorganisms.

The application of biotechnology to agriculturally important crop species has traditionally involved the use of *selective breeding* to bring about an exchange of genetic material between two parent plants to produce offspring having desired traits such as increased yields, disease resistance, and enhanced product quality. The exchange of genetic material through conventional breeding requires that the two plants being *crossed* (bred) are of the same, or closely related, species. Such active plant breeding has led to the development of superior plant varieties far more rapidly than would have occurred in the wild due to random mating. However, traditional methods of gene exchange are limited to crosses between the same or very closely related species; it can take considerable time to achieve desired results; and frequently, characteristics of interest do not exist in any related species. Modern biotechnology vastly increases the precision and reduces the time with which these changes in plant characteristics can be made and greatly increases the potential sources from which desirable traits can be obtained.

Methods

In the 1970s, a series of complementary advances in the field of molecular biology provided scientists with the ability to readily move DNA between more distantly related organisms. Today, this recombinant DNA technology has reached a stage where scientists can take a piece of DNA containing one or more specific genes from nearly any organism, including plants, animals, bacteria, or viruses, and introduce it into a specific crop species. The application of recombinant DNA technology frequently has been referred to as genetic engineering. An organism that has been modified, or transformed, using modern techniques of genetic exchange is commonly referred to as a *genetically modified organism* (GMO). However, the offspring of any traditional cross between two organisms also are "genetically modified" relative to the genotype of either of the contributing parents. Plants that have been genetically modified using recombinant DNA technology to introduce a gene from either the same or a different species also are known as *transgenic plants*, and the specific gene transferred is known as a *transgene*. Not all GMOs involve the use of cross-species genetic exchange; recombinant DNA technology also can be used to transfer a gene between different varieties of the same species or to modify the expression of one or more of a given plant's own genes, for example, to amplify the expression of a gene for disease resistance.

The application of recombinant DNA technology to facilitate genetic exchange in crops has several advantages over traditional breeding methods. The exchange is far more precise because only a single (or at most, a few), specific gene that has been identified as providing a useful trait is being transferred to the recipient plant. As a result, there is no inclusion of ancillary, unwanted traits that need to be eliminated in subsequent generations, as often happens with traditional plant breeding. Application of recombinant DNA technology to plant breeding also allows more rapid development of varieties containing new and desirable traits. Further, the specific gene being transferred is known so the genetic change taking place to bring about the desired trait also is known, which often is not the case with traditional breeding methods where the fundamental basis of the trait being introduced may not be known at all. Finally, the ability to transfer genes from any other plant or other organism into a chosen recipient means that the entire span of genetic capabilities available among all biological organisms has the potential to be genetically transferred or used in any other organism. This markedly expands the range of useful traits that ultimately can be applied to the development of new crop varieties. As a hypothetical example, if the genes that allow certain bacteria to tolerate high external levels of salt can serve the same purpose when transferred into crops such as potatoes, wheat, or rice, then the production of such improved food crops on marginally saline lands may be possible. Given that the acreage of such saline lands is estimated to be equal to 20 to 25 percent of the land currently under cultivation worldwide, this would be a significant contribution toward global food security.

Two primary methods currently exist for introducing transgenic DNA into plant genomes in a functional manner. For plants known as dicots (broad-leaved plants such as soybean, tomato, and cotton), transformation is usually brought about by use of a bacterium, *Agrobacterium tumefaciens*. *Agrobacterium* naturally infects a wide range of plants and it does so by inserting some of its own DNA directly into the DNA of the plant. By taking out the undesired traits associated with *Agrobacterium* infection and inserting a gene(s) of interest into the *Agrobacterium* DNA that will ultimately be incorporated into the plant's DNA, any desired gene can be transferred into a dicot's DNA following bacterial infection. The cells containing the new gene subsequently can be identified and grown using plant cell culture technology into a whole plant that now contains the new transgene incorporated into its DNA. Plants known as monocots (grass species such as maize, wheat, and rice) are not readily infected by *Agrobacterium*, so the external DNA that is to be transferred into the plant's genome is coated on the surface of small tungsten balls, and the balls are physically shot into plant cells. Some of the

Some Useful Definitions of Biotechnology and Its Component Technologies[1]

Biotechnology is any technique that uses living organisms or parts thereof to make or modify a product or improve plants, animals, or microorganisms for specific uses. All the characteristics of a given organism are encoded within its *genetic material*, which consists of the collection of deoxyribonucleic acid (DNA) molecules that exist in each cell of the organism. Higher organisms contain a specific set of linear DNA molecules called *chromosomes* and a complete set of chromosomes in an organism comprises its *genome*. Most organisms have two sets of genomes, one having been received from each parent. Each genome is divided into a series of functional units, called *genes*, there being 20,000 to 25,000 such genes in typical crop plants like corn and soybean. The collection of traits displayed by any organism (*phenotype*) depends on the genes present in its genome (*genotype*). The appearance of any specific trait also will depend on many other factors, including whether the gene(s) responsible for the trait is turned on (*expressed*) or off, the specific cells within which the genes are expressed, and how the genes, their expression, and the gene products interact with environmental factors.

The key components of modern biotechnology are as follows.

- *Genomics:* The molecular characterization of all the genes and gene products of a species.
- *Bioinfomatics:* The assembly of data from genomic analysis into accessible and usable forms.
- *Transformation:* The introduction of single genes conferring potentially useful traits into plants, livestock, fish, and tree species.
- *Molecular breeding:* The identification and evaluation of useful traits in breeding programs by the use of marker-assisted selection, for plants, trees, animals, and fish.
- *Diagnostics:* The more accurate and quicker identification of pathogens by the use of new diagnostics based on molecular characterization of the pathogens.
- *Vaccine Technology:* based on the use of modern immunology to develop recombinant DNA vaccines for improved disease control against lethal diseases.

DNA comes off of the balls and is incorporated into the DNA of the recipient plant. Those cells can also be identified and grown into a whole plant that contains the foreign DNA.

The ability to easily incorporate genetic material from virtually any organism into many different crop plants has reached the stage of commercial applicability. About 50 percent of the maize, soybeans, and cotton grown in the United States in 1999 had been modified using recombinant technologies. The major technical limitation on the application of recombinant DNA technology to improving plants is insufficient understanding of exactly which genes control agriculturally important traits and how they act to do so.

The study of genes involves the rapidly developing field of *genomics*, which refers to determining the DNA sequence and identifying the location and function of all the genes in an organism. It appears that many traits are conserved between species, that is, the same gene confers the same trait in different species. Thus, a gene for salt tolerance in bacteria may confer salt tolerance if it is transferred and expressed in rice or wheat. The advent of large-scale sequencing of entire genomes of organisms as diverse as bacteria, fungi, plants, and animals, is leading to the identification of the complete complement of genes found in many different organisms. This is dramatically enhancing the rate at which an understanding of the function of different genes is being achieved. From the standpoint of agricultural biotechnology, advances in genomics will lead to a rapid increase in the number of useful traits that will be available to enhance crop plants in the future.

Why Are Agricultural Biotechnology Products Being Developing?

New developments in agricultural biotechnology are being used to increase the productivity of crops, primarily by reducing the costs of production by decreasing the needs for inputs of pesticides, mostly in crops grown in temperate zones. The application of agricultural biotechnology can improve the quality of life by developing new strains of plants that give higher yields with fewer inputs, can be grown in a wider range of environments, give better rotations to conserve natural resources, provide more nutritious harvested products that keep much longer in storage and transport, and continue low-cost food supplies to consumers.

After two decades of intensive and expensive research and development in agricultural biotechnology, the commercial cultivation of transgenic plant varieties has commenced over the past three years. In 1999, it is estimated that approximately 40 million hectares of land were planted with transgenic varieties of over twenty plant species, the most commercially important of which were cotton, corn, soybean, and rapeseed.[2] The countries include several of the world's major producers and exporters of agricultural commodities: Argentina, Australia, Canada, China, France, Mexico, South Africa, Spain, and the United States. Approximately 15 percent of the area is in emerging economies. The value of the global market in transgenic crops grew from U.S. $75 million in 1995 to U.S. $1.64 billion in 1998.

The traits these new varieties contain are most commonly insect-resistance (cotton, corn), herbicide-resistance (soybean), and delayed fruit ripening (tomato). The benefits of these initial transgenic crops are better weed and insect control, higher productivity, and more flexible crop management. These benefits accrue primarily to farmers and agribusinesses, but there are also economic benefits accruing to consumers in terms of maintaining food production at low prices. The broader benefits to the environment and the community through reduced use of pesticides contribute to a more sustainable agriculture and better food security. Crop/input trait combinations presently being field-tested in emerging economies include virus-resistant melon, papaya, potato, squash, tomato, and sweet pepper; insect-resistant rice, soybean, and tomato; disease-resistant potato; and delayed-ripening chili pepper. There also is work in progress to use plants such as corn, potato, and banana as minifactories for the production of vaccines and biodegradable plastics.

Further advances in biotechnology will likely result in crops with a wider range of traits, some of which are likely to be of more direct interest to consumers, for example, by having traits that confer improved nutritional quality. Crops with improved output traits could confer nutritional benefits to millions of people who suffer from malnutrition and deficiency disorders. Genes have been identified that can modify and enhance the composition of oils, proteins, carbohydrates, and starch in food/feedgrains and root crops. A gene encoding beta-carotene/vitamin A formation has been incorporated experimentally in rice. This would enhance the diets of the 180 million children who suffer from the vitamin A deficiency that leads to 2 million deaths annually. Similarly, introducing genes that increase available iron levels in rice threefold is a potential remedy for iron deficiency that affects more than 2 billion people and causes anemia in about half that number.

The new developments in gene technology also may be useful to solve problems in human health care, agriculture, and the environment in poor countries, given the chance. So far, the major research and development efforts of the private sector in biotechnology have been directed at opportunities for introducing traits useful to producers in the markets in industrial countries, because this is where bioscience companies are able to recoup their investments. New modalities that mobilize both public and private resources are needed if poor people are not to be bypassed by the genetic revolution.

The recent report of the Nuffield Council on Bioethics in the United Kingdom[3] concluded that there is a compelling moral imperative to enable emerging economies to evaluate the use of new biotechnologies as tools to combat hunger and poverty. Creative partnerships between the developing countries, the international agricultural research centers, and the private sector could provide new means for sharing and evaluating these new technologies. Several emerging economies are making major investments of human and financial resources in biotechnology with the aim of using these new developments in science to improve food security and reduce poverty. These developments were discussed in detail at a conference in Washington, D.C., in October 1999 co-sponsored by the Consultative Group on International Agricultural Research (CGIAR) and the U.S. National Academy of Sciences.[4]

Applications of biotechnology in agriculture are in their infancy. Most current genetically engineered plant varieties are modified only for a single trait, such as herbicide tolerance or pest resistance. The rapid progress being made in genomics may enhance plant breeding as more functional genes are identified. This may enable more successful breeding for complex traits such as drought and salt tolerance, which are controlled by many genes. This would be of great benefit to those farming in marginal lands worldwide, because breeding for such traits has had limited success with conventional breeding of the major staple food crops.

Benefits and Risks of Agricultural Biotechnology

In assessing the benefits and risks involved in the use of modern biotechnology, there are a series of issues to be addressed so that informed decisions may be made as to the appropriateness of the use of modern biotechnology when seeking solutions to current problems in food, agriculture, and natural resources management. These issues include risk assessment and risk management within an effective regulatory system as well as the role of intellectual property management in rewarding local innovation and enabling access to technology developed by others. In terms of addressing any risks posed by the cultivation of plants in the environment, there are six safety issues proposed by the Organization for Economic Cooperation and Development (OECD) that need to be considered. These are gene transfer, weediness, trait effects, genetic and phenotypic variability, expression of genetic material from pathogens, and worker safety.[5]

In making value judgments about risks and benefits in the use of biotechnology, it is important to distinguish between technology-inherent risks and technology-transcending risks. The former include assessing any risks associated with food safety and the behavior of a biotechnology-based product in the environment. The latter emanate from the political and social context in which the technology is used and how these uses may benefit and/or harm the interests of different groups in society.

Technology-Inherent Risks

In terms of technology-inherent risks, the principles and practices for assessing these risks on a case-by-case basis are well established in most OECD countries and several emerging economies. These principles and practices have been summarized in a series of OECD reports published over the past decade or more. National, regional, and international guidelines for risk assessment and risk management provide a basis for national regulatory systems. Biosafety guidelines are available from several international organizations including the OECD, the United Nations Environment Program, the United Nations Industrial Development Organization, and the World Bank.

Principles, Practices, and Experience

Regulatory trends to govern the safe use of biotechnology to date, include undertaking scientifically based, case-by-case, hazard identification and risk assessments; regulating the end product rather than the production process itself; developing a regulatory framework that builds on existing institutions rather than establishing new ones; and building in flexibility to reduce regulation of products after they have been demonstrated to be of low risk.

The biosafety risk assessments conducted prior to thousands of experimental and field trials in the United States focus on the characteristics of the organism being assessed, including its novel traits, intended use of the organism, and features of the recipient environment. The concept of substantial equivalence between new and traditional products has been used as a basis for determining what safety tests are needed before commercialization of products derived from genetic engineering, and if product labeling is required, and if so, what information would be useful to consumers. Familiarity has emerged as a key biosafety principle in some countries. Although familiarity cannot be equated with safety, it has provided the basis for applying existing management practices to new products, and is premised upon case-by-case and step-by-step risk assessment and management of new products. This approach has been recommended by the OECD and is the basis of the U.S. regulatory system.[6]

A recent development, partly in response to negative public reactions to the growing use of genetically modified crops in agriculture in some countries, has been the introduction of measures in a number of countries, especially in Europe, and most recently Japan, to label some or all biotechnology-based products, with the aim of giving consumers more choice. There is also a view by some regulatory authorities for regulatory requirements relating to GMOs to be based on a more precautionary approach. This approach is based on the proposition that not enough may be known about long-term adverse effects of GMOs, and thus requires prior evidence of the safety of biotechnology-based products for human health and the environment. The current debate on labeling includes the issues of whether product labeling should be mandatory or voluntary, what information should be on the label so as to inform consumers as to their choice, and whether labeling is feasible in bulk commodities that may contain a mixture of GMO and non-GMO crops.

Toward an International Biosafety Protocol

During the negotiations to establish the Convention on Biological Diversity in the early 1990s, there was concern expressed by some governments that GMOs may pose a risk to biological diversity. Consequently, intergovernmental negotiations have been in progress over the past several years to negotiate a legally binding biosafety protocol under the Convention on Biological Diversity (CBD). The centerpiece of the draft protocol is an advance informed-agreement (AIA) procedure to be followed prior to the transboundary transfer of GMOs (called living modified organisms or "LMOs" in the protocol). LMOs that will come into contact with the environment of an importing country are to be covered under the AIA, to assess them for any potential adverse impacts on biodiversity. There is debate, however, as to which LMOs should be regulated by the protocol and for what purpose. Is the intention to provide international oversight of specific traits in LMOs that may adversely affect human health and the environment and/or impact on biodiversity, or is the AIA procedure to be focused on oversight of the gene-technology processes by which the LMOs were produced?

A key point of disagreement centers around whether LMOs, which are intended for food, feed, or processing rather than for use as seed in the importing country, should be covered under the AIA procedure. These LMOs, called "commodities," would include GM crops such as soya or corn, which form a growing component of the international agricultural-commodity trade in these crops. A group of major agricultural exporting countries (the Cairns group) argues that agricultural commodities should be excluded from the AIA procedure, because such LMOs are not intended for release into the environment and therefore cannot pose a threat to biological diversity. This is consistent with current trade in commodities, under existing international agreements, where seed contaminated with plant diseases can be marketed internationally for consumption but not for planting. The Cairns group also contend that providing detailed information on LMOs in bulk agricultural-commodity shipments is not feasible, given the commingling of genetically modified and conventional seed, as well as the lack of a direct business link between seed growers and exporters. Other countries are calling for all first-time transfers of LMOs, including commodities, to be covered by AIA, as the only way to monitor entry of such LMOs into a country. Some also believe that the protocol should allow for consideration of any human health impacts of LMOs as well as their environmental impact. These countries also point out that "intended use" of LMOs for processing (rather than planting into the environment) cannot always be guaranteed once these commodities are within a country's borders.

Another key dispute within the biosafety protocol negotiations is how decisions under AIA can be based on science and precaution. Those calling for sound science to be the basis for decision making note that reliance on an excessively precautionary approach could result in discriminatory or unjustifiable barriers to international trade in LMOs. Those favoring additional precautionary approaches note that unambiguous scientific evidence of harm relating to LMOs may not be forthcoming in the short term. The latter argue, therefore, for the need for precaution in the face of scientific uncertainty to ensure the safety of genetically modified products for human health and the environment. Countries also disagree about whether socioeconomic effects of LMOs, liability and compensation, and pharmaceutical products should be included in the protocol, although these topics fall outside the scope of the protocol, as set by the Conference of the Parties to the CBD in Jakarta in 1994 (Decision 2/5).

The final major issue is how a country's obligations under the CBD and any agreed biosafety protocol should relate to a country's rights and obligations under World Trade Organization (WTO) agreements. The next round of negotiations for the Biosafety Protocol are to be held in Montreal in January 2000.

Effects on Human Health

The health effects of foods grown from genetically modified crop varieties (sometimes called GM foods) depends on the specific content of the food itself and may have either potentially beneficial or occasional harmful effects on human health. For example, a GM food with a higher content of digestible iron is likely to have a positive health effect if consumed by iron-deficient individuals. Alternatively, transfer of genes from one species to another may also transfer allergic risk, and these risks need to be evaluated and identified prior to commercialization. Individuals allergic to certain nuts, for example, need to know if genes conveying this trait are transferred to other foods such as soybeans and would labeling be required if such crops were to be commercialized. There is also some concern as to the potential health risks from the use of antibiotic-resistance markers in GM foods, although there is no evidence of this.

Labeling also may be needed in some countries to identify other novel content resulting from genetic modification for cultural and religious reasons or simply because the consumers want to know what is the content of the food and how it was produced to make an informed choice, independent of any health risks.

Risks to the Environment

Among the potential ecological risks identified are increased weediness, due to cross-pollination whereby pollen from GM crops spreads to non-GM crops in nearby fields. This may allow the spread of traits such as herbicide-resistance from genetically modified plants to nontarget plants, with the latter potentially developing into weeds. This ecological risk may be assessed when deciding if a GMO with a given trait should be released into a particular environment, and if so, under what conditions. Where such releases have been approved, the monitoring of the behavior of GMOs after their release is a rich field for future research in crop ecology.

Other potential ecological risks stem from the widespread use of genetically modified corn and cotton with insecticidal genes from *Bacillus thuringiensis* (the Bt genes). This may lead to the development of resistance to Bt in insect populations exposed to the GM crops. An attempt to manage this risk is being done in the early plantings of GM crops by planting "refuge" sections of Bt-cotton fields with insect-susceptible varieties to reduce the opportunity of the insect population to evolve toward resistance to the plants having the Bt gene for resistance.[7] There also may be a risk to nontarget species, such as birds and butterflies, from the plants with Bt genes. The monitoring of these effects of new transgenic crops in the environment and the devising of effective risk-management approaches is an essential component of further research in risk management.

Technology-Transcending Risks

Technology-transcending risks include the social and ethical concerns that modern biotechnology may increase the prosperity gap between the rich and the poor, both internationally and within individual societies, and that it may contribute to a loss of biodiversity. There also are ethical concerns as to the moral dimensions of patenting living organisms and the cross-species movement of genes. These risks relate to the use of the technology, not the technology itself. The management of these risks requires policies and practices that give consumers choices while also promoting environmentally sustainable development through the judicious use of new developments in science and technology. The reduction of biodiversity is a technology-transcending risk. The reduction of biological diversity due to the destruction of tropical forests, conversion of more land to agriculture, overfishing, and the other practices to feed a growing world population is more significant than any potential loss of biodiversity due to the adoption of genetically modified crop varieties. This is not an issue restricted to transgenic crops. Farmers have adopted new commercially developed varieties in the past and will continue to do so when they perceive this to be to their advantage. On occasion, introduced varieties may enhance biological diversity, as for example for wheat in Turkey and corn in Mexico where new landraces are evolving by genetic introgression of genes from improved varieties into traditional landraces.

To slow the continuing loss of biodiversity, the main tasks are the preservation of tropical forests, mangroves and other wetlands, rivers, lakes, and coral reefs. The fact that farmers replace traditional varieties with superior varieties does not necessarily result in a loss of biodiversity. Varieties that are under pressure of substitution also can be conserved through in vivo and in vitro strategies. Improved governance and international support are necessary to limit loss of biodiversity. Actually or potentially useful biological resources should not be lost simply because we do not know or appreciate them at present.[8]

Regulatory Systems

Risks and opportunities associated with GM foods may be integrated into the general food-safety regulations of a country. The regulatory processes are a matter of continuing scrutiny and debate at the national and international levels as more products of biotechnology come close to market. A science-based, efficient, transparent regulatory system, which enjoys the confidence of the public and the business and farming communities, is essential in enabling the effective use of biotechnology. This system should be closely associated with existing regulatory arrangements for new pharmaceuticals, foods, and agricultural and veterinary products. National regulatory systems are complemented by international technical guidelines. National food-safety and biosafety regulations should reflect international agreements, a society's acceptable risk levels, the risks associated with not introducing modern biotechnology, as well as alternative means to achieve the desired goals.

Intellectual Property Management

Trade-related intellectual property rights (TRIPS) also will be an issue related to biotechnology and food at the forthcoming Seattle round of WTO negotiations. There is need for a fair system for intellectual property (IP) management that protects the interests of the inventors while promoting the safe use of the new biotechnologies. All countries who are signatories to the WTO have agreed to put in place a system for the protection of intellectual property rights, including protection of new plant varieties, although many have still to do so. These new IP systems need to include ways to reward not only the inventors of new technologies but also those farmers who have been traditional improvers of plant varieties over centuries. There also is a need to devise suitable systems for intellectual property protection that encourage and reward innovation at all levels and for all countries, not only for the technologically sophisticated.

Conclusion

The issues of major concern in relation to the future applications of biotechnology to crop improvement include the evaluation of any risks to human health and the environment; the need for mandatory and/or voluntary labeling of GM foods and/or agricultural commodities for international trade; the relationship between countries' responsibilities under the WTO; and international environmental treaties. These include the international protocol on biosafety being negotiated under the Convention on Biological Diversity, and whether this will provide oversight on traits and/or processes of genetic modification.

Governments and other responsible parties should effectively communicate with the public about the nature of new crop types and new crop varieties, about the unity of life processes in all organisms, and about the risks and benefits of agricultural biotechnology in their own country and internationally. There also is a need to continually improve the transparency and broad participation in the decision-making processes in relation to biotechnology, the release of genetically modified organisms into the environment, and the approval of genetically modified foods for commercial use.

Notes

1. Gabrielle J. Persley and John J. Doyle, overview to *Biotechnology for Developing Country Agriculture: Problems and Opportunities,* IFPRI 2020 Vision Focus 2 Briefs, October 1999 (Washington, D.C.: International Food Policy Research Institute, 1999).

2. Clive James, *Global Review of Commercialized Transgenic Crops 1999* (Ithica, N.Y.: International Service for the Acquisition of Agricultural Biotechnologists, 1999).

3. Nuffield Council on Bioethics, *Genetically Modified Crops: The Ethical and Social Issues* (London: Nuffield Council on Bioethics, 1999).

4. Consultative Group on International Agricultural Research (CGIAR), *Summary Report of a CGIAR/NAS International Conference on Biotechnology,* October 21–22, 1999 (Washington, D.C.: World Bank, 1999).

5. R. James Cook, "Toward Science-Based Risk Assessment for the Approval and Use of Plants in Agricultural and Other Environments," in *Proceedings of a CGIAR/NAS International Conference on Biotechnology,* October, 21–22, 1999 (Washington D.C.: World Bank, in press).

6. Calestous Juma and Aarti Gupta, "Safe Use of Biotechnology," brief no. 6 in *Biotechnology for Developing Country Agriculture: Problems and Opportunities,* IFPRI 2020 Vision Focus 2 Briefs, October 1999 (Washington, D.C.: International Food Policy Research Institute, 1999).

7. Fred Gould, "Sustainable Use of Genetically Engineered Crops in Developing Countries," in *Proceedings of a CGIAR/NAS International Conference on Biotechnology,* October, 21–22, 1999 (Washington D.C.: World Bank, in press).

8. Klaus M. Leisinger, "Disentangling Risk Issues," brief no. 5 in *Biotechnology for Developing Country Agriculture: Problems and Opportunities,* IFPRI 2020 Vision Focus 2 Briefs, October 1999 (Washington, D.C.: International Food Policy Research Institute, 1999).

The Tragedy of the Commoners

Biotechnology and Its Publics

Ambuj Sagar,
Arthur Daemmrich, and Mona Ashiya

In pluralistic societies, innovations offering new ways to promote independence or individuality are generally rapidly embraced, whereas those that appear to centralize power and authority are either outright attacked, or accepted only after contentious debates. Biotechnology, however, has repeatedly broken this mold, offering the potential for remarkably individualized products while also appearing to concentrate capabilities and power in the hands of a small number of global players. For example, biotechnology's promise to provide health care specific to individual needs goes hand in hand with giving corporations access to detailed genetic information. Likewise, its potential to fine-tune food production to local needs coincides with concentrating the capabilities to do so in a

Reprinted with permission from *Nature Biotechnology*, by Ambuj Sagar, Arthur Daemmrich, and Mona Ashiya, "The Tragedy of the Commoners: Biotechnology and Its Publics," vol. 18 (1 January 2000): 2–4. Copyright © 2000 Nature Publishing Group.

small number of institutions. Understandably, such scenarios have met with both support and opposition. The most recent, and perhaps the most contentious of the resulting controversies, has been the highly visible and ongoing debate on genetically modified (GM) agricultural products. Many observers seem to be surprised by the sudden rise of this issue, and its origin is often ascribed to a lack of public trust in regulatory institutions.

In one sense, the emergence of the GM issue should not be seen as surprising: it is only one of numerous debates over the applications of biotechnology. Scientists, politicians, and the public have also expressed concern about aspects of cloning, xenotransplantation, genetic testing, and gene prospecting, to name a few other prominent examples. We suggest that a major factor in the emergence of these controversies has been neglect of the needs, interests, and concerns of the primary stakeholders—the "commoners"—in the biotechnology arena. The sustainable development of biotechnology will require a renewed focus on stakeholders and their needs. This, in turn, demands a clearer understanding of public concerns as well as attention to issues of institutional structure and representation in decision-making processes. The framework presented in this paper suggests some steps in this direction. While many of the points raised in the following analysis are based on examples from agricultural biotechnology, our approach also applies to other areas where ongoing development hinges on proper process in the construction of new forms of governance.

The Complexities of Controversy

Observers of recent developments in the agricultural biotechnology arena often claim that different degrees of trust in regulatory institutions have shaped the lay public's response to the products of this new technology. Whereas protests have targeted GM foods in Europe, Americans have been comparatively quiet and accepting of such new foods. Some analysts have claimed that this difference emerges from greater regulatory transparency and an ability to modulate agency decisions (often through the courts) in the United States in contrast to the closed-door, elite decision-making procedures employed in Europe. They further suggest that experiences with BSE and dioxin-tainted foods have eroded the credibility of regulators in Europe to safeguard the public interest.

Whereas the above is true to some extent, a broader analysis of the factors governing debates over GM products shows limitations of reducing the controversy solely to degrees of trust in environmental and health safety agencies. A more textured account indicates that the public is also concerned with issues such as globalization and stratification of power, ethics, equity, and individual rights and choice. Such intertwined themes underlie not just the European public disquiet over GM products, they also cut across a number of issue areas in biotechnology (see figure 1 for a representative list). Put another way, concerns of "commoners" focus not just on the narrow aspects of applications of individual technologies, but also on the broader institutional and political context in which they are introduced. As sociologist Dorothy Nelkin has noted, "Controversies over science and technology are struggles over meaning and morality, over the distribution of resources, and over the locus of power and control."[1] Lay individuals, unhindered by disciplinary boundaries, often display a remarkable sensitivity to issues that policy experts fail to address. As a result, risks that are reduced to narrow variables and deemed amenable to traditional "scientific/technical" analyses by experts are not resolved easily once they become public controversies. In the current policy discourse, attention is paid mostly to specific issues such as GM products, genetic testing, or gene prospecting as they become contentious. There is little systematic analysis of underlying currents that lead to controversies.

Concerns of the Commoners

Public concerns must be understood as factors that shape the very discourse—the style and content of interactions—surrounding biotechnology around the globe. They determine the structure and boundaries of disputes and are key considerations for conflict resolution. Here we discuss two particularly important cross-cutting concerns to illustrate how they are germane to debates in biotechnology.

Risk and uncertainty. Recent public protests against GM foods are indicative of a divide between expert and lay perceptions of risk and uncertainty. Scholars of risk analysis have long noted that risk estimates vary significantly between experts and the public.[2] Public risk perception is influenced as much by social relations and feelings of power and powerlessness as by objective knowledge about the likelihood of large-scale accidents or individual harm. Furthermore, there is an increasing sense that the shift from an industrial era to the information age has produced a "risk society," characterized by self-inflicted dangers, increasing interdependence of human decisions, worldwide implications even of seemingly local events, and controversies that undermine the basis for calculating risk and insurance.[3] In such a society, regulatory institutions, corporations, and even scientists who base decisions regarding field trials or widespread marketing of

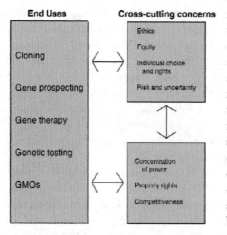

Figure 1. Concerns of "commoners" focus not just on the narrow aspects of applications of individual technologies, but also on the broader institutional and political context in which they are introduced.

GM foods on risk/benefit calculations fail to grasp the deeper ethical and social bases that shape public opinion.

Consequently, even when government officials in Europe, Asia, and Latin America respond to public concerns by instituting additional testing and regulatory requirements on biotechnology, they fail to account for the broader unquantifiable concerns of the "commoners." Attention has, instead, focused on environmental and human health risks alone. For example, this past June, the European Union's environment ministers adopted a decision imposing a de facto moratorium on new marketing approvals for GM organisms. Furthermore, ministers from Denmark, France, Greece, Italy, and Luxembourg independently insisted that they would block new applications to market GM seeds, plants, or foodstuffs until a new regulatory regime is put in place.[4] These announcements came two days after the Japanese Ministry of Agriculture, Forestry, and Fisheries (MAFF) stated it would tighten safety regulations on GM crops in the wake of a Cornell University study showing that pollen from insect-resistant Bt (Bacillus thuringiensis) maize is potentially toxic to monarch butterfly larvae. The MAFF will suspend approval of Bt crops in Japan until revised safety protocols are developed for GM crops. In a third example, the Brazilian state of Rio Grande do Sul has declared itself a "genetically modified free zone." This followed a Brazilian federal court decision in June barring the planting or distribution of GM soybeans in the country.

These events all illustrate the degree to which broad uncertainty is being examined in a traditional risk framework, making it amenable to testing and regulation under existing structures. But, in this process, moral and cultural unease as well as concerns about the unknown (or perhaps unknowable) consequences of some aspects of biotechnology are swept aside. Given this divide, the lay public is likely to increasingly demand "socially precautionary" principles in regulatory oversight. Decision makers should consider how they have represented the public and its concerns when making complex and often incommensurable calculations of economic benefits against ecological and health risks. All too often, key concerns are lost in the processes through which institutions model and represent stakeholders.

Globalization and its discontents. Current modes of exchanging knowledge, services, and consumer goods are rapidly reshaping the social, political, and economic structures of the world. This process of globalization is widening troubling gaps in access to knowledge and financial resources. In turn, these disparities are fueling criticism of new technologies—even within highly developed countries—based on several key features.

First among the issues to be considered are increasing income/resource disparities. The developing world, home to 80 percent of the world's population, generates less than 20 percent of the global gross domestic product, and there are few signs this divide will shrink in the foreseeable future. In 1997 the richest fifth of the world had an income seventy-four times that of the poorest fifth, more than double the inequality in 1960.[5]

Second, patterns of industrial ownership are changing. Recent mergers have led to the formation of transnational corporations of unprecedented size and wealth. Worldwide mergers in 1998 were worth $2.4 trillion and reflected a 50 percent increase over 1997, which itself was a record year.[6] In the biotechnology sector alone, mergers and acquisitions increased from $9.3 billion to $172.4 billion in the decade between 1988 and 1998.[7]

Third, there have been major shifts in the nature of financial flows to developing countries. Whereas public development aid dropped from $56.9 billion in 1990 to $47.9 billion in 1998, private flows to developing countries increased from $43.9 billion to $227.1 billion in the same period.[8] As a percentage of their combined gross national product, contributions from members of the OECD's Development Assistance Committee have fallen to their lowest level ever, from 0.33 percent in 1992 to 0.22 percent in 1997.[9]

Fourth, disparities are increasing in knowledge generation and utilization. North America, Europe, Japan, and newly industrialized countries accounted for 84.5 percent of the $470 billion spent on research and development worldwide in 1994 (the latest year for which data are available).[10] A similar trend is also evident when considering patent filings. Industrialized countries hold 97 percent of patents worldwide, and more than 80 percent of patents granted in developing countries belong to individuals or corporations based in industrialized countries.[11]

These disparities pose critical challenges to governments, particularly those of developing countries. In today's world where knowledge is considered the single "most important factor in determining the standard of living,"[12] the biotechnology pillar of the knowledge economy is often seen as increasing global divisions. For example, future applications of biotechnology may increasingly depend on genetic diversity. A substantial portion of the global genetic resources reside in the South, while the capabilities to commercially utilize them lie in the North. This raises serious concerns about appropriate benefit sharing. Furthermore, as developing countries become more dependent on private sources for capital, choices available to promote and expand research in biotechnology are constrained. Notably, even pragmatic analysts from the North are concerned that future innovations will be limited under an emerging industry structure where the top five biotechnology firms control more than 95 percent of gene-transfer patents.[13]

More broadly, critical reactions to emerging applications of biotechnology are based on control of research agendas, access to useful technologies, influence in decision-making forums, and debates over who will benefit from new technologies. Biotechnology is expected to give further impetus to the current process of globalization. Given the aforementioned concerns, it often serves as a lightning rod for expressing disquiet about the apparent future of the world. While such public concerns may not be easy to face or assuage, downplaying their relevance will only exacerbate the current polarization.

Who Speaks for the Stakeholders?

Given the substantial social and economic consequences of biotechnology, a broad swath of society has become a stakeholder in discussions on its applications. At the same time, a variety of institutions are involved in these debates. Figure 2 presents some major categories of relevant institutions and stakeholders.

Importantly, while some stakeholders may be able to participate directly in discussions about the adoption of new technologies, most are unable to do so. For this reason, channels to incorporate stakeholder views into debates take on critical importance. To some extent, such perspectives are inserted into discussions and decision making about biotechnology through participating institutions. However, these have to balance a number of interests and agendas, including those of the institutions themselves, of their internal constituents, and of stakeholders they explicitly represent. For example, the views of a biotechnology firm may be shaped by the personal ambitions of the company officers and scientists, the obligation to ensure returns to shareholders, and the need to satisfy customers. Similarly, nongovernmental organization (NGO) positions reflect the passions of activists, the concerns of their donors, and their public support base.[14] Such balancing acts may often give short shrift to the already marginalized voices of external stakeholders and raise serious issues about the authenticity of representation, even though many institutions derive their legitimacy from "representing" stakeholders.

Even the most well meaning organizations can view the interests of beneficiaries in a simplistic manner, or distort them unwittingly. Moreover, institutions may indulge strategically in a "virtual representation" of their stakeholders that offers both moral authority in defense of their own views and reduced costs of actual representation. The recent GM foods debate illustrates a situation where supporters and opponents of this technology have invoked their concern for the disenfranchised to defend their positions. Biotechnology firms, some government agencies, and other organizations cite the potential contribution of biotechnology to resolve world hunger, while NGOs highlight the potential adverse impacts of biotechnology on farmers and ecosystems in developing countries.[15]

Thus, some hard questions should be asked of any institution that participates in a given debate: Whose interests does it purport to represent and whose interests does it actually repre-

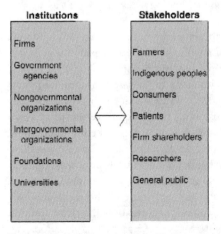

Figure 2. Major categories of institutions and stakeholders in biotechnology.

sent through its actions? How does it model and understand stakeholders' interests? Is the institution's present mode of representation consistent with past actions and stances in other policy arenas? More broadly, are the interests of all stakeholders represented equally? Finally, how is the institution accountable to its stakeholders? Such questions apply equally to industry, advocacy groups, and academia. Requiring institutions to provide answers should compel them to be more honest in their representation of stakeholders' views.

The issue of accountable representation is central to the biotechnology debate. Ultimately, stakeholders' interests should be given priority, not only for reasons of justice and fairness, but also because sustainable development of biotechnology will require widespread acceptance by the public of its goods and services. Accountable representation necessitates the integration of stakeholder interests into activities including basic and applied science, policy research, advocacy, and policy formulation and implementation.

Conclusion

The issue of trust certainly will continue to be of major significance for biotechnology in the coming years. But this issue is far more complex than generally portrayed, and not easily reduced to "trust in regulatory institutions." In our view, public trust—the ultimate arbiter of any technology in the marketplace—is based on perceptions of how a technology will influence the lives of various individuals, of how specific firms as well as the industry as a whole have represented public interests, and of the social, political, and economic landscape that serves as a backdrop for technological change. Thus, it becomes fundamentally important to pay attention to the "commoners" in the biotechnology debate, as well as to their needs and concerns.

Biotechnology's future ultimately relies on governing institutions listening and responding to the public, rather than discounting key stakeholders as irrational, scientifically illiterate, or technophobic. Institutions such as the biotechnology industry and government agencies stand to gain greater acceptance only by soliciting public input, implementing policies in a transparent and democratically representative fashion, and demonstrating their responsiveness to concerns raised by scientific experts, other organizations, and citizens and consumers around the world. New interactive approaches must be developed in order to bring stakeholders together and allow them to articulate cross-cutting concerns. Industry, given its stake in the debate, will need to go furthest by not merely recognizing "the public" as stakeholders, but by granting them some of the rights normally afforded citizens of modern states: open access to information, the opportunity to comment on proposed actions, the right to receive reasoned explanations, and above all, recognition that dissent can be bridged only through compromise.

Notes

1. Dorothy Nelkin, "Science Controversies," in *Handbook of Science and Technology Studies,* eds. S. Jasanoff et al. (Thousand Oaks, Calif.: Sage, 1995), 445.

2. Paul Slovic, "Beyond Numbers: A Broader Perspective on Risk Perception and Risk Communication," in *Acceptable Evidence: Science and Values in Risk Management,* eds. Deborah G. Mayo and Rachelle D. Hollander (New York: Oxford University Press, 1991), 48–65.

3. Ulrich Beck, *Risk Society: Towards a New Modernity* (London: Sage, 1992).

4. United Nations Development Program, *Human Development Report 1999* (New York: Oxford University Press, 1999).

5. Ibid.

6. "How to Merge: After the Deal," *Economist* (9 January 1999): 21–23.

7. United Nations Development Program, *Human Development Report 1999.*

8. World Bank, *Global Development Finance 1999* (Washington, D.C.: World Bank, 1999).

9. Organization for Economic Cooperation and Development (OECD), "Aid and Private Flows Fell in 1999," press release, 18 June 1998.

10. Rémi Barré, *World Science Report 1998* (Paris: UNESCO, 1998).

11. United Nations Development Program, *Human Development Report 1999.*

12. World Bank, *World Development Report 1999* (Washington D.C.: World Bank, 1999).

13. United Nations Development Program, *Human Development Report 1999.*

14. Of course, institutions and individuals are shaped by their past experiences. Institutional and personal experiences with disparate issues such as nuclear power, pesticides, globalization, or individual privacy often shape current perceptions in the debate on GM foods.

15. See, for example, Gabrielle J. Persley and John J. Doyle, overview to *Biotechnology for Developing Country Agriculture: Problems and Opportunities,* IFPRI 2020 Vision Focus 2 Briefs, October 1999 (Washington, D.C.: International Food Policy Research Institute, 1999); AgrEvo, "To Nourish Ten Million People," AgrEvo background paper, Berlin, 1998; Rural Advancement Foundation International, "Traitor Technology: The Terminator's Wider Implications," RAFI communiqué, Winnipeg, Manitoba, Canada, January/February 1999.

Risk Prevention and the Political Control of Genetic Engineering

Lessons from a Participatory Technology Assessment on Transgenic Herbicide-resistant Crops

Wolfgang van den Daele

Abstract

The review describes the political experiment of a participatory technology assessment which was organized in
Germany to test whether the endless battle over genetic
engineering could be transferred from the public arena to
a dialogue of rational argumentation. Transgenic herbicide-resistant crop plants were the topic of the technology
assessment. The claim that such plants pose particular risks
because they have been genetically modified could not be
defended in an exchange of arguments. The critics continued to reject the technology, but on different grounds,
arguing that there was no acceptable social need for herbicide-resistant plants, since better alternatives were available. This shift seemed to indicate that the real issue behind

Reprinted with permission from *AgBiotech News and Information*, 10, no.
11 (1998): 355N–58N. Copyright © 1998 CABI Publishing.

the conflict over genetic engineering is not the prevention of risk, but the quest for more democratic control of the dynamics of technological innovation. The critics refused, however, to ratify this finding as a result of the dialogue and thus avoided the redefinition of the conflict in the public agenda.

Putting Risk Arguments to Test

I n July 1998 the Swiss citizens rejected the claim that an unconditional ban on the release of genetically modified organisms into the environment should be included in the Swiss constitution. It remains to be seen whether, with this referendum, the long argument over the risks of genetic engineering which have dominated the public debate, particularly in German-speaking countries, will finally come to a halt. We had earlier put the arguments to the test of a participatory technology assessment organized by the Science Center for Social Research from 1991 to 1993 on transgenic herbicide-resistant crop plants. This technology assessment was specific for two reasons. First, it was *participatory*. It involved some fifty persons from environmental groups, industry, regulatory agencies, and the scientific community, giving full representation and a fair share of resources (for commissioning expert reports) to the critics of the technology. Second, it was *discursive*. Participants were expected to collect and discuss all available arguments in an ongoing process of communication and interaction. For that purpose, they attended a series of conferences which lasted more than ten days in all.

Discourse in such a social setting is remarkably different from the so-called public discourse pursued in mass communication. Participants in mass communication tend to use the rhetoric of arguments but rarely observe the discipline of argumentation. They normally confine themselves to the statement of their own strongest points, neglecting countervailing arguments or selecting for consideration only those which they can easily refute. In contrast, the participants in our technology assessment were bound to take the rules of argumentation seriously. The presence of advocates of opposing views guarantees that the full range of arguments and counterarguments are considered. Selectivity cannot be maintained. The participants may well be committed to restrictive positions and strategic interests, but as long as they participate in the process of communicative interaction, they can hardly ignore requests to substantiate reasons, to take objections into account, to present the empirical evidence for a statement, and to consider counterevidence.

Discussions over risks proceeded through various stages in our technology assessment:

- from recognizable risk with predictable consequence to hypothetical and unknown risks with unforseeable consequences

- from the isolated assessment of risks involved in genetically modified plants to a comparison of risk between transgenic and nontransgenic plants
- from the need to substantiate suspected risk to a reversal of the burden of proof; the absence of risk should be demonstrated
- from arguments over risks to arguments over social benefits; socioeconomic need should be a prerequisite for the introduction of new technology.

All these aspects had been raised in the public debate before. What the technology assessment showed was that there is a logical order or pattern of transformation, to which the criticism of genetic engineering will be submitted if put to the test of argumentation.

Comparing the Risks of Transgenic and Nontransgenic Plants

Basically, no risks from transgenic herbicide-resistant plants were recognizable which were not already known from nontransgenic plants. Recognizable risks were "normalized" through comparison. They no longer appeared dramatic if compared to the risks which are accepted with conventional agricultural crops and practices. The participants in the technology assessment agreed that it was not enough to consider risks which can be described and tested. The real issue with transgenic plants might well be that we do not know the risks. However, it was pointed out that in conventional breeding, too, one can neither foresee nor control what the physiological impact of new genes might be, given the genetic background of the host plant. Unexpected and undesirable side effects are abundant and must be coped with through testing and selection in the further development of new varieties. Thus, comparison with conventional plants not only "normalized" the recognizable risks of transgenic plants, but also the uncertainties and the hypothetical risks which might be implied by the fact that we have limited foresight of the possible consequences of such plants.

The critics had two arguments as to why it was not legitimate to compare conventional breeding and the construction of transgenic plants. More serious side effects should be expected with the latter because (1) transgenes insert at random and may cause insertional mutations in the host genome which can trigger changes in the transformed plant that are unrelated to the information coded in the transgene; and (2) genes can be transferred across species barriers and introduce metabolic pathways which have never belonged to the host plant species. Both arguments refer to hypothetical risks. There is no empirical evidence yet that more serious side effects do in fact occur in transgenic plants, nor is it possible to anticipate such effects theoretically in any detail. It is, however, held that, as one critic put it, one can infer from the "specific quality" of genetic engineering that transgenic plants present a "specific type of uncertainty."

In our technology assessment, the first of the above arguments was invalidated through comparison, by pointing out that insertional mutations (and pleitropic effects) are not specific to genetic engineering. They also occur with conventional breeding techniques and when natural transposable elements, which are known to exist in most plants, and also insert at random, jump around in the plant genome. The second argument was considered as valid in principle, but again weekend through comparison. While it may be true that the probability of side effects is theoretically higher in transgenic plants, because (and if) new metabolic pathways are transferred, it can also be argued that the probability of side effects is theoretically lower in transgenic plants, because with genetic engineering a single, identifiable gene product is transferred, whereas with crossing techniques an uncontrolled number of undetermined genes may be exchanged, all of which can interact with the existing metabolism. Thus, the assumption that transgenic plants will have more unexpected side effects than nontransgenic plants seemed as only as good in theory as the contrary assumption that transgenic plants have less unexpected side effects than nontransgenic plants.

The critics of genetic engineering finally retreated to the argument that even if our present knowledge does not warrant the assumption that transgenic plants involve specific and more severe risks than nontransgenic plants, it is still theoretically possible that such risks do in fact exist and may become apparent later. This argument could only be effectively turned against new technology if the rule that risk assumptions have to be substantiated was abandoned and the burden of proof shifted from those who claim risks to those who claim safety. At this point of the debate, our technology assessment proceeded beyond the established framework of risk prevention and raised fundamental issues of the politics of innovation.

Socioeconomic Need and the Quest for Democratic Control of Innovation

It became quickly apparent in our discussions that a full reversal of the burden of proof is not an operational rule. The unsubstantiated assumption that there may be unknown risks can easily be raised against any new technology and can hardly be refuted. No innovation would survive under such a rule. Consequently, the critics of genetic engineering made the next step and demanded that only those innovations should be admitted for which there is a clear socioeconomic need. They concluded that transgenic herbicide-resistant plants had to be prohibited under this test, because such plants represented no ecological advantage and little if any agronomic use; efficient weed control could be achieved in almost all cases using available selective herbicides, and nonchemical methods of weed control would be the ecologically preferable alternative, anyhow. Arguments as to whether transgenic herbicide-resistant plants are useful and satisfy a proper need were accepted as a necessary and legitimate topic of inquiry (and controversy) in the technology assessment. The proposal, however, to make "socioeconomic need" a legal prereq-uisite in the regulation of the technology was rejected by most participants. It was argued that this would replace market mechanisms with political decision-making and that the decline of the socialist countries had demonstrated that no model for an efficient economy exists in which decisions on innovation and investment are the domain of politics. The critics conceded the problems but insisted that never-theless some revisions of the established institutional balance between market mechanisms and democratic control of innovation were necessary and that the question of whether we really need a new technology must be put on the polit-ical agenda. The controversy over this point remained as unresolved in this tech-nology assessment as it is in the rest of the society.

The Limits of Participatory Technology Assessment

The exchange of arguments did not achieve a final consensus in our technology assessment. However, the dissent at the end of the discussions was not the same as at the beginning. The issues of debate had been transformed. They had shifted from arguments about risk prevention to arguments about the reform of political institutions and the future development of society. This shift seems to indicate that no conclusive reasons against the use of transgenic herbicide-resistant plants could be formulated within the framework of established risk regulation, and that the real issue behind the conflict over genetic engineering is the quest for democratic control of the process of technological innovation.

It would have been a real accomplishment if we had been able to publicize this transformation of issues, and hence a refiguration of the landscape of political controversy in Germany, as a finding of our participatory technology assessment. The critics of genetic engineering were not, however, prepared to take this step. They withdrew their participation in the technology assessment at the beginning of the final conference, at which they were expected either to accept the proposed conclusions or to reject them, giving additional reasons why they considered them incorrect. Apparently, it would have been difficult for them to declare explicitly that the conflict was not about risks, but about social goals and political reform, after they had committed themselves categorically to the rhetoric of risk, and used it successfully in the mobilization of the general public. As long as one claims risks, it is easy to argue that the dominant policy is irresponsible and offends against gen-erally accepted values. When risk arguments no longer play a role, it seems more legitimate to apply majority decisions in the choice of conflicting goals.

Participatory technology assessments are not a procedural fix to resolve polit-ical conflicts over technological innovation. They provide a forum for rational dis-course in which controversial arguments will not only be exchanged but also examined. Such discourse implies learning. However, while the learning may easily be accepted by the observing public (including parliaments, administrations, and the courts), the participants who represent the parties of the political conflict may refuse to adapt to what has been learned. Participatory technology assessment constitutes a limited context of cooperation and it operates at a distance from the real political arena. It remains a small island of argumentation in a large sea of strategic battle. Thus, in terms of "realpolitik," it must be expected that arguments which have been refuted within the technology assessment will continue to be used outside the technology assessment, as long as they can still impress the public. Even then, participatory technology assessments may be a valuable contribution to the political culture. They will not lead to consensus; political decisions will still have to be taken in dissent. However, procedures that give the critics a fair chance and submit controversial issues to the discipline of rational argumentation will contribute to the legitimacy of decision making in dissent.

References

For a full acount of the technology assessment see: Wolfgang van den Daele, Alfred Pühler, and Herbert Sukopp, "Grüne Gentechnik im Widerstreit" (Weinheim: Verlag Chemie, 1996). An English summary report is available as WZB-discussion paper FS II 97-302: Wolfgang van den Daele, Alfred Pühler, and Herbert Sukopp, "Transgenic Herbicide-Resistant Crops. A Participatory Technology Assessment," 1997. Requests for papers should be addressed to the author: Wolfgang van den Daele, Wissenschaftszentrum Berlin für Sozialforschung, Reichpietschufer 50, D-10785, Berlin. [Fax: (49) 30 25491-219, e-mail: daele@medea.wz-berlin.de]

Part 8

Precautionary Principle and Genetically Modified Foods

Introduction

The objectives of the Convention on Biological Diversity are stated in its first article:

> The conservation of biological diversity, the sustainable use of its components, and the fair and equitable sharing of the benefits arising out of the utilization of genetic resources, including by appropriate access to genetic resources and by appropriate transfer of relevant technologies. . . .[1]

The convention made explicit that conservation of biological diversity should be proactive: "Where there is threat of significant reduction or loss of biological diversity, lack of full scientific certainty should not be used as a reason for postponing measures to avoid or minimize such a threat." This stance on scientific uncertainty was formalized into the *precautionary principle* in the Cartagena Protocol:[2]

> In accordance with the precautionary approach contained in Principle 15 of the Rio Declaration on Environment and Development, the objective of this Protocol is to contribute to ensuring an adequate level of protection in the field of the safe transfer, handling and

use of living modified organisms resulting from modern biotechnology that may have adverse effects on the conservation and sustainable use of biological diversity, taking also into account risks to human health, and specifically focusing on transboundary movements. (Art. 1)

The precautionary principle is now a cornerstone of biotechnology policy, and a controversial cornerstone at that. Some who are opposed to GM foods would see the principle used to halt completely the development and release of transgenic organisms into the environment. The introduction of new technologies often elicits the view that we should be very cautious just in case we are opening a Pandora's box. Strong analogies are made between the introduction of nuclear energy and genetically modified foods, a connection that receives its proper elaboration in this section's first essay by Florence Dagicour. The analogy with nuclear energy suggests that a strongly precautionary attitude toward new technology is inevitable, and, as it more or less has in the case of nuclear energy, concerns fade with time, but we can nevertheless learn from these worries.

Perhaps they will, but only once we have determined how real GM food risks are for us and the environment. How one would go about applying the precautionary principle as these risks are being evaluated is the subject of Indur Goklany's paper. Goklany first develops a framework for applying the principle, and then anticipates its application in a number of cases. Far from supporting an outright ban on genetically modified crops, Goklany concludes that it might be imprudent in many cases to invoke the precautionary principle if the benefits of the technology truly outweigh the risks.

Henry Miller and Gregory Conko argue in the final selection that the precautionary principle reflects only the technophobic voice of a radical minority. If extreme precautions were taken with respect to all new technological introductions, we would live in very diminished circumstances. Risk is a part of life, they argue, and so what we need are better tools to manage risk, not idle bureaucrats picking up on the scare tactics of the antiprogressive few.

Notes

1. United Nations, *Convention on Biological Diversity*, article 1, United Nations, Rio, 1992.

2. Secretariat of the Convention on Biological Diversity, *Cartagena Protocol on Biodiversity to the Convention on Biological Diversity*, Secretariat of the Convention on Biological Diversity, Montreal, 2000.

27

Protecting the Environment

From Nucleons to Nucleotides

Florence Dagicour

Introduction

While the development of new technologies is seen as a necessary step for the economy, it also raises many concerns on an important contemporary issue: the protection of the environment. To manage the risks associated with new technologies, the precautionary principle has recently emerged in international environmental law to deal with so-called scientific uncertainty. The cornerstone of international environmental law, the 1992 Rio Declaration, provides states with some guidelines to implement such a principle: "Where there are threats of serious or irreversible damage, lack of full scientific certainty shall not be used as a reason for postponing cost-effective measures to prevent environmental degradation." While a precautionary approach is clearly required, its practical implementation remains uncertain even though authors have been carefully trying to define it.

In contrast with the prevention principle, which deals with known risks, the precautionary principle calls for caution with respect to new or very complex technologies where risks are unknown or very uncertain. The current and controversial area of biotechnology presents an ideal context to implement, and as a result, to better define the precautionary principle. Unknown or uncertain risks associated with biotechnology are numerous. For example, genetically modified (GM) plants and animals risk displacing natural species. Another source of unknown or uncertain risk is xenotransplantation. By transferring animal cells, tissues, and organs to human patients, xenotransplantation creates the risk of disease transmission. It may afford easy passage for animal viruses to infect human recipients and consequently enter the human population.[1] Unfortunately, the method for detecting infection is not clear-cut[2] and the risk still remains largely unquantified.

Some may regret that the development of such hazardous activities has come at a time when the precautionary principle is not yet fully understood. Authors have identified three conditions necessary to implement the principle: scientific uncertainty, risk of harm, and level of harm; however, these conditions are subject to various interpretations.[3] For example, how can we evaluate the risk of harm and, more important, the level of harm when the risks are unknown? The lack of a clear understanding of the precautionary principle could therefore be seen as jeopardizing the regulation of a technology, such as biotechnology, since it would appear that we hardly know how to deal with unknown risks. In the long term, this could endanger the environment because proper measures to prevent uncertain risks will not be taken. One can argue, however, that many technologies were developed before the precautionary principle came into existence. Nevertheless, the international community approached these technologies with a precautionary attitude in order to address concerns about human health and the protection of the environment. We can apply the lessons learned from taking this precautionary attitude to develop what is now being formalized under the name "precautionary principle." By doing so, we can learn from the lessons of previously developed, hazardous, and still controversial activities before questioning the real implications of the precautionary principle on biotechnology.

Nuclear energy presents an ideal foundation for carrying out this analysis. Not only did the civil use of nuclear energy start more than forty years ago, but it also went through the unfortunate experience of two major accidents: the 1979 Three Mile Island and the 1986 Chernobyl accidents. These two facts provide us with a good starting point to ask some fundamental questions regarding biotechnology, but the comparison between nuclear energy and biotechnology is even deeper. While the nature of the two technologies is obviously very different from one another, nuclear energy and biotechnology do share many similarities. A closer look may help us appreciate them.

Similarities between Nuclear Energy and Biotechnology

First of all, both biotechnology and nuclear energy were introduced in the hope of creating a better and more environmentally friendly world. Although nuclear energy first aimed at providing cheap electricity to everyone and promoting economic development, it is now seen as a tool to avoid the greenhouse effect. Unlike coal, natural gas, and oil-fired power plants, nuclear energy does not emit either carbon dioxide (partially responsible for global warming) or other noxious gases such as sulphur dioxide, mercury, and nitrogen oxide. In the context of global warming or of what President George W. Bush calls an "energy crisis," nuclear energy could be the cleanest source available to satisfy the world's energy requirements at a reasonable rate.[4] Biotechnology also brings its set of hopes. It will help us to produce purer and cheaper medications through genetically modified organisms (GMOs), to develop new ways of delivering vaccines to children in the developing world, and to increase our understanding of how diseases occur and ways to diagnose and prevent them. With biotechnology, we also hope to feed the planet with cheaper and more nutritious food, while ensuring the protection of the environment. GM fish can grow ten times faster than non-GM fish, which could help feed the developing world at a low cost. Finally, biotechnology may also help to clean the environment through bioremediation.

Both nuclear energy and biotechnology thus represent an opportunity for the developing world to promote its own development. With a clean and cheap source of energy, and the technology to improve its agricultural and medical sectors, the developing world will possess the fundamental tools to significantly increase its economic development and, ultimately, to become more independent from the rest of the world.

Despite this noble goal, an accident involving one of these two technologies could result in dramatic consequences from health, environmental, and economic viewpoints, not only at a national level but also at the international level. The Ukrainian Chernobyl accident demonstrates the disastrous potential of nuclear technology. While studies on the consequences of the Chernobyl accident vary significantly,[5] common conclusions show that hundreds of people died at the time of the accident and an increasing number of people are affected by thyroid cancer and congenital malformations.[6] The environment surrounding the nuclear plant, both in the Ukraine and in Belorussia, has been so highly contaminated that thousands of hectares of land cannot be inhabited or used for agricultural purposes. Scandinavian countries have been alarmed by the radioactivity of their lakes and rivers. Shortly after the accident, Germany banned all commercialization of agricultural products, such as milk, vegetables, and so on, because of the fear of radioactivity.[7] Besides the economic costs of the accident itself, many people have lost their jobs, and had to move to other regions and find new careers. While it is difficult to predict the consequences of a major accident in the biotechnology industry, we can easily expect transboundary damage to the environment. As recently discovered, if a GM salmon escapes from a farm, it may endanger the non-GM salmon by tainting the gene pool. To avoid the contamination, GM fish must be infertile; however, ensuring that one hundred percent of the fish are sterile remains very difficult. The risk of reproduction between GM and non-GM salmon is quite real.[8]

The risks associated with these two technologies also share the common characteristic of dual purposes, since both can be used for civil as well as military purposes. Everyone remembers the first public recognition of the utility of nuclear energy during the Second World War with the Hiroshima bomb. The fight against nuclear proliferation is now a permanent priority for the international community and the few states that own nuclear bombs. Nonproliferation seeks to prevent new countries, and specifically unstable countries, from acquiring nuclear weapons. The most recent example of military use of nuclear substances is the controversy over the use of depleted uranium in Kosovo and the Gulf War.[9] Like nuclear technology, biotechnology can easily be used as an weapon during conflicts. One author has indicated that "[w]hile nuclear proliferation and missile and submarine sales have received the most attention, by far the worst problem is [chemical and biological warfare] proliferation."[10] The use of biological warfare was seriously feared during Operation Desert Storm. The seriousness of chemical and biological warfare is also stressed by the adoption of the 1972 international convention on the prohibition of the development, production, and stockpiling of bacteriological (biological) and toxin weapons and on their destruction,[11] although it has been largely disregarded.

While both technologies rely on very detailed scientific expertise, certain areas of these technologies remain scientifically uncertain, making it very difficult to identify and evaluate potential risks to human health and the environment. While highly radioactive emissions are undoubtedly hazardous to human health and the environment, it is still unclear how dangerous low doses are.[12] The risks associated with GM plants or animals and with xenotransplantation are also still unclear, since no scientific evidence exists. Because little is known about risks, speculation significantly increases.

Finally, both technologies face strong public opposition. Antinuclear movements slowly appeared in the 1970s in response to the 1973 French nuclear tests in the Pacific. It was, however, the two major accidents, Three Mile Island and Chernobyl, that considerably strengthened public opposition. Later on, the still-unsolved final management of highly radioactive wastes continued to feed this opposition. Certainly unexpected for the Americans, the difficulty in accepting GMOs has been made clear by the Europeans, who refuse to buy any genetically modified products. In addition to the "genetically-modified free product" advertisements used by some European supermarkets, a new phenomenon symbolized by José Bové has occurred: the purposeful destruction of a McDonald's restaurant in France. His act illustrates not only the attachment of the French, but also Europeans in general, to what they call "natural" food, meaning food that has not been industrially or genetically produced but, instead, originates in the *terroir* (the land).

Based on these similarities, biotechnology policymakers should look at the forty years of experience with nuclear energy and extract the lessons learned from this experience. We now have the opportunity to adapt the nuclear approach to biotechnology without duplicating its mistakes. While a precautionary approach was one tool available to protect the environment from nuclear risks, it was not the only one.

Lessons to Learn

Understanding the Consequences of Public Opposition

While it is arguable that public opposition to technology is a secondary issue, we should look at the nuclear energy experience to reconsider this opinion. Although nuclear energy developed with the support of the public—the industry was, after all, creating new and lucrative jobs—all the public needed to change its mind was one accident, even a remote one. The Three Mile Island accident started the process of opposition faster than expected. Seven years later, the Chernobyl accident convinced most of the remaining people that nuclear activity was too hazardous. The few military operations using radioactive materials—such as the French nuclear tests or the use of depleted uranium during conflicts—significantly helped the antinuclear movements to make their point even clearer.

Public opposition to nuclear energy is now so strong that any minor incident is widely covered by the media and commented on by the academic world. It has even led Germany to adopt policies to progressively abandon its national production of nuclear energy over the next twenty years,[13] which will lead Germany to be more dependent on an outside energy supply. Public opposition has also significantly disturbed the transportation of radioactive wastes between Germany and France in the context of reprocessing arrangements.[14] Transportation of nuclear wastes was suspended for about three years between 1998 and 2001. Events in Sweden also demonstrated how strong the antinuclear movement has become. Following the Three Mile Island accident, the antinuclear movement forced the government to hold a referendum in 1979 on the future of the nuclear industry in Sweden. As a result, the government initiated a long-term program of denuclearization, and Sweden began planning to dismantle its nuclear power plants. While there are some arguments that legitimately justify this opposition to nuclear power, it remains surprising that public opposition is still so strong at a time when nuclear energy presents the tremendous benefit of avoiding global warming. The nuclear experience demonstrates that the public can take a long time to become opposed to an industry—after all, it took it more than fifteen years to react against the nuclear industry—but once opposition has taken root, it is even more difficult to go back to a neutral position. As Cynthia Picot et al. noted, "Policies that lack public support are policies that risk failure."[15]

Policymakers should remember this lesson before it is too late and quickly address the increasing public opposition to biotechnology. Opposition to biotechnology, however, presents different characteristics from that of nuclear energy. Indeed, while European governments have an economic incentive to proceed with biotechnology in general, and GMOs in particular, in order to remain competitive and encourage their economic growth, the public does not necessarily share these concerns and perceives biotechnology as a social, ethical, moral, and cultural concern. This is the fundamental difference between nuclear energy and biotechnology. While the public fears nuclear energy because of its danger to human health and the environment—the management of the highly radioactive waste issue illustrates it—biotechnology in general, and GMOs in particular, touch a very sensitive area—culture—that the international community and countries should be aware of. No matter how much GMOs improve the quality of food, genetically modified food will remain, for most of the European audience and especially for the French, an "unnatural" product because it does not come from the notion of *terroir*.[16]

To address this public-opposition concern, industry ought to demonstrate a strong motivation and good faith to work not only for the public, but also with it. While the public is willing to accept risks from new technologies in recognition of their benefits, it is no longer satisfied with industry's statement on the safety of a technology. The public now wants to know, understand, and participate in the process used to ensure this safety.

Transparency

The lack of transparency has also clearly been one factor fueling public opposition to the nuclear industry. For a long time, incidents were not published, in order to preserve public faith in the industry, and safety measures were kept confidential for industrial competition and to preserve the secrecy surrounding military operations. While this absence of transparency may be justifiable and legitimate, the public also needs to be properly informed when its safety and health are at stake. Restraining the release of information may be perceived by the public as purposely hiding bad news. Some nongovernmental organizations (NGOs) also have a good appetite for and actively look for any incidents in the nuclear industry to build support for themselves and their campaigns. Greenpeace has clearly defined the nuclear industry as its primary target, whereas others industries, such as the chemical industry, are equally or even more dangerous to human health and the environment.[17] Again, once the public—generously helped by NGOs—is opposed to an industry, convincing it of the good and safe operation of that industry requires a lot of time and effort, and a nonnegligible part of its budget. Biotechnology should avoid this situation by being as transparent as possible with the public by informing the public of the risks associated with activities, safety measures used to minimize potential negative impacts, accidents that have occurred, and so forth.

Strong International Cooperation

With the increasing globalization, strongly demonstrated by the Summit of Americas in April 2001, the public is not only concerned about hazardous activities operating within its own boundaries but also—and maybe mainly—about those activities operating outside of its boundaries.

One of the most remarkable approaches of the nuclear industry and the international community was very strong cooperation between states and national industries from the very beginning of nuclear development. Even though the 1946 proposal of an international control regime on nuclear installations was rejected, international cooperation dealing with scientific and technical matters, such as exchange of engineers, with financial as well as regulation issues strongly helped to ensure safe management of this hazardous activity. While this cooperation was clearly intended to promote the development of nuclear technology by ensuring its safe operation, it also resulted in protection of the environment, as environmental protection was becoming an important issue.

This cooperation is illustrated by the following three examples, which should be considered by the biotechnology industry. Early on, an international institution dealing specifically with nuclear energy—the International Agency for Atomic Energy (IAEA)—was created under the United Nations system.[18] Another institution was similarly created at the regional level for European countries, although Canada, the United States, and Japan are now members of the Agency of the Organization for Economic Cooperation and Development for Nuclear Agency (AEN/OECD).[19] These institutions encourage countries to work together and to exchange technical and scientific information through the exchange of engineers and reports. One interesting aspect of these institutions is that they are authorized to establish standards of safety for protecting health, minimizing danger to life, and so on. While these standards are nonbinding—meaning that states are not obliged to implement them—they do provide guidelines for member countries' operations. As a matter of fact, these guidelines have often been integrated into domestic laws and regulations.

Second, international cooperation also helps to promote and implement principles and instruments to protect the environment around the world. The sea dumping of radioactive wastes provides us with a good example. As early as the beginning of the 1960s, several nuclear countries, such as United States and the United Kingdom, recognized sea dumping of radioactive wastes as the best solution to avoid spreading radioactive emissions in the course of radioactive-waste management. Soon after, the IAEA and AEN/OCDE studied the impact of such dumping of radioactive waste from a scientific, technical, and legal viewpoint. A committee was created within the AEN/OCDE to study the problems related to the dumping of these wastes and to provide the necessary structure to exchange information on all aspects of the dumping. The result of this involvement led not only to the conduct of an environmental study, which can now be assimilated into an environmental-impact assessment, but also to a multilateral mechanism of consultation and control for sea dumping of radioactive wastes. From 1977 on, the dumping of these wastes was only undertaken under the supervision of the AEN/OCDE, and was finally suspended in 1982. This suspension was renewed in 1992 up to 2005. The sea-dumping experience calls for two major remarks. Not only did it initiate the process of environmental-impact assessment, which is now a fundamental instrument of international and domestic environmental law, but it demonstrated a precautionary approach, since sea dumping is now suspended while further studies are being conducted.

Based on this experience, an international forum or institution could provide the biotechnology industry with a structure to conduct research, consultation, or monitoring, which might be ignored or considered financially infeasible by one single country. Large-scale projects could be undertaken within this forum, while ensuring its safe operation and monitoring through a precautionary approach. It could also provide members with the incentive to develop new tools or instruments to protect the environment on the model of the environmental study used in the 1960s. Although prevention and precautionary approaches are necessary for both technologies—nuclear and biotechnology—they are particularly necessary in the biotechnology industry in which a specific damage could imply a nonreturn situation. The introduction of the South American cane toad into Queensland in 1935 illustrates this point. Cane toads were introduced to control two cane beetle species that were damaging the local sugar cane harvest. Unfortunately, not only did the toads fail to control the beetles, but the cane toads themselves became a significant pest. Now, researchers are looking into viruses and fungi that can be used to control the cane toads.[20] Unlike biotechnology, nuclear energy has the advantage of the "radioactivity decreasing" phenomena, which means that the radioactivity of a substance decreases with time. An intermediate radioactive waste will become low-radioactive waste after several years before becoming non-radioactive. Prevention has, however, been recognized over the years as a fundamental approach to protecting the environment from radioactive hazards, moving away from the initial approach of compensating for damages.

The third benefit that international cooperation can bring to the biotechnology industry is a procedure and eventually an obligation to provide early notification of accidents. This procedure finds its roots in nuclear law. Based on the experience of Chernobyl—where the former USSR did not notify anyone of the accident but, instead, Scandinavian countries discovered the accident by noting unexpectedly high levels of radioactivity—the international community rapidly responded by establishing an obligation to notify possibly affected countries or the

IAEA as early as possible of an accident so that they could take necessary measures to prevent or to reduce the consequences of the accident. This early-notification procedure supplements the report of incidents that members of international organizations initially committed to provide. The reporting system ensures a clear understanding of previous incidents around the world, and depending on how comprehensive the system is, can help other countries to avoid similar incidents. Although this reporting system has been operating on a voluntary basis for a long time by the IAEA and the AEN/OCDE, the convention on early notification has transformed it into an obligation.[21] An international reporting system and early notification of accidents should also be integrated within the regulation of biotechnology to ensure a better understanding of biotechnology and the risks associated with it.

The creation of an institution is not the solution to risks posed by biotechnology, neither should it become an automatic response to any new technology. However, such an intergovernmental forum will definitely help the international community to deal with biotechnology for two main reasons. First of all, as mentioned above, the biotechnology industry partially justifies its activities as a way to help developing countries. If so, a forum including developed and developing countries should be set up to encourage the better understanding of cultures from different members or regions, to identify and appreciate the real needs of developing countries, and to evaluate and assess how developed countries can ethically help the developing world while ensuring the development of this economic sector. Second, such international cooperation will also help industry to understand the public opposition around the world to this new technology, and especially in Europe. An international forum will not necessarily aim at understanding cultures from different continents, but will help the biotechnology industry and governments to be aware of the differences, and as a result, to better address the issue.

Like the IAEA, an international forum for biotechnology could be created under the United Nations system. Its overall goal should be to encourage technical and scientific cooperation in biotechnology around the world. It should be authorized to establish some standards regarding safety, to provide its members with effective assistance in the use of biotechnology, and to provide its members with a database on previous incidents, worldwide recognized experts, and so on.

Liability Regime

While the nuclear industry clearly offers interesting insights to duplicate in other industries, it also provides experiences that should be carefully considered when developing hazardous technologies. Aware of the risks associated with nuclear energy, the international community adopted a convention dealing with a specific civil liability regime in 1963 ensuring compensation for and protection of victims in case of nuclear accidents. Another convention on this matter was also signed in 1960 at the European level. This liability regime reflects the initial approach of environmental law, which was compensation. The prevention approach appeared later in the 1970s. While the liability regime was a welcomed step, analysis demonstrates that the regime is more protective of the nuclear industry than of the victims. The nature of the nuclear industry and its risks led competent authorities to develop a new and specific liability regime. To better protect victims, only the operator of the nuclear facility where an accident occurs is held liable for damages, whether his conduct was responsible for the accident or not. Under this absence of fault or "objective liability" system, the simple occurrence of a nuclear accident is enough to hold the operator liable. The channeling of liability onto the operator, as well as the objective liability, spares victims the painful investigation of who can be sued for compensation. These two aspects of the nuclear liability regime are, however, undermined by the need to prove injury. Due to "scientific uncertainty"—such as the effects of low doses of radiation—proving the relationship between injury and an activity becomes a very difficult burden, which as a rule of law, falls on the victim. An example of the burden of proof can be illustrated with the case of a person affected by thyroid cancer. While radioactive emissions can be a factor influencing the development of thyroid cancer, they are not a necessary condition. Other reasons—such as a person's genetic background—have to be taken into consideration. Indeed, a person living near a nuclear plant may never be affected by the disease, whereas another person living far from such a plant may be affected. Because a direct causal relationship between the activity and the injury is very difficult to prove, many victims have seen their legal actions denied for absence of proof. Not only is the burden of proof very difficult for the victim to establish, but it is also very expensive because experts are often required. Further weaknesses of this regime are the limitation of the operator's liability in both amount and time. Solid cancers caused by radiation often do not manifest themselves until well after the ten-year limitation period (although one recent amendment to one convention has extended the limitation period to thirty years for death). Compensation limits have also been very low. While this regime has clearly affected the public perception of the nuclear industry, the Chernobyl accident has demonstrated its inapplicability in the case of major accidents. The assessment of all damages was not feasible, the compensation limit was too small to cover all damages, and finally, the Soviet government was unable to financially compensate victims. Facing this situation, no state sued the USSR for international responsibility. It is also worth noting that environmental damages were not covered by any of the conventions up to the 1997 amendments of the international convention.

This brief description of the nuclear liability regime highlights the necessity for the biotechnology industry not only to soon adopt a liability regime, but also to adopt a regime that clearly protects victims as well as the industry against any futile legal actions. Although it is tempting to adapt the nuclear liability regime, as did the oil industry, the biotechnology industry particularly needs to consider the current public opposition before choosing its liability regime. Biotechnology needs to regain public support. One way to do so is to adopt a strong liability regime capable of ensuring proper compensation of victims and mitigation of environmental damage. The nuclear liability regime may also be difficult to implement within the biotechnology industry. For example, channeling of liability could be a serious legal issue in biotechnology because many actors intervene.

Strong International and Domestic Regulations

The public also wants strong regulations to ensure the safe operation of hazardous activities. However, national governments as well as international institutions have strongly favored "soft law" regulation instead of a "hard law" approach to regulate the nuclear industry. "Soft law" only includes nonbinding instruments, such as recommendations, guidelines, standards, and so forth, as opposed to "hard law," which designates binding instruments, such as conventions, treaties, laws, and regulations. The use of nonbinding instruments was justified for three main reasons. First of all, the predominant and legitimate expectation of industry and governments was the desire to promote nuclear energy. As mentioned above, nuclear energy sounded very promising at the beginning of its development, and governments were anxious to develop it despite its potential risks. Second, the complexity of the technology—strengthened by the different types of reactors—was seen as restraining its translation into legal terms and binding instruments. The nuclear industry called for flexibility, allowing a case-by-case approach to safety measures. Finally, the very strong cooperation among states seemed to satisfy the international community as well as national regulators that necessary safety measures were taken care of. Whereas these nonbinding measures can be beneficial—we demonstrated earlier than they are in the context of international safety standards—they are not enough, as such, to ensure the safety of a hazardous technology. They need to be supplemented by binding instruments. The nuclear industry failed to do so, although it is now slowly changing. Biotechnology should therefore develop its own "soft law" instruments, while also working on binding ones. This strategy has the advantage of ensuring that fundamental safety measures for biotechnology activities are implemented while negotiations on an international convention are going on. Due to divergences of interests among the international community, the adoption of a convention as well as its ratification may take years. Although the official negotiations of the 1994 convention on nuclear safety went fairly well—even though divergences existed—more than five years were necessary before the convention came into force and became binding. This strategy of "soft law" supplemented by "hard law" also implies that conventions should be negotiated as early as possible in the development of an activity to ensure regulation of that activity. In any case, it should not be a response to an accident. This is, unfortunately, what characterizes nuclear conventions. Before the 1986 Chernobyl accident, the only existing conventions were those dealing with the liability regime. With the occurrence of the two accidents and the increasing public opposition, the international community realized that binding international nuclear law was required to convince the public of the safe operation of nuclear plants around the world. Although the adoption of five new conventions and amendment of a previous one are a significant step toward a more reliable, transparent, and well–regulated industry, this move came too late. The public had already lost its confidence, and been disappointed by the nuclear industry. Biotechnology ought to be proactive to obtain the support of the public. Although the 1992 Convention on Biological Diversity and its 2000 Cartagena Protocol on Biosafety were adopted, these instruments touch a very small part of the range of biotechnological activities. The convention's objectives deal with the conservation of biological diversity, the sustainable use of its components, and the fair and equitable sharing of benefits arising out of the use of genetic resources, whereas the protocol establishes a procedure for the control of transboundary movements, transit, handling, and use of all living modified organisms. Others instruments should deal with the safety, early notification of accidents, liability, and so forth.

The specific nature of the nuclear industry also led competent authorities to create a very specific set of international and national regulations. Beside the specific liability regime described earlier, a different regulation applicable to the movements of radioactive wastes, as opposed to dangerous wastes, has been established. General international conventions dealing with the environment often exclude nuclear activities from their scopes. The result of this "soft law" and specific approach, which has been highly criticized by academics, authors, and environmental groups who call for stronger regulation, is both the loss of confidence and skepticism of the public. Without public support, biotechnology will have a hard time proving that its ultimate goal is to create a better world.

Conclusion

While the nuclear industry has developed in the absence of the precautionary principle, this too-short analysis has, I hope, demonstrated that nuclear activities have been operating with a precautionary approach, sometimes creating new instruments to protect the environment. Although the precautionary principle could become a promising tool to protect the environment, its current undeveloped stage should not be seen as an obstacle to develop new technologies. Although further studies are required to investigate the lessons to learn for the nuclear industry, this industry has already provided a few tools with which to better protect the environment. For a long time, the nuclear industry tended to neglect the unexpected development of "civil society," which is now modeling new rules for industry and governments. While the nuclear industry is paying the price of this change in society, biotechnology should not repeat past mistakes. With or without the precautionary principle, precautionary measures need to be implemented to satisfy the public.

Notes

1. A prime example is the HIV-1 virus, which probably came from chimpanzees. The worldwide AIDS epidemic may have been started by a single cross-species event or by the use of chimpanzee kidneys in Africa to create batches of poliovirus vaccine. Robin A. Weiss, "Xenografts and Retroviruses," *Science* 285, no. 5431 (1999): 1221–22.

2. Robin A. Weiss, "Science, Medicine, and The Future: Xenotransplantation," *British Medical Journal* 317, no. 7163 (1998): 931–34.

3. Owen McIntyre and Thomas Mosedale, "The Precautionary Principle as a Norm of Customary International Law," *Journal of Environmental Law* 9, no. 2 (1997): 221–42; Frank B. Cross, "Paradoxical Perils of the Precautionary Principle," *Washington and Lee Law Review* 53, no. 3 (1996): 851; James Cameron and Juli Abouchar, "The Precautionary Principle: A Fundamental Principle for the Protection of the Global Environment," *Boston College International and Comparative Law Journal* 14, no. 1 (winter 1991): 1–27; Bernard A. Weintraub, "Science, International Environmental Regulation, and the Precautionary Principle: Setting Standards and Defining Terms," *New York University Environmental Law Journal* 1, no. 1 (1992): 173; David Vander Zwag, "The Precautionary Principle in Environmental Law and Policy: Elusive Rhetoric and First Embraces," *Journal of Environmental Law and Practice* 8 (1999): 355; John Moffet, "Legislative Options for Implementing the Precautionary Principle," *Journal of Environmental Law and Practice* 7 (1997): 157; David Freestone, "The Precautionary Principle," in *International Law and Global Climate Change*, eds. Robin Churchill and David Freestone (London: Graham & Trotman Martinus Nijhoff, 1991); Pascale Martine Bidou, "Le principe de precaution en droit international de l'environment," (The principle of precaution in international law of the environment) *Revue Générale de Droit International Public* 103, no. 3 (1999): 36.

4. Patrice Hill, "Energy Needs May Spur Rebirth of Nuclear Power," *Washington Times,* 19 March 2001, A1.

5. William J. Schull, "The Genetic Effects of Radiation: Consequences of Unborn Life," *Nuclear Europe Worldscan* 4 (November–December 1998): 35–37.

6. Yves Marignac, "Le choix honteux de la communicanté internationale—Les populations oubliées de Tchernobyl," *Le Monde Diplomatique,* July 2000, 15.

7. "Development and Harmonization of Intervention Levels in Case of a Nuclear Accident," *Nuclear Law Bulletin* 45 (June 2000): 24–25; "The Accident at Chernobyl—Economic Damage and Its Compensation in Western Europe," *Nuclear Law Bulletin* 39 (June 1987): 58–65.

8. ABCnews, "Risks to Engineering Fish? Critics Say 'GM' Fish Could Mate with Unaltered Population," 11 April 2001 [online], http://abcnews.go.com; "Moratorium on Alteration of Salmon," *New York Times,* 9 May 2001 [online], http://www.nytimes.com/2001/05/09/science/09FISH.html.

9. Robert James Parsons, "Des mensonges couverts par les Nations Unies–Loi du silence sur l'uranium appauvri," *Le Monde Diplomatique,* February 2001, 22–23; Robert James Parsons, "'Depleted Uranium': A Tale of Poisonous Denial," *San Francisco Examiner,* 1 May 2000.

10. Joseph D. Douglas Jr., "Chemical and Biological Warfare Unmasked," *Wall Street Journal,* 2 November 1995.

11. Convention on the prohibition of the development, production, and stockpiling

of bacteriological (biological) and toxin weapons and on their destruction, London and Moscow, 10 April 1972, entry into force, 26 March 1975.

12. Jacques Lochard and Marie-Claude Grenery-Boehler, "Optimizing Radiation Protection: The Ethical and Legal Basis," *Nuclear Law Bulletin* 52 (December 1993): 9; Klau Becker, "How Dangerous Are Low Doses? The Debate about Linear versus Threshold Effects," *Nuclear Europe Worldscan* 4 (November–December 1998): 29–31.

13. "L'Allemage renonce au nucléaire," *Le Monde,* 15 June 2000; "L'Allemagne à la recherche de solutions de rechange," *Le Monde,* 29 July 2000. However, talk of abandoning nuclear energy first arose after the Chernobyl accident, Norbert Pelzer, "Current Problems of Nuclear Liability Law in the Post-Chernobyl Period," *Nuclear Law Bulletin* 39 (June 1987): 66–76.

14. "Derrière un convoi spectaculaire une activité économique en crise," *Le Monde,* 27 March 2001; "Que faire des déchets nucléaires?" *Le Monde,* 27 March 2001.

15. Cynthia Picot et al., "Nuclear Energy and Civil Society," *Nuclear Energy Agency News* 18, no. 2 (2000).

16. Marian Burros, "U.S. Plans Long-Term Studies on Safety of Genetically Altered Foods, *New York Times,* 14 July 1999, A18.

17. "Environmental organizations claim that more than 100,000 chemical substances have been on the European market since 1980, of which seventeen have undergone risk assessments," from "Bonn Calls for Tough New Rules on Chemicals Testing," *European Voice* 5, no. 17 (5 April–5May 1999).

18. Statute of the International Agency for Atomic Energy, 23 October 1956.

19. Statute of the Agency for the OECD for Nuclear Energy, 1958.

20. E. Richard Gold, "Hope, Fear, and Genetics: Judicial Responses to Biotechnology," *Judicature* 83, no. 3 (November–December 1999): 132.

21. Convention on early notification of a nuclear accident, Vienna, September 1986.

28

Applying the Precautionary Principle to Genetically Modified Crops

Indur M. Goklany

Executive Summary

The precautionary principle has often been invoked to justify a ban on genetically modified (GM) crops. However, this justification is based upon a selective application of the principle to the potential public health and environmental benefits of such a ban, while ignoring a ban's potential downside. This is due principally to the fact that the precautionary principle itself provides no guidance on its application in situations where actions (such as a ban on GM crops) could simultaneously lead to uncertain benefits and uncertain costs to public health and the environment.

Accordingly, I first develop a framework for applying the principle in cases where the final outcome is ambiguous because both costs and benefits are uncertain. Then, based on a brief survey of the public health and environmental costs and benefits of GM crops, I apply this framework to the broad range of consequences of a ban on

From *Center for the Study of American Business*, by Indur M. Goklany, "Applying the Precautionary Principle to Genetically Modified Crops," Policy Study Number 157, August 2000, pp. 1–35. Copyright © 2000 Weidenbaum Center. Reprinted with permission.

GM crops. This application of the framework indicates that by comparison with conventional crops, GM crops would increase the quantity and nutritional quality of food supplies. Accordingly, GM crops ensure that—despite the expected increases in human population—the world's progress in improving public health, reducing mortality rates, and increasing life expectancies during the twentieth century should be sustained into the twenty-first.

Plant and animal genes have always been part and parcel of the human diet, and consumption of these genes has not modified human DNA. The public health benefits from GM crops, therefore, are likely to be larger in magnitude and more certain than the adverse public health effects from the ingestion of any genes that may be transferred from various organisms into GM crops.

With respect to environmental effects, cultivation of GM, rather than conventional, crops would be more protective of biological diversity and nature. By increasing productivity, GM crops reduce the amount of land and water that would otherwise have to be converted to mankind's needs. Reductions in land conversion to agriculture would reduce soil erosion, conserve carbon stores and sinks, and improve water quality. GM crops also could help limit environmental damage by reducing reliance on synthetic fertilizers and pesticides, and increasing no-till cultivation, which would further reduce soil erosion, water pollution, and greenhouse gas emissions.

A comprehensive application of the precautionary principle indicates that a GM crop ban, contrary to the claims of its advocates, would *increase* overall risks to public health and to the environment. Thus it would be more prudent to research, develop, and commercialize GM crops than to ban such crops, provided reasonable caution is exercised.

Introduction

A popular formulation of the precautionary principle is contained in the Wingspread Declaration: "When an activity raises threats of harm to human health or the environment, precautionary measures should be taken even if some cause and effect relationships are not established scientifically."[1]

In this study, I attempt to apply this formulation of the principle to devise precautionary policies for one of the most contentious environmental issues facing the globe, namely, the issue of bioengineered crops. In this analysis, I will examine and apply the precautionary principle comprehensively to a broader set of public health and environmental consequences of banning GM crops.

A Framework for Applying the Precautionary Principle under Competing Uncertainties

Few actions are either unmitigated disasters or generate unadulterated benefits, and certainty in science is the exception rather than the rule. How, then, do we formulate precautionary policies in situations where an action could simultaneously lead to uncertain benefits and uncertain harms? Before applying the precautionary principle, it is necessary to formulate hierarchical criteria on how to rank various threats based upon their characteristics and the degree of certainty attached to them. Consequently, I offer six criteria to construct a precautionary "framework."

The first of these criteria is the *public health criterion*. Threats to human health should take precedence over threats to the environment. In particular, the threat of death to any human being outweighs similar threats to members of other species.

However, in instances where an action under consideration results in both potential benefits and potential costs to public health, additional criteria have to be brought into play. These additional criteria are also valid for cases where the action under consideration results in positive as well as negative environmental impacts unrelated to public health. I identify five such criteria as follows:

- The *immediacy criterion*. All else being equal, more-immediate threats should be given priority over threats that could occur later. Support for this criterion can be found in the fact that people tend to partially discount the value of human lives that might be lost in the more distant future.2 While some may question whether such discounting may be ethical, it may be justified on the grounds that if death does not come immediately, with greater knowledge and technology, methods may be found in the future to deal with conditions that would otherwise be fatal which, in turn, may postpone death even longer. For instance, if an HIV-positive person in the United States did not succumb to AIDS in 1995, because of the advances in medicine there is a greater likelihood in 2000 that he would live out his "normal" life span. Thus, it would be reasonable to give greater weight to premature deaths that occur sooner. This is related to, but distinct from, the *adaptation criterion* noted below.
- The *uncertainty criterion*. Threats of harm that are more certain (have higher probabilities of occurrence) should take precedence over those that are less certain if otherwise their consequences would be equivalent. (I will, in this study, be silent on how equivalency should be determined for different kinds of threats.)
- The *expectation value criterion*. For threats that are equally certain, precedence should be given to those that have a higher expectation value. An action resulting in fewer expected deaths is preferred over one that would result in a larger number of expected deaths (assuming that the "quality of lives saved" are equivalent). Similarly, if an action poses a greater risk to biodiversity than inaction, the latter ought to be favored.
- The *adaptation criterion*. If technologies are available to cope with, or adapt to, the adverse consequences of an impact, then that impact can be discounted to the extent that the threat can be nullified.
- The *irreversibility criterion*. Greater priority should be given to outcomes that are irreversible, or likely to be more persistent.

In the following pages, I first will outline the potential benefits and costs to public health and the environment from research and development and commercialization of GM crops. I will then apply the relevant criteria to determine the appropriate policy pursuant to a comprehensive precautionary principle.

Potential Benefits of Bioengineered Crops

Environmental Benefits

Agriculture and forestry, in that order, are the human activities that have the greatest effect on the world's biological diversity.[3] Today, agriculture uses account for 37 percent of global land area,[4] 70 percent of water withdrawals, and 87 percent of consumptive use worldwide.[5] It is also the major determinant of land clearance and habitat loss worldwide. Between 1980 and 1995, developing countries lost 190 million hectares (Mha) of forest cover mainly because their increase in agricultural productivity was exceeded by growth in food demand. Developed countries increased their forest cover by 20 Mha because their productivity outpaced demand. Agriculture and, to a lesser extent, forestry also affect biodiversity through water pollution and atmospheric transport due to such things as release of excess nutrients, pesticides, and silt.[6]

Demand for agricultural and forest products will almost certainly increase substantially. The world's population is expected to grow from about 6 billion today to between 10 and 11 billion in 2100, an increase of 70 to 80 percent. The average person is also likely to be richer, which ought to increase agricultural demand per capita. Accordingly, the predominant future environmental and natural resource problem for the globe is likely to be the challenge of meeting the human demand for food, nutrition, fiber, timber, and other natural resources while maintaining, if not improving, biological diversity.[7]

The question is whether biotechnology is more likely to help or to hinder reconciling these often-opposing goals.[8] Although most of the following discussion focuses on agriculture, with particular emphasis on developing countries, much of it is equally valid for other human activities that use land and water, for example, forestry, and in developed countries as well.

Decrease in Land and Water Diverted to Human Uses

The United Nations's latest most-likely estimate is that global population, which hit 6 billion in 1999, will grow to 8.9 billion in 2050.[9] Figure 1, based on the methodology outlined by Indur Goklany,[10] provides estimates of the additional land that would need to be converted to cropland from 1997 to 2050, as a function of the annual increase in productivity in the food and agricultural sector per unit of land. This figure assumes that global crop production per capita will grow

Figure 1

Net Habitat Loss to Cropland v. Increase in Agricultural Productivity, 1997 to 2050

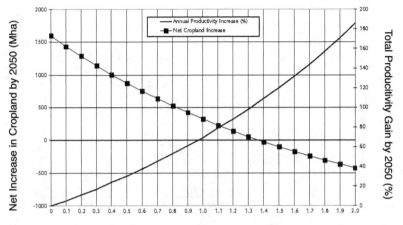

Annual Productivity Increase (%)

Source: Food and Agriculture Organization, FAO Databases [online], apps.fao.org [12 January 2000], per Indur M. Goklany, "Saving Habitat and Conserving Biodiversity on a Crowded Planet," *BioScience* 48 (November 1998): 941–53 and "Meeting Global Food Needs: The Envionmental Trade-offs Between Increasing Land Conversion and Land Productivity," in *Fearing Food: Risk, Health and Envionment*, ed. Julian Morris and Roger Bate (Oxford, U.K.: Butterworth-Heinemann, 1999).

at the same rate between 1997 and 2050 as it did between 1961-63 and 1996-98, and that new cropland will, on average, be as productive as existing cropland in 1997 (an optimistic assumption).

If the average productivity in 2050 is the same as it was in 1997—hardly a foregone conclusion[11]—the entire increase in production (106 percent under the above assumptions of growth in population and food demand) would have to come from an expansion in cropland. This would translate into additional habitat loss of at least 1,600 million hectares (Mha) (see figure 1) beyond the 1,510 Mha devoted to cropland in 1997.[12] Much of that expansion would necessarily have to come at the expense of forested areas.[13] It would lead to massive habitat loss and fragmentation, and put severe pressure on the world's remaining biodiversity.

On the other hand, a productivity increase of 1 percent per year, equivalent to a cumulative 69 percent increase from 1997 to 2050, would reduce the amount of new cropland needed to meet future demand to 325 Mha. Such an increase in productivity is theoretically possible without resorting to biotechnology. It would require large investments in human capital, research and development, extension services, infrastructure expansion (to bring new lands, where needed, into pro-

Figure 2

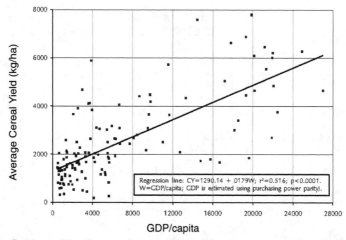

Cereal Yields as a Function of National per Capita
Gross Domestic Product (GDP), 1995

Source: Goklany, *The Future of the Industrial System*, International Conference on Industrial Ecology and Sustainability, University of Technology of Troyes, Troyes, France, 22–25 September 1999.

duction and integrate them with the rest of the world's agriculture system), inputs such as fertilizers and pesticides, and acquisition and operation of technologies to limit or mitigate environmental impacts of agriculture.[14]

A 1 percent a year increase in the net productivity of the food and agricultural sector (per unit area) is within the bounds of historical experience given that it increased 2 percent a year between 1961–63 and 1996–98.[15] More important, there are numerous existing but underused opportunities to enhance productivity in an environmentally sound manner. They are underused largely due to insufficient wealth—one reason why cereal yields are usually lower in poorer nations (see figure 2). Merely increasing the 1996–98 average cereal yields in developing and transitional nations to the level attained by Belgium-Luxembourg (the country grouping that had the highest average yield, that is, the yield ceiling, YC) would have increased global production in these years by 141 percent,[16] while increasing the average global cereal yield (2.96 tons per hectare [T/ha] in 1996–98)[17] to YC (7.80 T/ha in 1996–98) would increase global cereal production by 163 percent. Notably, the theoretical maximum yield is 13.4 T/ha or 350 percent greater than the average global cereal yield in 1996–98.[18]

Several conventional (that is, nonbioengineering) methods could be used to increase net productivity in the food and agricultural sector from farm to mouth. These methods include (1) further limiting preharvest crop losses to pests and diseases, which currently reduce global yields by an estimated 42 percent;[19] (2) increasing fertilizer use; (3) liming acidic soils; (4) adapting high yielding varieties to specific locations around the world (many scientists believe, however, that opportunities to further increase yields through conventional breeding techniques are almost tapped out);[20] and (5) reducing postharvest and end-use losses,[21] which are estimated at about 47 percent worldwide.[22] Improvements in yields based upon conventional technologies will depend, in large part, on the ability of developing nations to afford and operate (through economic development and growth of human capital) the necessary technologies.

Productivity improvements could come much more rapidly and more surely if biotechnology is used. Biotechnology could more easily reduce current gaps between average yields and yield ceilings, and between yield ceilings and the theoretical maximum yield, as well as push up the theoretical maximum yield. This begs the question of whether environmental costs of such productivity increases will also increase and whether yield increases are sustainable in the long run. This issue will be discussed in greater detail later.

If through biotechnology the annual rate at which productivity can be increased sustainably rises from 1 percent a year to 1.5 percent a year, then cropland could actually be reduced by 98 Mha rather than increased by 325 Mha, relative to 1997 levels (see figure 1). At the same time, the increased food demand of a larger and richer population could be met. If productivity could be doubled to 2 percent a year, then 422 Mha of current cropland could be returned to the rest of nature or made available for other human uses. Boosting annual productivity from 1 percent per year to 1.5 percent per year implies a net improvement in agricultural productivity of 30 percent due to biotechnology alone, while a 2 percent per year productivity increase corresponds to an overall improvement of 69 percent.

Several biotechnological crops, currently in various stages between research and commercialization, could put more food on the table per unit of land and water used in agriculture. Such crops, which could be particularly useful in developing nations, include:[23]

- *Cereals that are tolerant of poor climate and soil conditions.* Specifically, cereals that are tolerant to aluminum (so they can grow in acidic soils), drought, high salinity levels, submergence, chilling, and freezing are being developed.[24] The ability to grow crops in such conditions could be critical for developing countries. Forty-three percent of tropical soils are acidic.[25] More cropland is lost to high salinity than is gained through forest clearance. Salinity has rendered one-third of the world's irrigated land unsuitable for growing crops.[26] Moreover, if the world warms, the ability to tolerate droughts, high salinity, submergence, and acidity could be especially important for achieving global food security. In Kasuga et al.'s experiments,[27] 96 percent of genetically modified (GM) plants survived freezing, compared to less than 10 percent for the wild-type plant. Corresponding numbers for drought were 77 percent versus 2 percent and, for salinity stress, 79 percent versus 18 percent.

- *Rice that combines the best traits of the African and Asian varieties.* This bioengineered rice combines the ability of African rice to shade out weeds when young (which, however, inhibits photosynthesis later in the "pure" African variety) with the high yield capacity of the Asian variety.[28] In addition, the GM variety is highly resistant to drought, pests, and diseases. This could be particularly useful for Africa because its increases in rice yields have so far lagged behind the rest of the world's. This lag is one reason why malnourishment in sub-Saharan Africa has increased in the past several decades, in contrast to improved trends in nutrition elsewhere.[29]

- *Rice with the property of being able to close stomata more readily.*[30] Rice with this characteristic ought to increase water-use efficiency and net photosynthetic efficiency. Both aspects will be useful under dry conditions—conditions that, moreover, may get more prevalent in some areas if global warming continues.

- *Rice with the alternative C4 pathway for photosynthesis.* This trait could be especially useful if there is significant warming because the C4 pathway is more efficient at higher temperatures.[31] In addition, efforts are underway to try to reengineer RuBisCO—an enzyme critical to all photosynthesis—by using RuBisCO from red algae, which is a far more efficient catalyst for photosynthesis than that found in crops.[32]

- *Maize, rice, and sorghum with resistance to Striga, a parasitic weed that could decimate yields in sub-Saharan Africa.*[33]

- *Rice with the ability to fix nitrogen.*[34]

- *Rice and maize with enhanced uptakes of phosphorus and nitrogen.* Rice and maize production together account for 20 percent of global cropland use.[35]

- *Rice, maize, potato, sweet potato, and papaya with resistance to insects, nematodes, bacteria, viruses, and fungi.* For instance, papaya, which had been ravaged in Hawaii by the papaya ringspot virus, has now made a comeback due to a bioengineered variety resistant to that virus.[36]

- *Cassava, a staple in much of Africa, with resistance to the cassava mosaic virus and including a gene with an enzyme (replicase) with the ability to disrupt the life cycles of a number of other viruses.* This bioengineered cassava could, it is claimed, increase yields tenfold.[37] Also, because cassava naturally contains substances that can be converted to cyanide, it has to be carefully prepared before consumption. Work is proceeding on a genetically modified cassava that would be less toxic.[38]

- *Spoilage-prone fruits bioengineered for delayed ripening, thereby increasing their shelf life and reducing postharvest losses.* These fruits include bananas and plantains (important sources of food for many African nations),[39] and melons, strawberries, and raspberries.[40]

- *Crops bioengineered to reduce the likelihood of their seed pods shattering.* Shattering seed pods reduce yields of crops such as wheat, rice, and canola. It is estimated that genetically modifying canola in this way could increase canola yields by 25 to 100 percent.[41]

- *High-lysine maize and soybeans, maize with high oil and energy content, and forage crops with lower lignin content.* These alterations ought to improve livestock feed and reduce the overall demand for land needed for livestock.[42]

If the methods and genes used to bioengineer the above crops can be successfully adapted and transferred to other vegetables, tubers, fruits, and even trees, that would help reduce future land and water needs for feeding, clothing, and sheltering humanity.

Reduction in the Release of Nutrients, Pesticides, Silt, and Carbon into the Environment

The above GM crops, by increasing crop yields and reducing the amount of cultivated land necessary, would also reduce the area subject to soil erosion from agricultural practices. Reducing soil erosion, in turn, would limit associated environmental effects on water bodies and aquatic species and would reduce loss of carbon sinks and stores into the atmosphere. Furthermore, many of the same GM crops could also directly reduce the amount of nutrients and pesticides released into the environment.[43] These bioengineered crops include:

- *Nitrogen-fixing rice, and rice and maize bioengineered with the ability to increase uptakes of phosphorus and nitrogen from the soil.* In Europe and the United States, only 18 percent of the nitrogen and 30 percent of the phosphorus in fertilizers are incorporated into crops. Between 10 and 80 percent of the nitrogen and 15 percent of the phosphorus end up in aquatic ecosystems. Much of the remainder accumulates in the soil, to be eroded later into aquatic systems.[44] These nitrogen-fixing crops would reduce reliance on fertilizers. As a result, they would reduce ground- and surface-water pollution, risks of chemical spills, and atmospheric emissions of nitrous oxide. Nitrous oxide is a greenhouse gas that, pound for pound over a hundred-year period, is 310 times more potent a greenhouse gas than is carbon dioxide.[45]
- *Crops resistant to viruses, weeds, and other pests. Striga-resistant maize, rice, and sorghum are examples.* Various Bt crops, which contain genes from the *Bacillus thuringiensis* bacterium that has been used as a spray insecticide for four decades, also are being developed. One evaluation of Bt cotton in the United States estimates its planting on 2.3 million acres in 1998 reduced chemical-pesticide use by over a million pounds, increased yields by 85 million pounds and netted farmers an added $92 million compared to the performance of conventional cotton seed.[46] The usage of Bt maize that was planted on 14 million acres in the United States reduced pesticide spraying

on 2 million of those acres. The reduction in pesticide use would have been greater but for the fact that many farmers do not normally spray for the European maize borer, the target of the Bt toxin.[47]

Developing countries also can reduce pesticide usage by using pest-resistant crops. India is the world's third largest producer of cotton. The crop is grown on only 5 percent of India's land, yet cotton farmers buy about 50 percent of all pesticides used in the country.[48] In 1998, the devastation caused by pests reportedly contributed to five hundred suicides among Indian cotton farmers whose crops had failed. Field trials of Bt cotton at thirty locations in India show a 14 to 38 percent yield increase despite suspension of any spraying.[49]

- *Low phytic acid corn and soybean and phytase feed.* These altered crops help livestock better digest and absorb phosphorus. As a result, they can reduce phosphorus in animal waste and decrease runoff into streams, lakes, and other water bodies, mitigating one of the major sources of excess nutrients in the environment.[50] These GM crops would also reduce the need for inorganic phosphorus supplements in feed.

- *Crops tolerant of various herbicides, so that those herbicides can be used to kill weeds, but not the crop itself.* Herbicide-tolerant crops are among the most common applications of biotechnology today. One commercially available example is "Roundup Ready" soybeans, which are engineered to be tolerant to glyphosate. There are several potential benefits associated with such crops. They could help reduce the amount, toxicity, and persistence of herbicides employed. So far results from the field are mixed. Planting these crops seems to have reduced application of more hazardous and longer-lasting herbicides (for example, acetochlor), although overall herbicide use may have increased.[51]

Such herbicide-tolerant crops also would increase yields, while facilitating no-till cultivation. No-till cultivation is a highly effective method of stemming soil erosion and, thus, preserves future agricultural productivity. Erosion from cultivated land can be particularly damaging to the environment because the eroded particles can transport fertilizers and pesticides into aquatic systems and into the atmosphere. Finally, as noted, soil erosion releases stored carbon into the atmosphere.

Other Environmental Benefits of Bioengineered Plants and Trees

Crops can also be engineered to directly clean up environmental problems. For instance, GM plants can be used for bioremediation. Crops can be engineered to selectively absorb various metals and metal complexes such as aluminum, copper, and cadmium from contaminated soils.[52] Such plants could, for instance, detoxify methyl mercury in soils, thereby removing it from the food chain.

Researchers have also genetically modified aspen trees to produce 50 percent less lignin and 15 percent more cellulose. Lignin, a component of all wood, must be chemically separated from cellulose to make the pulp used in paper production. The GM tree has half the normal lignin to cellulose ratio of one to two. Overall, 15 percent more pulp may be produced from the same amount of wood. Moreover, the GM trees are 25 to 30 percent taller. Thus, the land, chemicals, and energy used to make a given quantity of paper ought to be reduced substantially and result in significantly lower environmental impacts at every stage, from tree farming to paper production.[53]

Other potential applications of biotechnology that could reduce environmental impacts include production of biodegradable plastics using oilseed rape and production of colored cotton, which could reduce reliance on synthetic dyes.[54]

Public Health Benefits

Having sufficient quantities of food is often the first step to a healthy society.[55] The increase in food supplies per capita during the last half-century is a major reason for the worldwide improvement in health status during that period. Between 1961 and 1997, food supplies per capita increased 23 percent.[56] Thus, despite a 40 percent increase in population between 1969–71 and 1994–96, chronic undernourishment in developing countries dropped from 35 to 19 percent of their population.[57] Improved nourishment helped lower global infant-mortality rates from 156 per 1,000 live births to 57 per 1,000 live births between 1950–55 and 1998. Life expectancies increased from 46.5 years to 65.7 years between 1950–55 and 1997. Better nourishment also enabled the average person to live a more fulfilling and productive life.[58]

Despite unprecedented progress during the last century, billions of people still suffer from undernourishment, malnutrition, and other ailments due, in whole or part, to insufficient food or poor nutrition. Table 1 lists the current extent and consequences of some of these food- and nutrition-related problems, and a qualitative assessment of the likelihood that using GM, rather than conventional, crops could reduce their numbers.

As shown in table 1, about 825 million people currently are undernourished, that is, cannot meet their basic needs for energy and protein.[59] Reducing these numbers over the next half-century while also reducing pressures on biodiversity in the face of anticipated population increases of 1.3 to 4.7 billion[60] requires increasing the quantity of food produced per unit of land and water. As discussed above, GM crops could help in this struggle.

Increasing food quantity is not enough; improving the nutritional quality of food is just as important. The diets of nearly half the world's population are deficient in iron, vitamin A, or other micronutrients (see table 1). Such deficiencies can lead to disease and premature death.[61] About 2 billion people do not have enough iron in their diets, making them susceptible to anemia. Another 260 mil-

Table 1
**Current Extent of Public Health Problems Partly or Wholly Caused by
Insufficient Food or Poor Nutrition, and the Likelihood That They
Could Be Alleviated Using GM, Rather Than Conventional, Crops**

Problem	Current Extent (Year)	Likelihood that GM crops would reduce problem
Undernourishment	825 million people (1994–96)	Very high
Malnutrition	6.6 million deaths per year in children under 5 years old (1995)	Very high
Stunting	200 million people (1995)	High
Iron-deficiency anemia	2,000 million people (1995)	High
Vitamin A deficiency	260 million people (1995)	High
Ischemic and cerebrovascular diseases	2.8 million deaths per year in high-income countries (1998) 9.7 million deaths per year in low/mid-income countries (1998) (includes those due to smoking)	Moderate
Cancers	2.0 million deaths per year in high-income countries (1998) 5.2 million deaths per year in low/mid-income countries (1998) (includes those due to smoking)	Moderate

Sources: World Health Organization, *The World Health Report 1999* (Geneva: World Health Organization, 1999); Food and Agriculture Organization, FAO Databases, at apps.fao.org; id., "The State of Food Insecurity in the World," at www.fao.org/FOCUS/E/SOFI/home-e.htm, 1999.

lion suffer from subclinical levels of vitamin A deficiency, which causes clinical xerophthalmia which, if untreated, may lead to blindness, especially in children. Vitamin A is also crucial for effective functioning of the immune system.[62] Through the cumulative effects of these deficiencies, in 1995, malnutrition was responsible for 6.6 million or 54 percent of the deaths worldwide in children under five years of age. These nutritional shortfalls also resulted in stunting in 200 million children, and clinical xerophthalmia in about 2.7 million people.[63]

In addition to helping to ensure that adequate quantities of food are available, bioengineering could also help reduce many of these micronutrient deficiencies. For instance, Swiss scientists have developed "golden rice" that is rich in beta-carotene, a precursor to vitamin A, and have crossed it with another bioengineered strain rich in iron and cysteine, which allows iron to be absorbed in the digestive tract. Two-thirds of a pound of this rice, an average daily ration in the tropics, will provide the average daily vitamin A requirement while reducing iron deficiency. An ancillary benefit is that golden rice would reduce the need for meat—one of the primary sources for dietary iron. As a result, overall demand for livestock feed, and the land, water, and other inputs necessary to produce that feed could be reduced to some degree.[64]

Scientists are also working on using bananas and other fruits as vehicles to deliver vaccines against the Norwalk virus, *E. coli*, hepatitis B, and cholera.[65] This could eventually lead to low-cost, efficient immunization of whole populations against common diseases with broader coverage than likely with conventional needle delivery.

Bioengineered crops can also help battle the so-called diseases of affluence, namely, ischemic heart disease, hypertension, and cancer. According to the World Health Organization, these diseases accounted for 4.8 million or 60 percent of the total deaths in high-income countries, and 14.9 million or 32 percent of deaths in the low- and middle-income countries in 1998 (see table 1).[66]

Several GM crops can help reduce this toll. For instance, genetically enhanced soybeans that are lower in saturated fats are already on the market. The International Food Information Council also notes that biotechnology could make soybean, canola, and other oils and their products, such as margarine and shortenings, more healthful.[67] Bioengineering could produce peanuts with improved protein balance, tomatoes with increased antioxidant content, potatoes with higher starch than conventional potatoes (reducing the amount of oil absorbed in French fries and potato chips), fruits and vegetables fortified with (or containing higher levels of) vitamins such as C and E, and higher-protein rice, using genes transferred from pea plants.

Moreover, levels of mycotoxins, which apparently increase with insect damage in crops, are lower on Bt corn. Some mycotoxins, such as fumonisin, can be fatal to horses and pigs, and may be human carcinogens. Thus, Bt corn, whether used as food for humans or feed for livestock, may be a healthier product than conventional corn.[68]

GM plants may also be able to save life and limb. It may sound like science fiction, but there is some speculation that plants can be engineered to biodegrade explosives in land mines and in ordnance at abandoned munitions sites.[69]

Finally, to the extent pest-resistant GM plants can reduce the amount, toxicity, and/or persistence of pesticides used in agriculture, accidental poisonings and other untoward health effects on farm workers or their immediate families could be reduced.

Potential Costs of Bioengineered Crops

Adverse Environmental Consequences

The major environmental concerns regarding GM crops are those related to crops that are designed to be resistant to pests and tolerant of herbicides. One potential risk is that target pests will become resistant to toxins produced by pest-resistant GM crops, such as Bt corn or Bt cotton. Although this is a possibility even if Bt is delivered via conventional sprays on non-GM plants, it is of greater concern with Bt plants. Under conventional spraying, target pests are exposed to Bt toxins only for brief periods, whereas currently available Bt crops produce toxins throughout the growing season. Thus, genetically engineered Bt crops could increase the chances of developing Bt-resistant pests.[70] Some laboratory studies suggest that target pests may evolve resistance more rapidly than previously thought possible.[71]

Strategies used to address pest resistance due to conventional pesticide spraying can, and should, be adapted for GM crops. Such strategies include ensuring that plants deliver high doses of Bt, while simultaneously maintaining refuges for non-Bt crops to ensure pest populations remain susceptible to Bt. The EPA has established the requirement that Bt corn farmers plant 20 percent of their land in non-Bt corn as refuges. For Bt corn grown in cotton areas, farmers must plant at least 50 percent non-Bt corn. In addition, the EPA requires expanded monitoring to detect any potential resistance.[72] Other strategies to delay development of pesticide resistance include crop rotation,[73] developing crops with more than one toxin gene acting on separate molecular targets,[74] and inserting the bioengineered gene into the chloroplast since that ought to express Bt toxin at higher levels.[75] Clearly, farmers have an economic stake in implementing such adaptive strategies so that their crop losses to pests are kept in check in the long term as well as the short term.

Another source of risk is that Bt from pest-resistant plants could harm, if not kill, nontarget species. This could happen if, for instance, Bt-laden pollen were to drift away from the field or if the toxin were to leak through the roots and be consumed by nontarget organisms.[76] Losey, Rayor, and Carter showed in a laboratory study that the mortality rate of monarch butterfly larvae fed for four days with milkweed dusted with Bt maize pollen was 44 percent, compared to zero for the control case, which used milkweed dusted with ordinary pollen.[77]

The extent to which, or even whether, the monarch butterfly population would be affected in the real world is a matter of debate.[78] One study suggests that under a worst-case scenario as much as 7 percent of the North American monarch population (estimated at 100 million) may die, although the real-world effect would probably be smaller.[79] Some have also argued that the major threat to monarchs is the habitat loss in their wintering grounds in Mexico.[80] Perhaps more important, the inadvertent effects of Bt crops due to pollen dispersal or root leakage could be virtually eliminated by bioengineering genes into the chloroplast rather than into nuclear DNA.[81]

Bt could also enter the food chain through root leakage or if predators prey on target pests. For instance, studies have shown that green lacewing larvae, a beneficial insect, that ate maize borers fed with Bt maize were more likely to die.[82] Again, the real-world significance of this experimental finding has been disputed based on the long history of Bt spraying on crops and other studies that showed beneficial insects essentially unharmed by such spraying.[83]

There is also a concern that bioengineered genes from herbicide- or pest-tolerant crops might escape into wild relatives leading to "genetic pollution" and creating "superweeds." Such a development would have an adverse economic impact on farmers, reducing crop yields and detracting from the very justification for using such GM crops.[84] Clearly, farmers have a substantial incentive for preventing weeds from acquiring herbicide tolerance and, if that fails, to keep such weeds in check.

Gene escape is possible if sexually compatible wild relatives are found near fields planted with GM crops. This is a possibility in the United States for sorghum, oats, rice, canola, sugar beets, carrots, alfalfa, sunflowers, and radishes.[85] However, the most common GM crops, namely soybeans and corn, have no wild U.S. relatives.[86] Moreover, centuries of conventional breeding have rendered a number of important crops, for example, maize and wheat, "ecologically incompetent" in many areas,[87] although that is no guarantee of safety.[88] Despite the use of conventionally bred, herbicide-tolerant plants, there has been no upsurge in problems due to herbicide-tolerant weeds.[89] If any weeds develop such tolerance, available crop-management techniques (such as another herbicide) can be used to control them.

Gene escape from GM crops to wild relatives is also an environmental concern. It has been argued that herbicide-tolerant "superweeds" could invade natural ecosystems. It is unclear why such a weed would have a competitive advantage in a natural system unless that system is treated with the herbicide in question. But if it is so treated, does it still qualify as a natural system? Regarding ecosystem function and biodiversity, the significance of genetic pollution, per se, is unclear. Would gene escape affect ecosystem function negatively? Does gene escape diminish or expand biodiversity?[90]

Genes also may escape from GM crops to non-GM crops of the same species. Such an escape would be unpopular with organic farmers, who are afraid it might "adulterate" their produce. Of course, producers of GM seeds and farmers planting GM seeds are not eager to have someone else profit from their investments, either. The chances of such gene escape can be reduced by maintaining a buffer between the two crops. The Royal Society also notes that because more crops (including corn, sorghum, sugar beets, and sunflowers) are now grown from hybrid seeds, this provides a measure of built-in security against such gene transfers.[91]

Gene escape could be limited further if the GM plant was engineered to be sterile or prevented from germinating using, for instance, "terminator technology." An alternative approach would be to insert the gene into the chloroplast, which would preclude its spread through pollen or fruit, as well as prevent root leakage.[92]

Finally, there is a concern that in the quest to expand yields, GM plants will work too well in eliminating pests and weeds, leading to a further simplification of agricultural ecosystems and further decreasing biodiversity. This concern, in conjunction with the other noted environmental concerns, needs to be weighed against the cumulative biodiversity benefits of reduced conversion of habitat to cropland, and decreased use of chemical inputs.

Adverse Public Health Consequences

A major health concern is that the new genes inserted into GM plants could be incorporated into a consumer's genetic makeup. However, there is no evidence that any genes have ever been transferred to human beings through food or drink despite the fact that plant and animal DNA has always been a part of the daily human diet.[93] In fact, an estimated 4 percent of the human diet is composed of DNA,[94] and an average adult Briton consumes 150,000 kilometers of DNA in an average meal.[95] It is unclear, for instance, why consuming beans which have been modified with genes from a pig would pose a greater risk to public health than consuming a dish of non-GM pork and beans.

Another concern is that genes transferred from foods to which many people are allergic could trigger allergies in unsuspecting consumers of such GM crops. Between 1 and 3 percent of adults and 5 and 8 percent of children in the United States suffer from food allergies. Each year, food allergies cause 135 fatalities and 2,500 emergency room visits.[96]

This concern regarding allergic reactions to GM foods can be traced to pre-commercialization tests of a GM soybean conducted by Pioneer Hi-Bred. The tests showed that a soybean that had been bioengineered to boost its nutritional quality using a gene from the Brazil nut was, in fact, allergenic.

Although this example shows that GM foods can be tested prior to commercialization for their allergic potential, opponents of GM foods have used this as an argument against bioengineered crops. Several databases of known allergens could be used to help identify problematic GM products before they are developed.[97] In fact, because bioengineering allows more precise manipulation of genes than does conventional plant breeding, it could be used to render allergenic crops nonallergenic.[98]

Yet another potential negative effect on public health is that antibiotic-resistant "marker" genes that are used to identify whether a gene has been successfully incorporated into a plant could, through consumption of the antibiotic gene by humans, accelerate the trend toward antibiotic-resistant diseases. However, compared to the threat posed by the use of antibiotics in feed for livestock and their overuse as human medicines, the increased risk due to such markers is slight.[99]

Applying the Precautionary Principle

The above discussion indicates there are risks associated with either the use or the nonuse of GM crops. Here I will apply the various criteria outlined in the framework presented previously for valuing actions that could result in uncertain costs and uncertain benefits. Ideally, each criterion should be applied individually to the human mortality, the non–mortality public health, and the non–public-health-related environmental consequences of GM crop use. However, because the severity and degree of uncertainty associated with the various costs and benefits for each of these sets of consequences are not equivalent, I will apply several criteria simultaneously.

Public Health Consequences

Population could increase 50 percent between 1998 and 2050 (from 5.9 billion to 8.9 billion, according to the United Nations's best estimate). Hence, by 2050, undernourishment, malnutrition, and their consequences on death and disease might also be expected to increase by 50 percent worldwide, if global food supply increases by a like amount and all else remains equal. Unless food production outstrips population growth significantly over the next half-century, billions in the developing world may suffer annually from undernourishment, hundreds of millions may be stunted, and millions may die from malnutrition. Based on the sheer magnitude of people at risk, and the degree of certainty of the public health consequences, one can state with confidence that limiting GM crops will increase death and disease, particularly among the poor.

GM crops could also reduce or postpone deaths due to diseases of affluence. The probability of reducing these deaths by moving forward with GM crops is lower than that of reducing deaths due to hunger and malnutrition, of course. But the expected number of deaths postponed could run into the millions. For instance, a 10 percent decrease in the 15 million deaths due to cancer, ischemic, and cerebrovascular diseases in low- and middle-income nations translates to 1.5 million lives saved (see table 1).

By contrast, the negative public health consequences of ingesting GM foods are speculative (for example, the effects due to ingesting transgenes) or relatively minor in magnitude (for example, a potential increase in antibiotic resistance or increased incidence of allergic reactions). Moreover, it is possible to reduce, if not eliminate, the effects of even those minor impacts.

As noted previously, the likelihood of allergic reactions can be reduced by checking various databases of known allergens prior to developing a GM crop and by testing food from such crops prior to commercialization. With respect to the risk of increasing antibiotic resistance, Novartis has developed a sugar-based alternative to antibiotic-resistant marker genes that has been used to develop about a dozen GM crops, including maize, wheat, rice, sugar beet, oilseed rape, cotton, and sunflowers.[100] With additional research, marker genes may be devised for other crops. Alternatively, practical methods of removing or repressing antibiotic-resistant marker genes may be developed.[101]

Thus, with respect to human mortality and morbidity, and employing the "uncertainty," "expectation value," and "adaptation" criteria outlined in the framework developed previously, the precautionary principle *requires* that we continue to research, develop, and commercialize (with appropriate safeguards, of course) those GM crops that would increase food production and generally improve nutrition and health, especially in the developing world.

Some have argued that many developed countries are "awash in surplus food."[102] Thus, goes this argument, developed countries have no need to boost food production. However, this argument ignores the fact that reducing those surpluses would be almost as harmful to public health in developing countries as curtailing their food production directly.

At present, net cereal imports of the developing countries exceed 10 percent of their production. Trade (and aid) voluntarily moves the surplus production in developed countries to developing countries suffering from food deficits. Without this movement of developed nations' food surpluses, food supplies in developing countries would be lower, food prices would be steeper, undernourishment and malnutrition would be higher, and associated health problems, such as illness and premature mortality, would be greater. As already noted, developing countries' food deficits are only expected to increase in the future because of high population growth rates and, possibly, could be further worsened by global warming. Therefore, developed countries' food surpluses will at least be as critical for future food security in developing countries as they are today.

The above argument against GM crops also assumes that such crops will produce little or no benefits for the inhabitants of developed countries. But, as noted, GM crops are also being engineered to improve nutrition in order to combat diseases of affluence afflicting populations in developed, as well as developing, nations. As noted, these diseases are among the major causes of premature death in the developed countries—approximately 4.8 million a year currently (see table 1).[103] Similarly, the health benefits of a GM crop like "golden rice," for instance, do not have to be confined to developing countries; developed countries, too, could avail themselves of its benefits. Thus even for developed countries, the potential public health benefits of GM crops far outweigh in magnitude and certainty the speculative health consequences of ingesting GM foods.

Hence, the "expectation value" and "uncertainty" criteria applied to public health require developed countries to develop, support, and commercialize yield-increasing and health- and nutrition-enhancing GM crops in order to improve public health worldwide.

Environmental Consequences

Another, related argument against using GM foods to increase food production is that there is no shortage of food in the world today, that the problem of hunger and malnutrition is rooted in poor distribution and unequal access to food because of poverty. Therefore, the argument goes, it is unnecessary to increase food production; ergo, there is no compelling need for biotechnology.[104]

But even if everyone had equal access—an unlikely proposition, at best—finite levels of food, fiber, and timber would still have to be produced to meet the demand. A figure similar to figure 1 could be developed for any level of food demand whether it is, say, half that of today (perhaps because of a perfect, cost-free distribution system and a magical equalization of income) or whether it is four times that (possibly due to runaway population growth). Thus, regardless of the level of demand, limiting GM crops would lower crop and forest yields per unit of land and water used. To compensate for the lower yields, more land and water would have to be pressed into humanity's service, leaving that much less for the rest of nature.[105]

Moreover, if bioengineering succeeds in improving the protein and micronutrient content of vegetables, fruits, and grains, it might persuade many more people to adopt and persevere with vegetarian diets, thereby reducing the additional demand that meat-eating places on land and water. Giving up GM crops would further increase pressures on biodiversity due to excess nutrients, pesticides, and soil erosion. Reduced conversion of habitat and forest to crop and timber land coupled with reduced soil erosion due to increased no-till cultivation of bioengineered crops also would help limit losses of carbon reservoirs and sinks (thereby potentially reducing global warming).

Arrayed against these benefits to ecosystems, biodiversity, and carbon stores and sinks from deploying GM crops are the environmental costs from widespread planting of pest-resistant and herbicide-tolerant GM crops *minus* the environmental costs of conventional farming practices. These costs include a potential decrease in the diversity of the flora and fauna associated with, or in the immediate vicinity of, GM crops if they reduce more nontarget pests and weeds than conventional farming practices, and the possible consequences of gene escape to weeds and non-GM crops.

Hence, with respect to the environmental consequences, one must still conclude, based on the "uncertainty" and "expectation value" criteria, that the precautionary principle requires the cultivation of GM crops. On net, bioengineered crops should conserve the planet's habitat, biodiversity, and carbon stores and sinks, provided due caution is exercised, particularly with respect to herbicide-tolerant and pest-resistant GM crops.

It may be argued that if genes escape and are established in "natural" ecosystems, this may lead to irreversible harm to the environment; thus, under the "irreversibility" criterion, GM crops ought to be banned. However, increased habitat clearance and land conversion resulting from such a ban may be at least as irreversible, particularly if it leads to species extinctions.

It is worth noting that the precautionary principle supports using terminator-type technology because it would minimize the possibility of gene transfer to weeds and non-GM plants. Some of the same groups that profess environmental concerns about genetic pollution are the most vehement critics of terminator technology.[106] Clearly, in the policy calculus of these groups, the presumed negative economic consequences to farmers due to their inability to propagate GM crops from sterile seeds (and the antipathy of these organizations toward multinationals' profits) outweighs the environmental benefits of GM crops.

Conclusion

The precautionary principle often has been invoked to justify a prohibition of GM crops.[107] However, this policy is based upon a selective application of the precautionary principle to a limited set of consequences of such a policy. Specifically, the justification for a ban considers the potential public health and environmental benefits of a ban on GM crops, but ignores the probable public health and environmental benefits that would necessarily be foregone.

By comparison with conventional crops, GM crops would increase the quantity and nutritional quality of food supplies. Bioengineered crops hold the promise of improving public health and reducing mortality rates worldwide. In addition, cultivation of GM, rather than conventional, crops would be more protective of habitat, biological diversity, and carbon stores and sinks. Crops are being genetically modified to increase productivity, thus reducing the amount of land and water that would otherwise have to be diverted to humanity's needs. GM crops could also reduce the environmental damage resulting from the use of synthetic fertilizers and pesticides, and from soil erosion.

A ban on GM foods, contrary to the claims of its proponents, would be imprudent rather than precautionary. The precautionary principle—properly applied, using a broader consideration of the public health and environmental consequences of a ban—argues instead for a sustained effort to research, develop, and commercialize GM crops, provided reasonable caution is exercised during testing and commercialization of these crops.

In this context, an action, precondition, or restriction regarding testing or commercialization is "reasonable" if its public health benefits are not likely to be negated by reductions in the quantity or quality of food that would otherwise be available. Halting testing in this instance would increase food costs and reduce broader access to higher quality food, particularly for the poorer and most vulnerable segments of society. Also, the environmental gains flowing from a "reasonable" precaution should more than offset the environmental gains that could otherwise be obtained. In other words, a reasonable precaution is one that does not kill the goose that lays the golden egg, as a ban on GM crops would do.

Notes

1. Carolyn Raffensperger, Joel Tickner, and Wes Jackson, eds., *Protecting Public Health and the Environment: Implementing the Precautionary Principle* (Washington, D.C.: Island Press, 1999).

2. Maureen L. Cropper and Paul R. Portney, "Discounting Human Lives," *Resource* 108 (1992): 14.

3. Indur M. Goklany, "Saving Habitat and Conserving Biodiversity on a Crowded Planet," *BioScience* 48 (November 1998): 941–53.

4. Food and Agriculture Organization, FAO databases [online], http://www.apps. fao.org [12 January 2000].

5. United Nations Commission on Sustainable Development, 5th session, document E/CN.17/1997/9 [online], gopher://gopher.un.org:70/00/esc/cn17/1997/off/ 97-9.EN [23 July 1999].

6. David S. Wilcove et al., "Quantifying Threats to Imperiled Species in the United States," BioScience 48 (August 1998): 607–15.

7. Goklany, " Saving Habitat and Conserving Biodiversoty on a Crowded Planet"; Goklany, "The Importance of Climate Change Compared to Other Global Changes," in Proceedings of the Second International Specialty Conference: Global Climate Change—Science, Policy, and Mitigation/Adaptation Strategies, Crystal City, Virginia, 13–16 October 1998 (Sewickley, Penn.: Air and Waste Management Association, 1998); Goklany, "Meeting Global Food Needs: The Environmental Trade-offs betweem Increasing Land Conversion and Land Productivity," in Fearing Food: Risk, Health, and Environment, eds. Julian Morris and Roger Bate (Oxford, U.K.: Butterworth-Heinemann, 1999); and Goklany, "Potential Consequences of Increasing Atmospheric CO_2 Concentration Compared to Other Environmental Problems," Technology 7, supplement 1(2000): 189.

8. Goklany, "Meeting Global Food Needs."

9. United Nations Population Division, Long-Range World Population Projection: Based on the 1998 Revision, report ESA/P/WP.153 (New York: United Nations, 1999).

10. Goklany, "Saving Habitat"; Goklany, "Meeting Global Food Needs."

11. Goklany, "Saving Habitat."

12. Food and Agriculture Organization, FAO Databases.

13. Goklany, "Saving Habitat."

14. Goklany, "Meeting Global Food Needs."

15. Food and Agriculture Organization, FAO Databases.

16. Ibid.

17. Ibid.

18. Hans Linnemann et al., MOIRA: Model of International Relations in Agriculture: Report of the Project Group "Food for a Doubling World Population" (Amsterdam: North Holland Publishing, 1979).

19. E. C. Oerke et al., "Conclusion and Perspectives," in Crop Production and Crop Protection: Estimated Losses in Major Food and Cash Crops, eds. E. C. Oerke et al. (Amsterdam: Elsevier Science, 1994).

20. Gordon Conway and Gary Toenniessen, "Feeding the World in the Twenty-first Century," Nature 402 (1999 supplement): C55–C58; Charles C. Mann, "Crop Scientists Seek a New Revolution," Science 283, no. 5400 (1999): 310–14.

21. Indur M. Goklany and Merritt W. Sprague, An Alternative Approach to Sustainable Development: Conserving Forests, Habitat, and Biological Diversity by Increasing the Efficiency and Productivity of Land Utilization (Washington, D.C.: Office of Program Analysis, Department of the Interior, 1991).

22. William H. Bender, "An End-Use Analysis of Global Food Requirements," Food Policy 19, no. 4 (1994): 381–95.

23. Indur M. Goklany, "Biotechnology and Biodiversity: The Risks and Rewards of Genetically Modified Crops," Bioscience (2000): in review.

24. Juan Manuel de la Fuente et al., "Aluminum Tolerance in Transgenic Plants by Alteration of a Na^+/H^+ Antiport in Arabidopsis," Science 285, no. 5431 (1999): 1256–58;

Mie Kasuga et al., "Improving Plant Drought, Salt, and Freezing Tolerance by Gene Transfer of a Single Stress-inducible Transcription Factor," *Nature Biotechnology* 17, no. 3 (1999): 287–91; M. S. Swaminathan, "Genetic Engineering and Food: Ecological and Livelihood Security in Predominantly Agricultural Developing Countries," (paper presented at the joint Consultative Group on International Agricultural Research/National Academy of Scence Biotechnology conference, Washington, D.C., October 1999) [online], http://www.cgiar.org/biotech/swami.htm [11 November 1999]; Conway and Toenniessen, "Feeding the World in the Twenty-first Century"; Anne S. Moffat, "Crop Engineering Goes South," *Science* 285, no. 5426 (1999): 370–71; Qifa Zhang, "Meeting the Challenges of Food Production: The Opportunities of Agricultural Biotechnology in China" (paper presented at CGIAR/NAS Biotechnology conference, Washington, D.C., October 1999) [online], www.cgiar.org/biotech/zhang.htm [11 November 1999]; Elisabeth Pennisi, "Plant Biology: Transferred Gene Helps Plants Weather Cold Snaps, " *Science* 280, no. 5360 (1998): 36; see also, Kirsten Jaglo-Ottosen et al., "Arabidopsis CBF-1 Overexpression Induces COR Genes and Enhances Freezing Tolerance," *Science* 280, no. 5360 (1998): 104–106.

25. World Bank, "New and Noteworthy in Nutrition," issue 24, 13 October 1994 [online], http://www.worldbank.org/html/extdr/hnp/nutrition/nnn/nnn24.htm [5 January 2000].

26. Wolf B. Frommer, Uwe Ludewig, and Doris Rentsch, "Taking Transgenic Plants with a Pinch of Salt," *Science* 285, no. 5431 (1999): 1222–23.

27. Kasuga et al., "Improving Plant Drought, Salt, and Freezing Tolerance."

28. Conway and Toenniessen, "Feeding the World in the Twenty-first Century."

29. Goklany, "Saving Habitat."

30. Mann, "Crop Scientists Seek a New Revolution."

31. Maurise S. B. Ku et al., "High-Level Expression of Maize Phosphoenolpyruvate Carboxylase in Transgenic Plants," *Nature Biotechnology* 17, no. 1 (1999): 76–80; Gerry Edwards, "Tuning up Crop Photosynthesis," *Nature Biotechnology* 17, no. 1 (1999): 22–23; Conway and Toenniessen, "Feeding the World in the Twenty-first Century."

32. Charles C. Mann, "Genetic Engineers Aim to Soup up Crop Photosyntheses," *Science* 283, no. 5400 (1999): 314–16.

33. Charles C. Mann, "Biotech Goes Wild," *Technology Review* (July/August 1999) [online], http://www.techreview.com/articles/July99/fulltext.htm [21 February 2000]; Colin MacIlwain, "Access Issues May Determine Whether Agri-Biotech Will Help the World's Poor," *Nature* 401, no. 6760 (1999): 341–45; Conway and Toenniessen, "Feeding the World in the Twenty-first Century."

34. International Rice Research Institute, "Nitrogen-fixing Rice Moves closer to Reality," IRRI Science online, 21 December 1999, http://www.iclaim.org.Science.htm [1 February 2000].

35. Conway and Toenniessen, "Feeding the World in the Twenty-first Century"; C. S. Prakish, "A First Step Toward Engineering Improved Phosphate Uptake," ISB New Report, May 1998 [online], http://www.Isb.vt.edu/news/1998/news98.may.html [15 January 2000]; Inside Purdue, "Ragothama: Phosphorous Uptake Gene Discovered," 13 January 1998 [online], http://www.purdue.edu/PER/1.13.98.IP.html [19 January 2000].

36. Conway and Toenniessen, "Feeding the World in the Twenty-first Century"; Dan Ferber, "Risks and Benefits: GM Crops in the Cross Hairs," *Science* 286, no. 5445 (1999): 1662–66.

37. Anne S. Moffat, "Engineering Plants to Cope with Metals," *Science* 285, no. 5426 (1999): 369–70.

38. Conway and Toenniessen, "Feeding the World in the Twenty-first Century."

39. Ibid.

40. Peggy G. Lemaux, "Plant Growth Regulators and Biotechnology," (paper presented at the Western Plant Growth Regulator Society, Anaheim, California, January 1999) [online], http://www.plantbio.berkeley.edu/~outreach/REGULATO.htm [19 January 2000].

41. Sarah J. Lijiegren et al., "Shatterproof MADS-box Gene Control Seed Dispersal in Arabidopsis," *Nature* 404, no. 6779 (2000): 766–70.

42. Barbara Mazur, Enno Krebbers, and Scott Tingey, "Gene Discovery and Product Development for Grain Quality Traits," *Science* 285, no. 5426 (1999): 372–75; Conway and Toenniessen, "Feeding the World in the Twenty-first Century."

43. Goklany, "Biotechnology and Biodiversity: Risks and Rewards."

44. Stephen Carpenter et al., "Nonpoint Pollution of Surface Waters with Phosphorous and Nitrogen," *Issues in Ecology* 3 (summer 1998) [online], esa.sdsd.edu/carpenter.htm [10 February 2000].

45. Intergovernmental Panel on Climate Change, *Climate Change 1995: The Science of Climate Change,* eds. J. T. Houghton et al. (Cambridge: Cambridge University Press, 1996).

46. Ferber, "Risks and Benefits."

47. Ibid.

48. C. S. Prakash, "Relevance of Biotechnology to Indian Agriculture," May 1999 [online], http://www.terlin.org/discuss/biotech/abstracts.htm [15 January 2000].

49. Hindu Business Line, "Bt Cotton Trials Show Yield Rise," 18 January 2000 [online], http://www.indisserver.com/bline/2000/01/19/stories/071903al.htm [18 January 2000].

50. Grabau Laboratory, "Improving Phosphorous Utilization in Soybean Meal Through Phytase Gene Engineering," Grabau Laboratory [online], www.biotech.vt.edu/plants/grabau/projects.html [9 February 2000]; Libby Mikesell, "Ag Biotech May Help Save the Bay," Biotechnology Industry Organization [online], www.bio.org/food&ag/cbf.html [21 February 2000]; CeresNet, "Environmental Benefits of Agricultural Biotechnology," 2 February 1999 [online], wvvw.ceresnet.org/Cnetart/990202 Environ_Benefits-A gBio.txt [10 February 2000].

51. Ferber, "Risks and Benefits."

52. Anne S. Moffat, "Engineering Plants to Cope with Metals."

53. Michigan Technological University, "New Aspen Could Revolutionize Pulp and Paper Industry," 11 October 1999 [online], www.admin.mtu.edu/urel/breaking/1999/aspen.htm [10 January 2000].

54. Eleanor Lawrence, "Plastic Plants," *Nature* science update, 24–30 September 1999 [online], www.helix.nature.corn /nsu/990930/990930-5.html [11 January 2000].

55. World Health Organization, *The World Health Report 1999* (Geneva: World Health Organization, 1999); Indur M. Goklany, "The Future of the Industrial System" (paper presented at the International Conference on Industrial Ecology and Sustainability, University of Technology of Troyes, Troyes, France, September 1999).

56. Food and Agriculture Organization, FAO Databases.

57. Food and Agriculture Organization, "The State of Food Insecurity in the World" [online], www.fao.org/FOCUS/E/SOFI/home-e.htm [12 January 2000].

58. World Health Organization, *The World Health Report 1999*; United Nations Population Division, *Long-Range World Population Projections*; Goklany, "The Future of the Industrial System."

59. Food and Agriculture Organization, "The State of Food Insecurity in the World."

60. United Nations Population Division, *Long-Range World Population Projections.*

61. World Health Organization, *The World Health Report 1999*; Food and Agriculture Organization, "The State of Food Insecurity in the World."

62. World Health Organization, "About WHO Nutrition," 21 September 1999 [online], www.who.int.aboutwho/en/promoting/nutrition.htm [5 January 2000].

63. World Health Organization, "Malnutrition—The Global Picture," 2 November 1999 [online], www.who.int/nut/malnutrition_worldwide.htm [5 January 2000].

64. Trisha Gura, "New Genes Boost Rice Nutrients," *Science* 285, no. 5430 (1999): 994–95; Mary Lou Guerinot. "The Green Revolution Strikes Gold," *Science* 287, no. 5451 (2000): 241–43; Xudong Ye et al., "Engineering the Provitamin A (Beta-Carolene) Biosynthetic Pathway into (Carotenoid-Free) Rice Endosperm," *Science* 287, no. 5451 (2000): 303–305; see also F. Onto et al., "Iron Fortification of Rice Seed by the Soybean Ferritin Gene," *Nature Biotechnology* 17, no. 3 (1999): 282–86.

65. Anne S. Moffat, "Engineering Plants to Cope with Metals"; Paul Smaglik, "Needle-Free Vaccines: Success of Edible Vaccine May Depend on Picking Right Fruit," *The Scientist* 12 (17 August 1998).

66. World Health Organization, *The World Health Report 1999.*

67. International Food Information Council, "Backgrounder—Food Biotechnology," updated April 1999 [online], www.ificinfo.health.org backgrnd/BKGR14.hlm [12 January 2000].

68. Gary P. Munkvold and Richard L. Hellmich, "Genetically Modified, Insect Resistant Corn: Implications for Disease Management" APSnet Plant Pathology online feature, 15 October–30 November 1999, http://www.scisoc.orgfeature/BtCorn/Top.html [19 February 2000].

69. Frederick Bolin, "Leveling Land Mines with Biotechnology," *Nature Biotechnology* 17, no. 8 (1999): 732; Christopher F. French et al., "Biodegradation of Explosives by Transgenic Plants Expressing Pentaerythritol Tetranitrate Reductase," *Nature Biotechnology* 17, no. 5 (1999): 491–94.

70. Fred Gould, "Sustaining the Efficacy of Bt Toxins," in *Agricultural Biotechnology and Environmental Quality: Gene Escape and Pest Resistance,* eds. Ralph W. Hardy and Jane B. Segelken, NABC Report 10 (Ithaca, New York: National Agricultural Biotechnology Council, 1998); see also Theo Wallimann, "Bt Toxin: Assessing GM Strategies," *Science* 287, no. 5450 (2000): 41c.

71. Liu Yong-Biao et al., "Development Time and Resistance to Bt Crops," *Nature* 400, no. 6744 (1999): 519; AgBiotechNet, "Hot Topic: Bt Plants: Resistance and Other Issues: New Findings on Insect Resistance to Bt Cotton," July 1999 [online], www.agbiotechnet.com/topics/hot.asp [4 February 2000].

72. Environmental Protection Agency, "Bt Corn Insect Resistance Management Announced for 2000 Growing Season," press release, 14 January 2000.

73. Fred Gould, "Sustaining the Efficacy of Bt Toxins."

74. Gordon Conway, "Food for All in the Twenty-first Century," *Environment* 42 (January/February 2000): 8–18.

75. Henry Daniell, "The Next Generation of Genetically Engineered Crops for Herbicide and Insect Resistance: Containment of Gene Pollution and Resistant Insects," AgBiotechNet1 (August 1999), ABN 024 [online], www.agbiotechnet.com/reviews/aug99/html/Daniell.htm [12 February 2000]; Madhuri Kota et al., "Overexpression of the *Bacillus Thuringiensis* (Bt) CRy2Aa2 Protein in Chloroplasts Confers Resistance to Plants against Susceptible and Bt-resistant Insects," in *Proceedings of the National Academy of Sciences USA* 96 (1999): 1840–45.

76. John E. Losey, Linda S. Rayor, and Maureen E. Carter, "Transgenic Pollen Harms Monarch Larvae," *Nature* 399, no. 6733 (1999): 214; Theo Wallimann, "Bt Toxin: Assessing GM Strategies"; Deepak Saxena, Saul Flores, and G. Stotzky, "Transgenic Plants: Insecticidal Toxin in Root Exudates from Bt Corn," *Nature* 402, no. 6761 (1999): 480.

77. Losey, Rayor, and Carter, "Transgenic Pollen Harms Monarch Larvae."

78. Ferber, "Risks and Benefits."

79. Ferber, "Risks and Benefits"; see also S. Milius, "New Studies Focus Monarch Worries," *Science News* 156 (1999): 391.

80. Ricki Lewis and Barry A. Palevitz, "GM Crops Face Heat of Debate," *The Scientist* 13 (1999): 1; see also Mary Beth Sheridan, "A Delicate Balancing Act in Mexico," *Los Angeles Times*, home edition, 29 February 2000, sec. A.

81. Madhuri Kota et al., "Overexpression of the *Bacillus Thuringiensis* (Bt) CRy2Aa2 Protein"; Susan E. Scott and Mike J. Wilkinson, "Low Probability of Chloroplast Movement from Oilseed Rape (*Brassica Napus*) into Wild *Brassica Rapa*," *Nature Biotechnology* 17, no. 4 (1999): 390–92; Dean Chamberlain and C. Neal Stewart, "Transgene Escape and Transplastomics," *Nature Biotechnology* 17, no. 4 (1999): 330–31.

82. Angelika Hilbeck et al., "Effects of Transgenic *Bacillus Thuringiensis* Corn-fed Prey on Mortality and Development Time of Immature *Chysoperla Carnea* (*Neuroptera: Chrysopidae*)," *Environmental Entomology* 27 (1998): 480–87.

83. Alan J. Gray and Alan F. Raybould, "Crop Genetics: Reducing Transgene Escape Routes," *Nature* 392, no. 6677 (1998): 653–54; see also C. L. Wraight et al., "Absence of Toxicity of *Bacillus thuringiensis* Pollen to Black Swallowtails under Field Conditions," *Proceedings of the National Academy of Sciences*, published online before print, 6 June 2000, object identifier 10.1073/pnas.130202097, http://www.pnas.org.

84. Gray and Raybould, "Crop Genetics: Reducing Transgene Escape Routes."

85. Mann, "Biotech Goes Wild"; Philip J. Regal, "Scientific Principles for Ecologically Based Risk Assessment of Transgenic Organisms," *Molecular Ecology* 3 (1994): 5–13 [online], www.psrast.org/pjrisk.htm [11 January 2000]; Lemaux, "Plant Growth Regulators and Biotechnology."

86. James R. Cook, "Toward Science-Based Risk Assessment for the Approval and Use of Plants in Agriculture and Other Environments," (paper presented at the joint CGIAR/NAS Biotechnology Conference, Washington, D.C., October 1999) [online], www.cgiar. org/biotechc/cook.htm [11 November 1999]; Mann, "Biotech Goes Wild."

87. Royal Society, "Genetically Modified Plants for Food Use," September 1998 [online], www.royalsoc.ac.uk/policy/index.html [11 January 2000].

88. Regal, "Scientific Principles for Ecologically Based Risk Assessment of Transgenic Organisms."

89. Royal Society, "Genetically Modified Plants."

90. See also Mark Sagoff, "What's Wrong with Exotic Species?" *Report from the Institute for Philosophy and Public Policy* 19 (fall 1999): 16–23; A. J. S. Rayl, "Are All Alien Invasions Bad?" *The Scientist* 14 (20 March 2000).

91. Royal Society, "Genetically Modified Plants."

92. Daniell, "The Next Generation of Genetically Engineered Crops"; Royal Society, "Genetically Modified Plants."

93. Royal Society, "Genetically Modified Plants."

94. Lisa Sheppard, "GMO Food Safety Risk Is Negligible," College of Agricultural, Consumer, and Environmental Sciences, University of Illinois at Urbana-Champaign, news release, 23 November 1999 [online], www.ag.uiuc.edu/news/artides/943382465.html [10 December 1999].

95. Lewis and Palevitz, "GM Crops Face Heat of Debate."

96. U.S. Senate Committee on Agriculture, Nutrition, and Forestry, "The Science of Biotechnology and Its Potential Applications to Agriculture," Hearing before the Committee on Foreign Relations, 106th Cong., 1st sess., 6 October 1999 [online], www.senate.gov/~agriculture/Hearings/Hearingsj999/buc99106.htm [11 January 2000].

97. Royal Society, "Genetically Modified Plants"; National Center for Food Safety and Technology, Biotechnology Information for Food Safety Database [online], www.iit.edu/~sgendel/fa.htm [11 January 2000].

98. U.S. Senate Committee on Agriculture, Nutrition, and Forestry, "The Science of Biotechnology and Its Potential Applications to Agriculture: Hearing"; Kathleen Scalise, "New Solution for Food Allergies Effective with Milk, Wheat Products, Maybe Other Foods, UC Researchers Discover," University of California, Berkeley, news release, 19 October 1997 [online], www.urel.berkeley.edu/urel_1/CampusNewsPressReleases/releas es/i 0_19_97a.html [5 January 2000].

99. Royal Society, "Genetically Modified Plants"; Food and Agricultural Organization/World Health Organization, "Biotechnology and Food Safety Special Issues," report of a joint FAO/WHO consultation, Rome, 30 September–4 October 1996 [online], http://www.fao.org/waicent/faoinfo/economic/esn/biotech/six.htm [12 June 2000]; Robert May, "Genetically Modified Foods: Facts, Worries, Policies, and Public Confidence," briefing from the chief science adviser, Sir Robert May FRS, February 1999 [online], www.gn. apc.org/pmhp/dc/genetics/cso-gmos.htm; Dan Ferber, "Superbugs on the Hoof," *Science* 288, no. 5467 (2000):792–94; "Corrections and Clarifications" (to Ferber, "Superbugs on the Hoof"), *Science* 288, no. 5472 (2000): 1751.

100. Andy Coghlan, "On Your Markers," *New Scientist*, 20 November 1999 [online], www.newscientist.com/nsplus/insight/gmworld/gmnews97.html [20 November 1999].

101. Royal Society, "Genetically Modified Plants"; Keith Harding, "Biosafety Reviews: Biosafety of Selectable Marker Genes," Biosafety Information Network and Advisory Service Online, United Nations Industrial Development Organization, www.bdt.org.br/binas/index.html [12 June 2000].

102. See, for example, Jonathan Williams, "Organic Farming in the Uplands of Mid Wales," statement at Earth Options, Second "Look Out Wales" Environmental Forum, May 1998 [online], www.wyeside.co.uk/expotec/Jonathan_williams.htm [19 March 2000[.

103. World Health Organization, *The World Health Report 1999.*

104. MacIlwain, "Access Issues."

105. Goklany, "Saving Habitat"; Goklany, "Meeting Global Food Needs."

106. Greenpeace, "Stop Monsanto's Terminator Technology," press release, 20 Sep-

tember 1998 [online], www.greenpeace.org/~geneng/highlights/pat/98 09_20.htm [12 January 2000]; Friends of the Earth, "FoE Supports Tory GM Moratorium Call: What about the Precautionary Principle Mr. Blair?" press release, 3 February 1999 [online], http://www.foe.co.uk/pubsinfo/infoteam/pressrel/ 1999/199902031 70456.html [15 May 2000].

107. Friends of the Earth, "FoE Supports Tory GM Moratorium Call"; Friends of the Earth, "FoE Remains Skeptical about Monsanto's Terminator Pledge," press release, 5 October 1999 [online], http://www.foeeurope.org/press/foe_remainssceptical.htm [21 February 2000].

29

Precaution without Principle

Henry I. Miller and Gregory Conko

Remember the admonition not to believe a bureaucrat who claims that "I'm from the government and I'm here to help you?" Well, government regulators now have a more subtle, updated version of that assertion: a wolf in sheep's clothing called the "precautionary principle." It has already laid waste to several industries and boasts a body count in the tens of thousands. It is now being used to cripple public-sector and academic researchers as well as the biotechnology industry.

Although a widely accepted definition of the "principle" does not exist, its thrust is that regulatory measures should prevent or restrict actions that raise even conjectural threats of harm to human health or the environment, although there may be incomplete scientific evidence as to their potential significance. Several European countries have used the precautionary principle to justify paralyzing restrictions on agricultural and food biotechnology, and the European Commission (EC) has invoked it to justify a moratorium on the approval of new recombinant DNA–

modified products.[1] Use of the precautionary principle is sometimes represented as "erring on the side of safety." But we believe the way it is typically applied to research and development and to commercial products can actually increase risk.

Potential risks should be taken into consideration before proceeding with any new activity or product, whether it is the choice of site for a power station or the introduction of a new drug into the pharmacy. But advocates of the precautionary principle focus primarily on the possibility that technologies could pose unique, extreme, or unmanageable risks. What is missing from the precautionary calculus is an acknowledgment that even when technologies introduce new risks, most confer net benefits; that is, their use reduces many other, far more serious hazards. Examples include blood transfusions, magnetic resonance imaging (MRI) scans, and automobile air bags, all of which offer immense benefits and only minimal risk.

The real danger of the precautionary principle is that it distracts consumers and policymakers from known, significant threats to human health and often diverts limited public health resources from those genuine and far greater risks. Consider, for example, the environmental movement's misguided crusade to rid society of all chlorinated compounds. By the late 1980s, environmental activists were attempting to convince water authorities around the world of the possibility that carcinogenic by-products of chlorination made drinking water a potential cancer risk. Peruvian officials caught in a budget crisis used this supposed threat to public health as a justification to stop chlorinating much of their country's drinking water. That decision contributed to the acceleration and spread of Latin America's 1991–1996 cholera epidemic, which afflicted more than 1.3 million people and killed at least 11,000.[2]

Anti-chlorine campaigners more recently have turned their attacks to phthalates, liquid organic compounds added to certain plastics to make them softer. These soft plastics are used for important medical devices, particularly fluid containers, blood bags, tubing, and gloves; children's toys, such as teething rings and rattlers; and household and industrial items, such as wire coating and flooring. Waving the banner of the precautionary principle, activists claim that phthalates could have numerous adverse health effects—even in the face of significant scientific evidence to the contrary.[3] Governments have taken these unsupported claims seriously, and several formal and informal bans have been implemented around the world. Industry has been stymied, consumers denied product choices, and doctors and their patients deprived of lifesaving tools. During the past few years, skeptics began more intensively to scrutinize the precautionary principle. In response to those assessments, the EC, a prominent user and abuser of the precautionary principle, last year published a formal communication to promote the legitimacy of the concept.[4] The EC resolved that, under its auspices, precautionary restrictions would be "proportional to the chosen level of protection," "nondiscriminatory in their application," and "consistent with other similar measures." The commission also avowed that EC decision makers would carefully weigh "potential benefits and costs." But all of these stipulations have been flagrantly ignored or abused in the commission's regulatory approach to recombinant DNA–modified—or in their argot, "genetically modified" (GM)—foods.

Dozens of scientific bodies, including the U.K.'s Royal Society, the U.S. National Academy of Sciences, the World Health Organization, and the American Medical Association, have analyzed the oversight that is appropriate for gene-spliced organisms and arrived at remarkably congruent conclusions: The newer molecular techniques for genetic improvement are an extension, or refinement, of earlier, far less precise ones; adding genes to plants or microorganisms does not make them less safe either to the environment or to eat; the risks associated with recombinant DNA–modified organisms are the same in kind as those associated with conventionally modified organisms; and regulation should be based upon the risk-related characteristics of individual products, regardless of the techniques used in their development.

Notwithstanding the EC's promises that the precautionary principle would not be abused, regulators treat recombinant DNA–modified plants and microorganisms in a discriminatory and inconsistent fashion, and without proportionality to risk. Both the fact and degree of regulation turn on the use of certain production methods—that is, on whether recombinant DNA techniques have been used—regardless of the level of risk posed by individual products.

For example, recombinant herbicide-tolerant crop plants, such as soybeans and canola, are subject to lengthy, hugely expensive mandatory testing and pre-market evaluation, whereas plants with virtually identical properties but developed with older, less precise genetic techniques are exempt from such requirements. In the United States, Department of Agriculture requirements for paperwork and field-trial design make field trials with gene-spliced organisms ten to twenty times more expensive than the same experiments with virtually identical organisms that have been modified with conventional genetic techniques.[5] The real-world impacts of this wholly disproportionate approach are instructive. If a student doing a school biology project takes a packet of "conventional," but genetically improved, tomato or pea seeds to be irradiated at the local hospital and plants them in his backyard in order to investigate interesting mutants, he need not seek approval from any local, national, or international authority. However, if the seeds have been modified by the addition of one or a few genes by recombinant DNA techniques, this would-be researcher (or equivalent highly skilled agricultural scientists) faces a mountain of bureaucratic paperwork and expense.

Not only does this discrimination flaunt the scientific consensus about the essential continuity between the traditional and molecular genetic improvement of plants, but it also ignores the fact that recombinant DNA technology is more precise and predictable and the modifications far better characterized than with other techniques. Logical application of the precautionary principle to situations of scientific uncertainty would dictate that greater precaution apply to the cruder, less precise, less predictable "conventional" forms of genetic modification. Instead, by torturing the precautionary principle, regulators have chosen to set the burden of proof far higher for recombinant DNA technology than for conventional plant breeding. And, as the EC's moratorium on new product approvals demonstrates, even when that extraordinary burden of proof is met through unprecedented amounts of testing and evaluation, regulators frequently declare themselves unsatisfied.

Remarkably, although the EC characterized its communication on the precautionary principle as an attempt to impart greater consistency and clarity, it specifically declined to define the principle, adding naively that "it would be wrong to conclude that the absence of a definition has to lead to legal uncertainty." Although reliance on regulatory agencies and courts to define and elaborate statutory policy is not unusual, this reluctance to define what purports to be a fundamental principle makes confusion and mischief inevitable, leaving innovators' legal rights and regulators' legal obligations subject to the wholly subjective and sometimes nefarious judgment of governments or even individual regulators.

As it is being applied, the precautionary principle provides neither evidentiary standards for "safety" nor procedural criteria for obtaining regulatory approval, no matter how much evidence has been accumulated. In effect, regulators are given carte blanche to decide what is "unsafe" and what is "safe enough," with no means to ensure that their decisions actually reduce overall risk or that they make any sense at all. Contrary to the claims of its supporters, the precautionary principle tends to make governments less accountable, not more so, because its lack of definition allows regulators to justify any decision. In spite of the assurance of the European Union and other advocates of precautionary regulation to the contrary, regulators of biotechnology applied to agriculture and food production seldom consider the potential risk-reducing benefits of new technologies. For example, the use of recombinant DNA–modified plants with enhanced pest or disease resistance has reduced farmers' use of chemical pesticides, reducing runoff into waterways, and the exposure of workers who manufacture, transport, and apply these chemicals. It has also permitted farmers to more widely adopt environment-friendly, no-till farming practices. And recently developed rice varieties enhanced with provitamin A and iron could drastically improve the health of hundreds of millions of the malnourished in developing countries. These are the kinds of tangible environmental and health benefits that have been given little or no weight in precautionary risk calculations. But benefits aside, the safety of this new technology is not really in doubt. Both theoretical and empirical evidence shows the extraordinary predictability and safety of gene-spliced organisms. Recombinant DNA–modified plants are now grown worldwide on more than 100 million acres annually, and more than 60 percent of processed foods in the United States contain ingredients derived from recombinant organisms. There has not been a single mishap resulting in injury to a single person.

For antibiotechnology activists, the deeper issue is not really safety at all. Often, the controversies over the testing and use of gene-spliced organisms—and in particular, the metastasis of the precautionary principle—stem from a social vision that is not just strongly antitechnology, but one that poses serious challenges to academic, individual, and corporate freedom.

In the Western democratic societies, we enjoy long traditions of relatively unfettered scientific research, except in the very few cases where bona fide safety issues are raised. (An example with contemporary relevance is the ban on research using live foot-and-mouth disease virus in the mainland United States.) Traditionally, we shrink from permitting small, authoritarian minorities to dictate our social agenda, including what kinds of research are permissible and which technologies and products should be available in the marketplace. Thus, for remarkably well behaved recombinant DNA technology, a refinement of earlier techniques, it is beside the point whether the purpose of investigating a new plant variety or microorganism is to test a scientific hypothesis or a marker gene, to produce a more elegant rose, to offer a marginal improvement for purposes of downstream processing, or to improve the lot of malnourished children.

It is precisely the antitechnology nature of the precautionary principle that makes it the darling of many nongovernmental organizations. Greenpeace, one of the principal advocates of the precautionary principle, wrote in its 1999 Internal Revenue Service filings that the organization's goal is not the prudent, safe use of recombinant DNA–derived foods or even their labeling; rather, they demand nothing less than these products' "complete elimination [from] the food supply and the environment."[6] Many of these groups do not merely proselytize for illogical and stultifying regulation or outright bans on product testing and commercialization; they advocate and carry out vandalism of field trials.

Carolyn Raffensperger, executive director of the Science and Environmental Health Network, a consortium of radical groups, asserts that the precautionary principle "is in the hands of the people," as illustrated, according to her, by violent demonstrations against economic globalization, such as those in Seattle at the 1999 meeting of the World Trade Organization.[7] "This is [about] how they want to live their lives," says Raffensperger.

In our view, it's really about how a small, vocal, violent group of radicals wants to dictate to the rest of us how we should live our lives. In other words, the issue here is freedom and its infringement by ideologues who disapprove, on principle, of a certain technology. But bullies should not be permitted to use untruths, conspiracy, and violence to oppose legitimate research into technologies that can improve our safety and well-being. We should no longer allow extremists to dictate the terms of the debate.

Notes

1. John Hodgson, "National Politicians Block GM Progress," *Nature Biotechnology* 18, no. 9 (2000): 918–19.

2. Christopher Anderson, "Cholera Epidemic Traced to Risk Miscalculation," *Nature* 354 (1991): 255.

3. William Durodié, *Poisonous Propaganda: Global Echoes of an Anti-vinyl Agenda* (Washington, D.C.: Competitive Enterprise Institute, 2000).

4. European Commission, "Communication from the Commission to the Precautionary Principle," COM, Brussels, 2 February 2000.

5. Suzanne L. Huttner, Henry I. Miller, and Peggy G. Lemaux, "U.S. Agricultural Biotechnology: Status and Prospects," *Technological Forecasting and Social Change* 50, no. 1 (1995): 25–39.

6. Greenpeace, Federal Income Tax Filing with the U.S. Internal Revenue Service (IRS), form 990, part 3, Statement of Program Service Accomplishments.

7. David Appell, "The New Uncertainty Principle," *Scientific American* 284, no. 1 (2001): 18–19.

Part 9

Developing Countries

Introduction

We saw that the reason for and promise of golden rice was that supposedly it could be used for the benefit of peoples in developing countries, or—less euphemistically—of peoples in nondeveloped countries. This section looks head-on at the whole question of the needs of such peoples in these countries and whether the prospect of GM foods is a boon or a burden.

Food biotechnology is often said to have the capacity to relieve the world's worst off. A sixth of the world's population is marginalized, and growth in biological knowledge and biotech is almost the exclusive property of industrialized countries where private biotechnology owns controlling interests. A situation like this should spell gloom and doom for developing countries, but many are optimistic that wise use of biotechnology will improve conditions in developing countries.

Robert Tripp argues that significant barriers stand in the way of the responsible use of agricultural biotechnology in the developing world. Tripp's solutions generally involve greater public involvement in what is largely a privately controlled biotech industry. Public-private partnerships could pave the road toward the development and distribution of biotechnology that would truly benefit developing countries. But, of course, the key is that the technologies developed must be appropriate for the nations and people who will receive them. Florence Wambugu, who has worked as an agriculture research scientist both in Kenya and at Monsanto in the United States, argues that industrialized nations' activists interfere with the development of these partnerships. Neither the biotech industry nor the groups claiming to represent the interests of those in developing countries reflect the concerns of—and here we return to Sagar's point—the "commoners." As Wambugu says, "Africans can speak for themselves."

Wambugu is actively improving African nations' biotechnology capacity because she believes that the locus of control must shift to developing countries if the right technologies are going to be developed. That developing countries should developed their own biotechnology runs contrary to the idea of foreign aid and donor technology. Calestous Juma and Karen Fang do not think that industrialized-world ownership and control of biological knowledge and biotechnology will lead to sustainable, context-sensitive development in developing countries. What is required, they argue, is local capacity strengthening developing countries' biotechnology industries. Intellectual property rights might have to change if industrialized countries are to contribute to this capacity strengthening, as will science and industry policies within developing countries. As Juma and Fang see it, the biotechnology is not in enough developing countries' food policy quivers. If this situation persists, opportunities to develop the sustainable wise-use of biotechnology may be missed.

'Twixt Cup and Lip

Biotechnology and Resource-poor Farmers

Robert Tripp

Although there seems no doubt that transgenic crops will find many applications in developing countries, their potential contribution to poverty reduction is not well understood. Many observers have correctly pointed to biotechnology's capacity for offering productivity gains to meet increasing food demand. What they discuss less frequently, however, are the challenges in allowing those gains to be realized by resource-poor farmers.

One of the most frequent points of comparison is the green revolution. It led to the widespread adoption of productive new varieties, but the impact was greatest in relatively favored environments, where markets were well established and inputs were available. Transgenic crops could circumvent such requirements. Engineered resistance to pests and disease could eliminate the need for expensive chemicals; changes in crop physiology could address limitations of poor soils or climate; nutritional enhancement can address dietary deficiencies caused by inadequate crop production. Transgenic crops could deliver benefits to resource-poor farmers within the seed. But real value will only accrue to such farmers if a number of largely nontechnical barriers can be overcome.

At least two infrastructural problems may significantly limit the poverty relevance of transgenic crops.

If biotechnology is to be directed toward poverty reduction, then public biotechnology research will have to address crops and areas that are unattractive to the private sector. Such research requires a significant investment of public resources. There is a natural tendency to direct such investments toward areas with high expected returns or where political pressure on the research system is most effective. The poorest farmers are usually without much political influence. In endeavoring to fulfill an intention to develop "pro-poor" technology, the significant countervailing forces against targeting marginalized farmers, especially by underfunded public research systems, need to be acknowledged and addressed.

A second infrastructural barrier is the seed industry in many developing countries. In many instances, liberalization has brought an end to inefficient public seed production without providing the incentives for an adequate private sector replacement. Where a commercial seed industry is in place, this offers an obvious pathway, but many farmers (such as those in most of sub-Saharan Africa) do not have access to such markets. Even where a commercial seed industry exists, its ability to serve resource-poor farmers depends on responsible and well-informed input retailers and some degree of consumer awareness. In the case of publicly developed varieties, there may be additional options for seed distribution including government-sponsored multiplication and distribution (relying on subsequent farmer-to-farmer diffusion), or small-scale seed projects. However, the larger programs may entail considerable expense, and the experience to date with small seed projects has not been encouraging.

There is a third factor, too: The adequacy of farmers' access to information about production problems and alternatives. This challenge is certainly not confined to biotechnology. However, the nature of many transgenic varieties exacerbates it.

Briefly, the problem is this. Many modern varieties, including those of the green revolution, rapidly diffuse to farmers. Such varieties often succeed because they offer radically different and easily distinguishable characteristics. Farmers learn about the management requirements of new varieties and their advantages and disadvantages, often through trial and error. They build a body of knowledge that guides them in choosing particular varieties to suit particular circumstances, and then managing them appropriately. However, in many areas where modern varieties are widely grown, it is not uncommon to find that farmers are uncertain about the identities of "second-generation" modern varieties (many of which offer precisely the disease- or pest-resistance envisioned for transgenic varieties). This identity confusion erodes the value of the associated knowledge, and it is directly relevant for the prospects of biotechnology.

The precision of genetic engineering, avoiding the trade-offs characteristic of conventional plant breeding by providing, for instance, disease-resistance without any other changes in a variety's appearance or performance, is a double-edged sword. If a new transgenic variety is not immediately distinguishable from conventional varieties, what are the chances that farmers will recognize and demand it? The answer in this case depends on the distribution and severity of the particular disease, but farmers may not be able to draw causal inferences from the variety's performance in fields where many other yield-limiting factors are probably in evidence.

Nutritionally enhanced transgenic crops may be similarly difficult to recognize. Even in cases of severe nutritional deficiency, farmers are unlikely to make a connection between the consumption of a particular variety and health status. If the new variety cannot be easily identified, then accompanying nutrition education is necessary to help farmers (and other consumers) recognize the appropriate variety and use it properly.

In those cases where a nutritionally superior variety can be recognized (as in the case of yellow, vitamin A–enriched rice), there may be the problem that the variety is seen as a low-status product, aimed at the poor. (For instance, any campaign to convince people who grow and consume white maize to switch to more nutritious yellow varieties would face tremendous opposition.)

There are thus several factors that suggest caution in making predictions about the poverty impact of transgenic crops. My purpose here is not to be unduly pessimistic, but to ask researchers to be realistic in their approach to biotechnology's potential contribution to agricultural development and poverty reduction. Biotechnology will only be effective if it is part of a package of broader changes that include the provision of adequate information and the development of seed delivery systems. First, public agricultural research must be better supported. Investments in biotechnology laboratories, without concomitant attention to developing researchers' capacities to interact with farmers, will be ineffective. Second, a clearer division of labor and better collaboration between public and private research is in order. Third, policies must be in place to strengthen the agricultural sector, to support a domestic seed industry, and to develop adequate markets.

These tasks are the responsibility of national governments, donor agencies, and private industry (which must contribute more to poverty reduction). They require a long-term commitment to building the institutions that support a productive and equitable agriculture.

Why Africa Needs Agricultural Biotech

Florence Wambugu

The public debate on transgenic crops in Europe is centred on fear and mistrust, quite possibly resulting from the experience over "mad cow disease." A recent report[1] from the Food Safety Authority of Ireland to address European Union concerns on genetically modified (GM) crops concluded that there is no evidence that transgenic foods are unsafe. The report, by a group led by Patrick Wall, the authority's chief executive, says that concern in Europe is based on ethical, socioeconomic, and anti-multinational issues; lack of knowledge or misinformation; environmentalism; food labeling; and consideration of the needs of developing countries.

Many of these concerns have nothing to do with food safety. Transgenic foods are eaten daily in the United States, Australia, Canada, Mexico, and elsewhere with no reported undue effects.[2] Nevertheless, the experts' advice does not seem to influence public opinion in Europe, probably because of a strong anti-biotechnology lobby that actively promotes misinformation and fear, and also because in some cases people have had good reason to distrust "expert" pronouncements.

One example of Europeans' concern for the Third

World is "terminator technology"—plants engineered to be sterile. But this technology is only a concept and is not being further developed. No products are planned for Africa or elsewhere. Critics of biotechnology have used the fear of this technology to promote serious anti-multinational attitudes—for example, crops in trials have been burned in some parts of the world.

Another concern promoted by critics of food biotechnology is that of toxins or allergies. An example is the case of the unpublished study by Arpad Pusztai, formerly of the Rowett Research Institute in Scotland, who suggested that rats fed with GM potatoes expressing a snowdrop lectin were slowly being poisoned. After an independent scientific review, these results were found to be misleading and to have been misinterpreted.[3] But the anti-biotechnology lobby is still using them strongly to advance its case in Europe, even though transgenic foods are rigorously tested for possible toxins and allergens before commercialization.

Surely there are parallels to be drawn with an antibiotic such as penicillin, which has continued to be used for many years despite many people being allergic to it because the benefits clearly outweigh the risks. Why is the same reasoning not applied to transgenic foods, where risks at even this low level are not proven? The anti-biotech lobby also cites as controversial the recombinant DNA processes used to develop transgenic foods. But the same processes are used to develop numerous pharmaceuticals for humans and animals, and many other industrial products. The public seems prepared to accept the application of GM techniques to new pharmaceutical products but not to food production. Why should there be different standards for crops and pharmaceuticals, particularly in Africa where the need for food is crucial for survival?

African Perspective

The critics of biotechnology claim that Africa has no chance to benefit from biotechnology, and that Africa will only be a dumping ground or will be exploited by multinationals.[4] On the contrary, small-scale farmers in Africa have benefited by using hybrid seeds from local and multinational companies, and transgenic seeds in effect are simply an added-value improvement to these hybrids. Local farmers are benefiting from tissue-culture technologies for banana, sugar cane, pyrethrum, cassava, and other crops. There is every reason to believe they will also benefit from the crop-protection transgenic technologies in the pipeline for banana, such as sigatoka, the disease-resistant transgenic variety now ready for field trials. Virus- and pest-resistant transgenic sugar-cane technologies are being developed in countries such as Mauritius, South Africa, and Egypt.

The African continent, more than any other, urgently needs agricultural biotechnology, including transgenic crops, to improve food production. African countries need to think and operate as stakeholders, rather than accepting the "victim mentality" created in Europe. Africa has the local germplasm, some of it well-characterized and clean, being held in gene banks in trust by centers run by the Consultative Group of International Agricultural Research. It also has the indigenous knowledge, local field ecosystems for product development, capacities, and infrastructure required by foreign multinational companies.

The needs of Africa and Europe are different. Europe has surplus food and has never experienced hunger, mass starvation, and death on the regular scale we sadly witness in Africa. The priority of Africa is to feed her people with safe foods and to sustain agricultural production and the environment.

Africa missed the green revolution, which helped Asia and Latin America achieve self-sufficiency in food production. Africa cannot afford to be excluded or to miss another major global "technological revolution." It must join the biotechnology endeavor. Transgenic food production increased from 4 million to 70 million acres worldwide from 1996 to 1998 with measurable economic gains and with sustainable agricultural production.[5] It would be a much higher risk for Africa to ignore agricultural biotechnology. Africa's crop production per unit area of land is the lowest in the world. For example, the production of sweet potato, a staple crop, is 6 tons per hectare compared to the global average of 14 tons per hectare. China produces on average 18 tons per hectare, three times the African average. There is the potential to double African production if viral diseases are controlled using transgenic technology.

The African continent imports at least 25 percent of its grain. The use of biotechnology to increase local grain production is far preferable to this dependence on other countries, particularly as the population growth rate exceeds food production. The inability to produce adequate food forces Africa to rely on food aid from industrialized nations when mass starvation occurs. Although biotechnology is not the only answer to this problem, Africa should certainly benefit in many ways from its use, for example, in improved seed quality and resistance to pests and diseases.

The average maize yield in Africa is about 1.7 tons per hectare compared to a global average of 4 tons per hectare. Some biotechnology applications can be used to reduce this gap, for example, in the case of the maize streak virus (MSV), which causes the loss of 100 percent of the crop in many parts of the continent. A biotechnology-transfer project is under way to develop MSV-resistant varieties. The project is brokered by the International Service for the Acquisition of Agri-Biotech Applications (ISAAA), and involves the collaboration of the Kenya Agricultural Research Institute (KARI), the University of Cape Town, the International Centre for Insect Physiology and Ecology in Kenya, and the John Innes Centre in the United Kingdom. Funding is coming from the U.S. Rockefeller Foundation, and Novartis in Europe has donated some technology to KARI.

Researchers at KARI are studying the mechanism of MSV resistance and trying to map the genes responsible. Advanced biotechnology skills, including the use of advanced agroinoculation techniques and molecular markers, is at the core of this effort. A priority in Kenya is also to produce high-yielding, drought-tolerant crop varieties to boost food production in the 71 percent of the country that is arid or semiarid.

Africa needs biotechnology to solve its environmental problems, and there is unlimited public demand for agricultural-biotechnology products and services. In Kenya, the demand for tree seedlings reaches 14 million per year, whereas the country can only supply 3 million, a clear indication of the need for tissue-culture and cloning techniques to curb deforestation and boost reforestation using indigenous species threatened with extinction. These technologies are being successfully used in South Africa, and the ISAAA has facilitated a project for application in Kenya. There are issues of intellectual property rights and patents that require hard work to develop or acquire, and advanced agricultural-biotechnology skills will be needed. There may also be a need to work out collaboration agreements with the private sector or with companies that already have patents.

Biotechnology in Africa is needs-based. After working at KARI for nearly a decade to help improve sweet-potato production using traditional breeding and agronomy methods, I made no progress. An opportunity to work in the private biotechnology sector abroad resulted in the development of a transgenic variety that is resistant to sweet-potato feathery mottle virus, which can reduce yields by 20 to 80 percent. Control of this disease will improve household food security for millions. This project involved collaboration between KARI, a project called Agricultural Biotecology for Sustainable Productivity, funded by the U.S. Agency for International Development, and Monsanto. The work by Kenyan scientists focuses on local varieties, and there will be a smooth and sustainable transfer of the technology, which will be shared with neighboring countries. Kenyan scientists have been trained in gene-technology techniques. The ISAAA has been asked to help with the transfer and licensing agreement. Similar projects are under way for bananas, sugar cane, and tropical fruits.

Remaining Problems

Needless to say, Africa has many problems—a shortage of skilled people (especially in biotechnology), poor funding of research, lack of appropriate policies, and civil strife. Nevertheless, countries such as South Africa, Egypt, Zimbabwe, and Kenya are taking practical steps to ensure that they can use biotechnology for sustainable development.

African countries need to avoid exploitation and to participate as stakeholders in the transgenic biotechnology business. They need the right policies and agencies, such as operational biosafety regulatory agencies, breeders' rights, and an effective local public and private sector, to interface with multinational companies that already have the technologies. Consumers need to be informed of the pros and cons of various agricultural-biotechnology packages, the dangers of using unsuitable foreign germplasm, and how to avoid the loss of local germplasm and to maintain local diversity. Other checks and balances are required to avoid patenting local germplasm and innovations by multinationals; to ensure policies on intellectual property rights and to avoid unfair competition; to prevent the monopoly buying of local seed companies; and to prevent the exploitation of local consumers and companies by foreign multinationals. Field trials need to be done locally, in Africa, to establish environmental safety under tropical conditions.

The main goal is to find a balanced formula for how local institutions can participate in transgenic product development and share the benefits, risks, and profits of the technology, as they own the local germplasm needed by the multinationals for sustainable commercialization. New varieties must not simply replace local ones. The removal of genes that were in the public domain into the private sector raises concern in Africa.

All these issues mean that Africa must strengthen its capacity to deal with various aspects of biotechnology, including issues of biosafety, creating and sustaining gene banks, and encouraging the emergence of a local biotechnology private sector. The great potential of biotechnology to increase agriculture in Africa lies in its "packaged technology in the seed," which ensures technology benefits without changing local cultural practices. In the past, many foreign donors funded high-input projects, which have failed to be sustainable because they have failed to address social and economic issues such as changes in cultural practice. The criticism of agribiotech products in Europe is based on socioeconomic issues and not food-safety issues, and no evidence so far justifies the opinion of some in Europe that Africa should be excluded from transgenic crops. Africans can speak for themselves.

Notes

1. www.fsai.ie.

2. Clive James, "Global Review of Commercialized Transgenic Crops," ISAAA brief no. 8, ISAAA, 1998; Food and Agriculture Organization, Biotechnology and Food Safety Food and Nutrition paper no. 61, FAO, Rome, 1996; "Genetically Engineered Food Production Gathers Pace," CSIRO media release 99/117, CSIRO, Guelph, Canada, 1999.

3. Natasha Loder, "Royal Society: GM Food Hazard Claim Is 'Flawed,'" *Nature* 399, no. 6733 (1999): 188.

4. Christian Aid et al., "Keeping Your Life and Environment Free of Genetically Modified Food," *GM-FREE* 1, no. 2 (June/July 1999); Nuffield Council on Bioethics, "Genetically Modified Crops" [online], http://www.nuffield.org/bioethics/publication/modified crops/index.html.

5. James, "Global Review of Commercialized Transgenic Crops."

Bridging the Genetic Divide

Calestous Juma and Karen Fang

Introduction

The emergence of biotechnology has invoked a major global controversy over the future of world agriculture. These debates have often reflected the interests of industrialized countries and paid little attention to the needs of developing countries, especially those related to food requirements of low-income populations. This paper argues that biotechnology represents an important technology option for meeting the long-term food needs of developing countries. However, current trends in technology development threaten to deny these countries the benefits of this emerging technology, thereby creating a "genetic divide" between countries.

Such a divide would affect international relations more profoundly than the "digital divide" has done because it would affect fundamental areas of human welfare such as agriculture, health, and environmental management. The paper bases its analysis on the view that biotechnology merely represents a set of tools available to society to solve technical problems and does not itself address social issues such as inequity. But the absence in improvements in methods of agricultural production can lead to economic maladjustment and worsening living conditions. It is in this respect that a focus on the role of biotechnology in the economies of developing countries is an important aspect of the development discourse in general and public policy in particular.

The first section of this paper traces the divergent approaches to biotechnology between the developed and developing countries. The second section reviews current biotechnology applications and technological trends and their relevance to developing countries. The third section examines the need to redirect existing technological efforts to address the needs of developing countries and the management demands associated with this process. The last section examines the public-policy implications arising from increased application of biotechnology to meet development goals.

Technology and Human Needs

The role of science and technology in globalization is currently at the center of a wide range of international controversies involving drugs, food, and software. These controversies are part of the ongoing challenge to globalization in general and the perceived global dominance of the United States in the post–Cold War period in particular.[1] These debates stem from the fact that science and technology is currently viewed in the industrialized countries as a major tool for international competitiveness. But this view has emerged concurrently with the growing recognition of the need to use emerging technologies to address the needs of the poor in developing countries. There is an apparent conflict between the way science and technology is currently used and the expectations of a large section of humanity that is not able to afford basic requirements such as essential drugs, basic nutrition, and energy.[2]

Social movements and groups around the world are therefore questioning the role of new technologies in meeting the needs of the poor. They contend that new technologies have helped to widen the gap between the rich and the poor within and between nations. Others have gone as far as suggesting that modern technological innovations are a source of social, economic, and ecological problems facing developing countries. There is a general mood of suspicion, cautiousness, and hostility toward technological innovation worldwide, often fueled by a wide range of social movements.[3] This questioning of technology is reflected in the global debates over access to essential drugs, genetically modified (GM) foods, and open-source software.

The debate over agricultural biotechnology has so far focused on international trade in GM foods. Attempts to resolve the growing concerns through the 1992 United Nations Convention on Biological Diversity (CBD) dealt largely with environmental aspects of living modified organisms (LMOs).[4] Issues related to human health are being negotiated under the Codex Alimentarius Commission administered by the Food and Agriculture Organization and the World Health

Organization.[5] Other bodies in the United Nations systems are handling different aspects of the biotechnology debate. For example, the United Nations Commission on Science and Technology for Development is examining issues related to biotechnology capacity building in developing countries. The United Nations Conference on Trade and Development (UNCTAD) is looking into the trade-related aspects of biotechnology.[6] Despite these efforts, the debate has so far focused largely on the risks associated with biotechnology and sidestepped its potential benefits in developing countries.

Ironically, the discussions over biotechnology in the late 1990s focused largely on its potential applications to meeting the needs of developing countries. Governments negotiated and signed the CBD based on detailed consideration of the potential role of biotechnology in development. Article 16(1) of the Convention on Biological Diversity provides the basis for international cooperation in the field of biotechnology: "Each Contracting Party, recognizing that technology includes biotechnology, and that both access to and transfer of technology among Contracting Parties are essential elements for the attainment of the objectives of this Convention, undertakes subject to the provisions of this Article to provide and/or facilitate access for and transfer to other Contracting Parties of technologies that are relevant to the conservation and sustainable use of biological diversity or make use of genetic resources and do not cause significant damage to the environment." Indeed, developing countries argued that they could use their biological resources to create new industries using the emerging technology.

In addition to the CBD, the role of biotechnology in development is clearly articulated in Chapter 16 of Agenda 21, the program of work of the 1992 United Nations Conference on Environment and Development. According to chapter 16 of agenda 21, biotechnology "promises to make a significant contribution in enabling the development of, for example, better health care, enhanced food security through sustainable agricultural practices, improved supplies of potable water, more efficient industrial development processes for transforming raw materials, support for sustainable methods of afforestation and reforestation, and detoxification of hazardous wastes. Biotechnology also offers new opportunities for global partnerships, especially between the countries rich in biological resources (which include genetic resources) but lacking the expertise and investments needed to apply such resources through biotechnology and the countries that have developed the technological expertise to transform biological resources so that they serve the needs of sustainable development. Biotechnology can assist in the conservation of those resources through, for example, ex situ techniques."

The promises of these international commitments remain unrealized largely because biotechnology has increasingly been defined in terms of its risks, and little consideration has so far been given to finding mechanisms that offer a balanced assessment of its potential.[7] Enterprises in developed countries have in turn been slow to engage in technological partnerships in developing countries because of concern over the lack of a policy environment that supports the use of emerging technologies. For example, they are concerned about the absence of effective biosafety and intellectual property protection systems. Efforts to place biotechnology in a risk frame have gone hand in hand with concerns over globalization and the growing dominance of the United States in international agricultural trade.[8] In other words, trade competition has provided a context in which international debates over biotechnology have been conducted.

It is therefore not a surprise that the first international regime regulating biotechnology—the Cartagena Protocol—deals mainly with transboundary movement of LMOs, which is a euphemism for "international trade." Arguing for the recognition of the trade-related aspects of these debates does not in any way seek to diminish the importance of environmental and human health concerns. These are indeed critical and need to be addressed in their own right and should be take seriously. But focusing only on these issues mystified more fundamental underpinnings of the debate and is unlikely to resolve the issues.[9]

The rules and decision procedures set out under the Cartagena Protocol are designed to regulate international trade and have inspired a wide range of analyses that examine its relationship with the World Trade Organization (WTO).[10] Follow-up negotiations on international regulation of biotechnology now focus on issues such as labeling and traceability, which are also trade-related issues. In recognition of the fact that these debates are partly rooted in international concerns over trade, the United States and the European Union have sought to use other mechanisms such as roundtables or forums to resolve their differences.

There are indeed a number of developing countries that are concerned about the trade implications of biotechnology, especially in the context of globalization. Developing countries fear that trade liberalization could undermine their commodity exports through product substitution using biotechnology.[11] Cheaper foods produced through the use of modern biotechnology are also perceived as a threat to existing commodity markets. The requirement to remove subsidies for agricultural production as part of the globalization processes threatens to alter the competitiveness of agricultural commodities from countries that do not have access to modern technology. The uncertainties associated with such changes in international agricultural trade are a source of resistance to market liberalization as well as the rapid introduction of new technologies.

Industrial-country consumers continue to express skepticism toward transgenic foods. This is partly because they have a wide range of foods from which to make the necessary choices. They therefore question the need to use new technologies to make incremental changes in their foods without offering tangible benefits, especially those that help to improve human welfare. Indeed, industrialized countries already face challenges associated with excessive production of food. Many of these countries, especially in Europe, have put in place policies that seek to link food production with environmental conservation. Corresponding institutional reforms that combine agricultural, environment, and consumer protection ministries illustrate a change in policy focus and public outlook.

Industry in the developed countries is looking into ways of producing foods that are relevant to the consumers. Fields such as "nutraceuticals" or "functional foods" are emerging as a response to the growing concern among consumers about their health and well-being in general. The success of such investments is still in doubt, but it is evident that the concerns in industrialized countries stem from the view that meeting food security is no longer the concern of consumers. Much of the consumer interest is shifting to the quality of the food they consume and its contributions to improved health.

The situation in many developing countries—especially in Africa—is different. Low-income families in these countries are faced with a wide range of challenges that include malnutrition, hunger, and the related illnesses. Addressing these challenges requires the deployment of the available technological options. The poor often rely on a limited range of food sources, and as ecological degradation continues, the capacity to meet their needs diminishes. Raising agricultural productivity while promoting sustainable land use becomes a key element. Indeed, in many poor regions of the world, agricultural production is done by women who also have other critical household responsibilities.

Responding to these challenges requires investment in technologies that are appropriate to the needs of low-income communities that live in diverse ecological zones often located in areas that are not served by major markets. Agricultural production in these areas will also need to be equally diverse and to reflect local needs and preferences. Genetic modification and the emerging techniques of genomics offer the possibility to design farming systems that are responsive to local needs and reflect sustainability requirements. In other words, genetic modification and genomics make it possible to design farming systems that are decentralized, responsive to local needs, and more productive than existing methods.

Divergent Technological Trajectories

The capacity to modify living organisms to perform new functions offers humanity the potential to make the transition from classical-farming methods to decentralized production systems that are consistent with ecological principles. It is because of this adaptive potential that developing countries have been particularly interested in building capacity in this field. Let us take the African region as an example. The green revolution only partially touched this continent. Advances in maize breeding helped to extend the scope of food production in many countries. Efforts in other fields showed dismal results. There are many reasons for this. First, the Cold War concerns that inspired the green revolution in Latin America and Asia took on a different character in Africa.[12] Raising food productivity was not a strategic way of responding to superpower competition in the region. As a result, promoting agricultural research was not given the kind of priority that it received in other regions.

The green revolution relied on a long history of prior research and accumulated knowledge on corn, wheat, and rice and focused on limited technical tasks such as raising yields using increased inputs. Africa's food-consumption patterns did not lend themselves to large-scale uses of these crops. Africa lacked the institutional foundations for research in these crops. Subsequent efforts to create research institutions that focused on tropical crops have not registered the same levels of productivity gains and impacts as wheat, rice, and corn. In fact, the fate of these institutions now hangs in the balance as international assistance to tropical agriculture declines and most international agricultural institutions fail to keep up with the demands of a changing global knowledge system.

There are, however, other ecological factors that set Africa apart from other continents. Much of the continent is arid or semiarid and marked by ecological diversity. These variations are associated with mosaics of productive activity with limited scope for the kind of mass agricultural production that has been promoted in regions of the world. Agricultural research to meet the needs of isolated rural populations was beyond the reach of classical plant-breeding institutions. Moreover, the sheer diversity of crops used in the region and the absence of large centralized markets undermines the feasibility of plant-breeding programs.

Today's technological capabilities in fields such as genomics make it possible to adapt crops to these diverse ecosystems in ways that are consistent with the principles of sustainable agriculture. Herbicide resistance, disease and stress tolerance, and other traits can be applied to promote sustainable agriculture in regions that do not support agriculture today. It is this technological flexibility and the creation of niche markets that developing countries hoped to use to improve their farming methods and reduce pressure on the environment. But the evolution of biotechnology has taken a different path with a focus on markets of the industrialized world or the temperate regions. This is partly because of the logic of technological agglomeration that favors the accumulation of knowledge in areas with previous investments in technological capabilities and supportive institutions. Developing countries that need biotechnology most are also the ones that are least involved in its development.

The use of transgenic crops has been expanding rapidly, but this diffusion has been in the temperate regions. In 2000, transgenic crops covered an estimated 44.2 million hectares, a twenty-five-fold increase over the 1996 figure. This rapid expansion occurred mainly in the United States, Canada, Argentina, and China, accounting for 99 percent of the coverage of transgenic crops. The bulk of this was in the United States (68 percent), with Argentina accounting for 23 percent, Canada 7 percent, and China 1 percent. Most of this coverage is in large farms where genetic modification has been used to introduce incremental changes in existing crops.[13] These incremental adjustments in crops explain why the distribution of transgenic crops is limited to geographical areas with similar ecological conditions.

Transgenic applications are currently limited to soybean, corn, canola, and cotton crops. Transgenic soybeans covered 25.8 million hectares in 2000, corn 10.3

million hectares, cotton 5.3 million hectares, and canola 2.8 million hectares.[14] The bulk of the crops contain traits for herbicide tolerance and disease resistance. These trends show that the early diffusion of transgenic crops has been largely in the temperate regions and has been limited to a few major commercial crops. The promise of biotechnology to meet the needs of low-income families in the developing world still remains a distant dream.

There are two main reasons why the promise has not been realized. First, crop development for low-income families has traditionally been carried out by the public sector. But biotechnology has emerged from the private sector that lacks the incentives to invest in crops for low-income families. Second, agricultural research in the public sector has been declining over the years, so little investment has gone into developing crops for low-income families. It is unlikely that the situation will change without a redirection of existing research priorities in private enterprises through the provision of appropriate incentives as well as a significant increase in public-sector funding for agricultural research. In addition, institutional arrangements will need to be created to facilitate closer cooperation between private- and public-sector institutions.

The divergence in technological evolution is likely to be reinforced by three factors. First, the continuing uncertainty over market access for GM products in Europe will reduce the pace of technological innovations in products intended for international markets. Indeed, in the short run, premium markets now exist for non-GM crops. This general trend could discourage investment in biotechnology research and as a consequence slow down technological development in developing countries. Second, developing countries that are engaged in biotechnology are likely to redirect their efforts toward meeting local needs. Indeed, many of the products being developed in developing countries are destined for local consumption partly because of urgency in this field and partly because of uncertainty in international markets. Third, a number of biotechnology firms in the industrialized world are willing to share their technology on the condition that it is used to address local food needs and not export crops. A number of enterprises have granted royalty-free uses of their inventions for the development of rice enhanced with vitamin A on the condition that the product is grown only by farmers earning less that $10,000 a year. Monsanto has licensed its technology to Kenya royalty-free for use in the development of virus-resistant sweet potatoes for local consumption.

Redirecting Technological Effort

Efforts to redirect biotechnology to address the needs of low-income families in developing countries should be placed in a large policy framework that addresses other social issues. More important, such strategies should be part of policies designed to use science and technology to achieve sustainable-development goals. In addition, biotechnology should be considered as one of the tools in a larger portfolio of technological options. In this regard, biotechnology is simply a set of tools and the embodied knowledge needed to solve specific problems. How this is done depends largely on the choice of problems and the nature of institutional arrangements in which the technology is issued.

This view does not imply that technology is neutral. The choice of technological trajectories often reflects the economic, social, and cultural context in which it emerges. This does not mean that its use always reproduces the same conditions that characterized its origins. Technologies are often modified and adapted to reflect new socioeconomic conditions depending on prevailing social goals. Indeed, it is the flexibility that is embodied in biotechnology techniques that make it possible for them to be applied under different farming systems. It is true that biotechnology is currently used mainly in large-scale agriculture in the United States. But it is also true that the same technology is being used in small-scale agriculture in China, South Africa, and Kenya. What matters is therefore the choice of farming systems.

The choice of technology should be driven by the determination of local needs. Many developing countries have already indicated their various agricultural-development priorities that could be addressed using genetic modification. Many African countries, for example, lie in regions where drought-tolerance, disease-resistance and crop-yield increases are priorities. Crops such as cassava, millet, yams, and sorghum are prime candidates for genetic modification. Modifications that seek to prolong the shelf life of foods could have a significant impact on reducing postharvest losses. The use of herbicide tolerance in low-till agriculture is another area of priority, especially in helping to lessen farm labor and providing farm workers—most of whom are women—with opportunities to engage in other activities.

Another potential area for biotechnology application is the development of livestock that is tolerant to tropical diseases and stresses. Modern methods such as genomics could be applied in this area without requiring transgenesis. Also related to agricultural production is the significance of revegetation in marginal areas. Investment in fast-growing plants could help facilitate ecological restoration in many denuded regions of the world. Such research could also add to the fodder requirements of these countries.

Redirecting global research and development efforts to focus on these challenges will entail considerable international cooperation, increases in public-sector funding, and incentives for private enterprises. It will also take the creation of an atmosphere that is tolerant to the use of emerging technologies in implementing sustainable-development goals. But where international cooperation is not possible, bilateral responses that might include realignments in international trade relations will become the only option open to countries that view biotechnology as strategic to their mutual interests. Such a scenario is already emerging under bilateral cooperation arrangements being signed between countries with strong biotechnology-based industries.

The use of biotechnology to address development problems needs to be conducted as part of a wider technology-management approach. There are three categories of risks that need to be addressed when considering the role of biotechnology for low-income families. These relate to health, environmental, and socioeconomic considerations. The advent of biotechnology demands that all countries put in place measures that ensure safety to human health and the environment. Such measures involve the judicious use of risk-assessment, risk-management, and risk-communication strategies. In addition, equity considerations also call for social policies that address the impact of new technologies on rural populations. Such policies should include how to create alternative livelihoods for farm workers displaced by new technological practices.

Many developing countries are reluctant to engage in biotechnology development because they fear some industrialized countries would erect barriers against their products. These concerns are real and have created an atmosphere of distrust that is likely to undermine not only the global trading system, but also the ability of developing countries to meet their human needs. The emerging trends are reflected in acrimonious debates in the World Trade Organization (WTO) and other international forums on issues such as access to essential drugs and trade in GM foods.

A final area of concern is the impact of intellectual property protection on the abilty of the developing countries to use biotechnology. There are two dimensions to this point. First, for international research institutions that undertake agricultural research and increasingly deal with intellectual property issues, ways need to be found to enable them to have access to technologies needed to meet the needs of low-income families. Second, national research institutes in developing countries face similar challenges. Biotechnology firms such as Monsanto have made public pledges to share technologies with developing countries. The realization of such pledges will require considerable institutional innovations to provide the required comfort among the providers and users of technology.

Emerging trends suggest that in the early phases of biotechnology, developing countries are likely to focus their attention on GM crops for local consumption rather than for international markets. This is partly because of the prevailing uncertainty over export markets, and the preference of biotechnology enterprises to limit the use of their technology to nonexport uses. Such a trajectory is helping to bring biotechnology in line with the initial expectations that these techniques would be used to meet the human needs. But the extent to which such a trajectory will make a significant difference will depend on other factors such as the availability of the technology-management capabilities needed to provide confirmation to those providing technology for such uses. So far only a small number of developing countries have such capabilities.

International Technology Cooperation

Science and technology are increasingly being recognized as a key force in shaping international relations. While the role of science and technology in international security and energy is widely acknowledged, it is only now that this pervasive role in international relations is being recognized.[15] But this phenomenon is evident in the character and content of a number of major international agreements adopted by governments in the last decade, especially in the environmental field.[16] Advances in a number of fields such as genetics have resulted in the call for new international regulatory instruments. Technological developments in other fields such as satellite imagery are starting to shape the prospects for monitoring compliance to international agreements. In other fields, advances in the sciences are opening up new opportunities for solving persistent problems in health, agriculture, and environmental management and thereby raising new issues about access to new knowledge for development.

These trends are starting to place new demands on the functioning of ministries of foreign affairs and other government organs that deal with international-development issues. Recent international negotiations surrounding the role of biotechnology in the international economy have highlighted the growing impact of science and technology in international diplomacy. Relations between countries are largely based on existing patterns of industrial and agricultural production as well as the associated trade. Technologies that change the patterns of production have the potential to create new trade relations and affect the nature of international cooperation.

Aid agencies and their counterparts in developing countries are increasingly dealing with scientific and technical issues involving knowledge of advances in fields such as molecular biology and ecology. For example, participating effectively in the global debate on GM foods requires an appreciation of the biological sciences as well as the related fields such as law, economics, ethics, and sociology. The technical nature of many international negotiations is favoring countries that use scientific knowledge to inform their positions. In addition, it is also shaping the way governments interact with industry and nongovernmental organizations (NGOs).

The creation of science-advice capacity in the U.S. Department of State is an example of the importance that governments are putting on science and technology in international diplomacy. This development not only serves as a source of inspiration for developing countries, but it also provides new opportunities for capacity building on science diplomacy in developing countries. The existence of such capacity in the foreign ministries of developing countries will help to create a basis for a common vocabulary on key international issues in areas such as biosafety, intellectual property protection, and overall scientific and technical cooperation. Helping developing countries to bring science and technology to their diplomatic activities will help improve cooperation and dialogue between nations.

Developing countries will also need to be creative in designing institutions that are adapted to a globalizing world. Take, for example, the case of the need to make effective use of a trained workforce. A number of developing countries—including several in Africa—have large pools of their nationals operating from the industrialized countries. For a long time, such overseas residence was considered as part of the "brain drain" and considerable effort was put to appeals to national instinct as well as the design of repatriation programs. There is no evidence that either of these strategies worked. But a few developing countries are now starting to establish research institutions in the industrialized countries that employ their nationals and others to work on problems of relevance to their economies. Similarly, others are taking advantage of communications technologies to promote linkages between their nationals in the diaspora and research institutions at home.

Conclusion

Promoting the responsible use of biotechnology to meet the needs of low-income countries will require fundamental policy adjustments in the developing and developed countries. Developing countries need to formulate policies that recognize the importance of science and technology in overall economic development and in agricultural production in particular. A review of existing agricultural policies in the developing countries is needed to accommodate the imperatives of emerging technologies, changing markets, shifting public perceptions about safety, and rising environmental concerns. Such a reexamination should be in line with the need to adopt strategies for sustainable agriculture that take into account the availability of new technological options.

Industrialized countries could play a key role by exhibiting greater sensitivity to the needs of developing countries. In addition, they particularly need to play a leading role in exploring how scientific and technological advances in general and biotechnology in particular could help the problems of low-income families. This will entail increases in public-sector funding, greater scientific and technical cooperation, and the creation of incentives that allow private enterprises to work on developing-country challenges. Intellectual property rights holders will need to show greater creativity in ensuring that those who work on meeting the needs of low-income families can operate without undue constraints imposed by the implementation of WTO rules.

On the whole, the final outcome will depend on the degree to which developing countries take a more active role in defining biotechnology as a strategic field. It is their actions that will determine the degree to which the "genetic divide" determines their relative position in the global economy. So far the available evidence suggests that only a handful of developing countries have serious science and technology policies that can help them become players in the biotechnology revolution. Many of them are hampered from making any major investments in this field because of their historical links with regions of the world that benefit from adopting a cautious approach toward biotechnology. Resolving the current impasse will take more than biosafety assurances; it will entail new international alliances defined by commitments to emerging technological opportunities for solving sustainable-development challenges.

Notes

An earlier version of this paper was published in the *International Journal of Biotechnology.*

1. These debates also reflect trade conflicts between the United States and Europe but are often articulated as global issues that involve all nations. Indeed, other nations are affected by any conflicts—however minor—that occur between the largest trading regions in the world.

2. United Nations Development Program, *Human Development Report* (Oxford: Oxford University Press, 2000).

3. Social movements working in the field of environment tend to attribute ecological degradation to technological change. Other groups concerned attribute cultural decay to technological change, and still others argue that technological change is a source of unemployment. Although there is some truth in these claims, the reality is more complex and defies simplistic attributions of the kinds that dominate popular discourse.

4. The Cartagena Protocol on Biosafety to the Convention on Biological Diversity was adopted on January 20, 2000, in Montreal and has so far been ratified by Bulgaria, Norway, and Trinidad and Tobago. Article I of the protocol states: "In accordance with the precautionary approach contained in Principle 15 of the Rio Declaration on Environment and Development, the objective of this Protocol is to contribute to ensuring an adequate level of protection in the field of the safe transfer, handling, and use of living modified organisms resulting from modern biotechnology that may have adverse effects on the conservation and sustainable use of biological diversity taking also into account risks to human health, and specifically focusing on transboundary movements."

5. The commission was set up in 1961 to administer food safety and quality standards.

6. See, for example, Simonetta Zarrilli, "International Trade in Genetically Modified Organisms and Multilateral Negotiations: A New Dilemma for Developing Countries," United Nations Conference on Trade and Development, Geneva, 2000.

7. Calestous Juma, *Science, Technology, and Economic Growth: Africa's Biopolicy in the Twenty-first Century* (Tokyo: United Nations University Press, 2000).

8. Robert Paarlberg, "The Global Food Fight," *Foreign Affairs* 79, no. 3 (2000): 24–38.

9. Indeed a trade-related conflict, consumers and company employees become soldiers and civil society and corporate leaders become generals to fight for the values they care about.

10. Article 2(4) of the Cartagena Protocol calls for consistency between the protocol and other international treaties: "Nothing in this Protocol shall be interpreted as restricting the right of a Party to take action that is more protective of the conservation and sustainable use of biological diversity than that called for in this protocol, provided that such action is consistent with the objective and the provisions of this Protocol and is in accordance with that Party's other obligations under international law."

11. "The commodities situation may worsen for developing countries should major consumer countries (in the North) develop laboratory substitutes for natural commodities through the use of biotechnology. There would be more displacement of the South's export commodities," Martin Khor, *Rethinking Globalization* (London: Zed Books, 2001).

12. John H. Perkins, *Geopolitics of the Green Revolution: Wheat, Genes, and the Cold War* (Oxford: Oxford University Press, 1997).

13. Clive James, *Global Trends in the Commercialization of Transgenic Crops* (Ithaca, N.Y.: International Service for the Acquisition of Agribiotech Applications, 2001).

14. Ibid.

15. Calestous Juma, "The UN's Role in the New Diplomacy," *Issues in Science and Technology* 17, no. 1 (2000): 37–38.

16. See, for example, Mostafa Tolba, *Global Environmental Diplomacy: Negotiating Environmental Agreements for the World, 1973–1992* (Cambridge: MIT Press, 1998).

Part 10

Assessing Environmental Impacts

Introduction

Suppose you modify a domesticated crop so that it is resistant to insect pests, and that you sow it and the crop hybridizes with a related weed. Could we now find ourselves facing a whole new generation of noxious super-weeds that are genetically resistant to pests, and hence of much greater danger to food security than they are now? This is the kind of issue to be faced in this final section of the collection. And here, as elsewhere, opinion is mixed, with some thinking that ecological disturbance caused by transgene movement is a real and present danger, and others much less inclined to panic. These latter point out that even if genetically modified organisms can do environmental damage, that damage is miniscule compared to the environmental impacts wrought by conventional agriculture—some of which might be lessened by GMOs.

Norman Ellstrand really is one who worries about the potential threats of GM plants. He has run experiments, putting such modified plants next to weeds, and he finds that the hybridization problem is real. He looked at sorghum, one of the world's most important crops, and johnsongrass, a major and persistent weed. They are very different species with different chromosome numbers, and johnsongrass normally fertilizes itself. Yet the two plants do spontaneously hybridize, and the offspring are fertile. Ellstrand stresses that this does not mean instant trouble, but that the potential is there and that when "problems are realized, they can be doozies." He ends his discussion on a cautious, but somewhat pessimistic note.

The most widely used genetically modified plants have transgenes that confer herbicide tolerance. The touted benefit of these genetically modified herbicide-tolerant (GMHT) crops is that they require less herbicide because applications can be timed to weed-emergence cycles. Herbicides are then delivered in bursts that would kill non-herbicide-tolerant crops. Over the course of a planting, these applications add up to less herbicide than conventional applications. Johnson and Hope cast doubt on biotechnology advocates' claim that agriculture is becoming less chemically dependent, and they also challenge the bigger claim that GMHT crops promote the enhancement of biodiversity in general.[1]

In the last selection, Anthony Trewavas pursues the issue of environmental concerns about GHMT crops with his usual vigor. Trewavas argues that the real threats to the environment are plants that fall beneath environmentalists' radar— the plants that stock cherished English gardens. More introduced species exist there and hybridize with native plants than in any crop. He shows little patience for environmentalists' concerns because he believes that the benefits of appropriately used genetic engineering in agriculture will far outweigh any attendant environmental risks. Environmentalists argue that a return to organic farming is the only environmentally defensible response to GM crops, yet organic farming causes environmental damage and poses its own risks to human health, its own unique risks that, in Trewavas' opinion, are greater than those posed by GM crops. What would Prince Charles say to that?

Note

1. It probably is too soon to tell, particularly with the effects of genetically modified crops on nontarget organisms such as the skylark. The effects of Bt corn on the monarch butterfly has caused considerable controversy in which evidence is lined up on both sides of the debate, without anything definitive on either. See Losey et al. (1999); Niiler (1999); Sears et al. (2000); Hansen and Obrycki (2000); Pimentel and Raven (2000); and ESA (2001) in Suggestions for Further Reading.

When Transgenes Wander, Should We Worry?

Norman C. Ellstrand

It is hard to ignore the ongoing, often emotional, public discussion of the impacts of the products of crop biotechnology. At one extreme of the hype is self-rightoeus panic, and at the other is smug optimism. While the controversy plays out in the press, dozens of scientific workshops, symposia, and other meetings have been held to take a hard and thoughtful look at potential risks of transgenic crops. Overshadowed by the loud and contentious voices, a set of straightforward, scientifically based concerns have evolved, dictating a cautious approach for creating the best choices for agriculture's future.

Plant ecologists and population geneticists have looked to problems associated with traditionally improved crops to anticipate possible risks of transgenic crops. Those that have been most widely discussed are (a) crop-to-wild hybridization resulting in the evolution of increased weediness in wild relatives, (b) evolution of pests that are resistant to new strategies for their control, and (c) the impacts on nontarget species in associated ecosystems (such as the unintentional poisoning of beneficial insects).[1]

Exploring each of these in detail would take a book, and such books exist.[2] However, let us consider the questions that have dominated my research over the last decade to examine how concerns regarding engineered crops have evolved. Those questions are How likely is it that transgenes will move into and establish in natural populations? And if transgenes do move into wild populations, is there any cause for concern? It turns out that experience and experiments with traditional crops provide a tremendous amount of information for answering these questions.

The possibility of transgene flow from engineered crops to their wild relatives with undesirable consequences was independently recognized by several scientists.[3] Among the first to publish the idea were two Calgene scientists, writing: "The sexual transfer of genes to weedy species to create a more persistent weed is probably the greatest environmental risk of planting a new variety of crop species."[4] The movement of unwanted crop genes into the environment may pose more of a management dilemma than unwanted chemicals. A single molecule of DDT [1,1,1,-trichloro-2,2-bis(p-chlorophenyl)ethane] remains a single molecule or degrades, but a single crop allele has the opportunity to multiply itself repeatedly through reproduction, which can frustrate attempts at containment.

In the early 1990s, the general view was that hybridization between crops and their wild relatives occurred infrequently, even when they were growing in close proximity. This view was supported by the belief that the discrete evolutionary pathways of domesticated crops and their wild relatives would lead to increased reproductive isolation and was supported by challenges breeders sometimes have in obtaining crop-wild hybrids. Thus, my research group set out to measure spontaneous hybridization between wild radish (*Raphanus sativus*), an important California weed, and cultivated radish (the same species), an important California crop.[5] We grew the crop as if we were multiplying commercial seed and surrounded it with stands of weeds at varying distances. When the plants flowered, pollinators did their job. We harvested seeds from the weeds for progeny testing. We exploited an allozyme allele (*Lap-6*) that was present in the crop and absent in the weed to detect hybrids in the progeny of the weed. We found that every weed seed analyzed at the shortest distance (1 m) was sired by the crop and that a low level of hybridization was detected at the greatest distance (1 km). It was clear, at least in this system, that crop alleles could enter natural populations.

But could they persist? The general view at that time was that hybrids of crops and weeds would always be handicapped by crop characteristics that are agronomically favorable, but a detriment in the wild. We tested that view by comparing the fitness of the hybrids created in our first experiment with their nonhybrid siblings.[6] We grew them side by side under field conditions. The hybrids exhibited the huge swollen-root characteristic of the crop; the pure wild plants did not. The two groups did not differ significantly in germination, survival, or ability for their pollen to sire seed. However, the hybrids set about 15 percent more seed than the wild plants. In this system, hybrid vigor would accelerate the spread of crop alleles in a natural population.

When I took these results on the road, I was challenged by those who questioned the generality of the results. Isn't radish probably an exception? Radish is outcrossing and insect pollinated. Its wild relative is the same species. What about a more important crop? What about a more important weed? We decided to address all of those criticisms with a new system. Sorghum (*Sorghum bicolor*) is one of the world's most important crops. Johnsongrass (*Sorghum halepense*) is one of the world's worst weeds. The two are distinct species, even differing in chromosome number, and sorghum is largely selfing and wind pollinated. Sorghum was about as different from radish as you could get.

We conducted experiments with sorghum paralleling those with radish. We found that sorghum and johnsongrass spontaneously hybridize, although at rates lower than the radish system, and detected crop alleles in seed set by wild plants growing 100 m from the crop.[7] The fitness of the hybrids was not significantly different from their wild siblings.[8] The results from our sorghum–johnsongrass experiments were qualitatively the same as those from our cultivated radish–wild radish experiments. Other labs have conducted similar experiments on crops such as sunflower (*Helianthus annus*), rice (*Oryza sativa*), canola (*Brassica napus*), and pearl millet (*Pennisetum glaveum*).[9] In addition, descriptive studies have repeatedly found crop-specific alleles in wild relatives when the two grow in proximity.[10] The data from such experiments and descriptive studies provide ample evidence that spontaneous hybridization with wild relatives appears to be a general feature of most of the world's important crops, from raspberries (*Rubus idaeus*) to mushrooms (*Aqaricus bisporus*).[11]

When I gave seminars on the results of these experiments, I was met by a new question: If gene flow from crops to their wild relatives was a problem, wouldn't it already have occurred in traditional systems? A good question. I conducted a thorough literature review to find out what was known about the consequences of natural hybridization between the world's most important crops and their wild relatives.

Crop-to-weed gene flow has created hardship through the appearance of new or more difficult weeds. Hybridization with wild relatives has been implicated in the evolution of more aggressive weeds for seven of the world's thirteen most important crops.[12] It is notable that hybridization between sea beet (*Beta vulgaris* subsp. *maritima*) and sugar beet (*B. vulgaris* subsp. *vulgaris*) has resulted in a new weed that has devastated Europe's sugar production.[13]

Crop-to-wild gene flow can create another problem. Hybridization between a common species and a rare one can, under the appropriate conditions, send the rare species to extinction in a few generations.[14] There are several cases in which hybridization between a crop and its wild relatives has increased the extinction risk for the wild taxon.[15] The role of hybridization in the extinction of a wild subspecies of rice has been especially well documented.[16] It is clear that gene flow from crops to wild relatives has, on occasion, had undesirable consequences.

Are transgenic crops likely to be different from traditionally improved crops? No, and that is not necessarily good news. It is clear that the probability of problems due to gene flow from any individual cultivar is extremely low, but when those problems are realized, they can be doozies. Whether transgenic crops are more or less likely to create gene-flow problems will depend in part on their phenotypes. The majority of the "first-generation" transgenic crops have phenotypes that are apt to give a weed a fitness boost, such as herbicide resistance or pest resistance. Although a fitness boost in itself may not lead to increased weediness, scientists engineering crops with such phenotypes should be mindful that those phenotypes might have unwanted effects in natural populations. In fact, I am aware of at least three cases in which scientists decided not to engineer certain traits into certain crops because of such concerns.

The crops most likely to increase extinction risk by gene flow are those that are planted in new locations that bring them into the vicinity of wild relatives, thereby increasing the hybridization rate because of proximity. For example, one can imagine a new variety that has increased salinity tolerance that can now be planted within the range of an endangered relative. It is clear that those scientists creating and releasing new crops, transgenic or otherwise, can use the possibility of gene flow to make choices about how to create the best possible products.

It is interesting that little has been written regarding the possible downsides of within-crop gene flow involving transgenic plants. Yet a couple of recent incidents suggest that crop-to-crop gene flow may result in greater risks than crop-to-wild gene flow. The first is a report of triple herbicide resistance in canola in Alberta, Canada.[17] Volunteer canola plants were found to be resistant to the herbicides Roundup (Monsanto, St. Louis), Liberty (Aventis, Crop Science, Research Triangle Park, North Carolina), and Pursuit (BASF, Research Triangle Park, North Carolina). It is clear that two different hybridization events were necessary to account for these genotypes. It is interesting that the alleles for resistance to Roundup and Liberty are transgenes, but the allele for Pursuit resistance is the result of mutation breeding. Although these volunteers can be managed with other herbicides, this report is significant because, if correct, it illustrates that gene flow into wild plants is not the only avenue for the evolution of plants that are increasingly difficult to manage.

The second incident is a report of the Starlink Cry9C allele (the one creating the fuss in Taco Bell's taco shells) appearing in a variety of supposedly nonengineered corn.[18] Although unintentional mixing of seeds during transport or storage may explain the contamination of the traditional variety, intervarietal crossing between seed-production fields could be just as likely. This news is significant because, if correct, it illustrates how easy it is to lose track of transgenes. Without careful checking, there are plenty of opportunities for them to move from variety to variety. The field release of "third-generation" transgenic crops that are grown to produce pharmaceutical and other industrial biochemicals will pose special challenges for containment if we do not want those chemicals appearing in the human food supply.

The products of plant improvement are not absolutely safe, and we cannot expect transgenic crops to be absolutely safe either. Recognition of that fact suggests that creating something just because we are now able to do so is an inadequate reason for embracing a new technology. If we have advanced tools for creating novel agricultural products, we should use the advanced knowledge from ecology and population genetics as well as social sciences and humanities to make mindful choices about to how to create the products that are best for humans and our environment.

Notes

This article was written while I was receiving support from the U.S. Department of Agriculture (grant no. 00-33120-9801). I thank Tracy Kahn for her thoughtful comments on an earlier draft of the manuscript and Maarten Chrispeels for his encouragement and patience.

1. Allison A. Snow and Pedro Palma, "Commercialization of Transgenic Plants: Potential Ecological Risks," *BioScience* 47 (1997): 86–96; Rosie S. Hails, "Genetically Modified Plants: The Debate Continues" *Trends in Ecology and Evolution* 15 (2000): 14–18.

2. For example, Jane Rissler and Margaret Mellon, *The Ecological Risks of Engineered Crops* (Cambridge: MIT Press, 1996); Scientists' Working Group on Biosafety, "Introductory Materials and Supporting Text for Flowcharts," pt. 1 of *Manual for Assessing Ecological and Human Health Effects of Genetically Engineered Organisms* (Edmonds, Wash.: Edmonds Institute, 1998); "Flowcharts and Worksheets," pt. 2 of *Manual for Assessing Ecological and Human Health Effects of Genetically Engineered Organisms.*

3. Rita E. Colwell et al., "Genetic Engineering in Agriculture," *Science* 229 (1985): 111–12; Norman C. Ellstrand, "Pollen as a Vehicle for the Escape of Engineered Genes?" in *Planned Release of Genetically Engineered Organisms*, eds. John Hodgson and Andrew M. Sugden (Cambridge: Elsevier, 1988), S30–S32; Philip J. Dale, "Spread of Engineered Genes to Wild Relatives," *Plant Physiology* 100 (1992): 13–15.

4. Robert M. Goodman and Nanette Newell, "Genetic Engineering of Plants for Herbicide Resistance: Status and Prospects," in *Engineered Organisms in the Environment: Scientific Issues,* eds. Harlyn O. Halvorson, David Pramer, and Marvin Rogul (Washington, D.C.: American Society for Microbiology, 1985), 47–53.

5. Terry Klinger, Diane R. Elam, and Norman C. Ellstrand, "Radish as a Model System for the Study of Engineered Gene Escape Rates via Crop Weed Mating," *Conservation Biology* 5 (1991): 531–35.

6. Terry Klinger and Norman C. Ellstrand, "Engineered Genes in Wild Populations: Fitness for Weed-Crop Hybrids of Radish, *Raphanus sativus* L," *Ecological Applications* 4 (1994): 117–20.

7. Paul E. Arriola and Norman C. Ellstrand, "Crop-to-Weed Gene Flow in the Genus *Sorghum* (*Poaceae*): Spontaneous Interspecific Hybridization between Johnsongrass, *Sorghum halepense,* and Crop Sorghum, *S. bibolor,*" *American Journal of Botany* 83 (1996): 1153–60.

8. Paul E. Arriola and Norman C. Ellstrand, "Fitness of Interspecific Hybrids in the Genus Sorghum: Persistence of Crop Genes in Wild Populations," *Ecological Applications* 7 (1997): 512–18.

9. For review, see Norman C. Ellstrand, Honor C. Prentice, and James F. Hancock, "Gene Flow and Introgression from Domesticated Plants into Their Wild Relatives," *Annual Review of Ecological Systems* 30 (1999): 539–63.

10. For review, see ibid.

11. Compare with ibid.

12. Ibid.

13. Ingrid M. Parker and Detlef Bartsch, "Recent Advances in Ecological Biosafety Research on the Risks of Transgenic Plants: A Transcontinental perspective," in *Transgenic Organisms: Biological and Social Implications*, eds. Jurgen Tomiuk, Klaus Wohrmann, and Andreas Sentker (Basel, Switz.: Birkhauser Verlag, 1996), 147–61.

14. Norman C. Ellstrand and Diane R. Elam, "Population Genetic Consequences of Small Population Size: Implications for Plant Conservation," *Annual Review of Ecological Systems* 24 (1993): 217–42; Gary R. Huxel, "Rapid Displacement of Native Species by Invasive Species: Effect of Hybridization," *Biological Conservation* 89 (1999): 143–52; Diana E. Wolf, Naoki Takebayashi, and Loren H. Rieseberg, "Predicting the Risk of Extinction through Hybridization," *Conservation Biology* (2002): in press.

15. Ernest Small, "Hybridization in the Domesticated-Weed-Seed Complex," in *Plant Biosystemics,* ed. William F. Grant (Toronto: Academic Press, 1984), 195–210.

16. Yun-Tzu Kiang, Janis Antonovics, and Lin L. Wu, "The Extinction of Wild Rice (*Oryza perennis formosana*) in Taiwan," *Journal of Asian Ecology* 1 (1979): 1–9.

17. Mary MacArthur, "Triple-resistant Canola Weeds Found in Alberta," *Western Producer* [online], http://www.producer.com/articles/20000210/news/20000210news01.html [10 February 2000].

18. Patricia Callahan, "Genetically Altered Protein Is Found in Still More Corn," *Wall Street Journal*, 22 Novermber 2000, B5.

34

GM Crops and Equivocal Environmental Benefits

Brian Johnson and Anna Hope

The United Kingdom and other European governments have domestic and international statutory obligations to conserve many farmland species, including a wide range of birds, and cannot do so simply by protecting isolated natural sites. This is because farmland-dependent species are wide-ranging, often spending different parts of the year in different habitats. In contrast to the United States, which has a wide range of large wilderness areas, European countries have little wilderness of use to species dependent on farmland ecosystems, and a high proportion of relatively intensively farmed land. If we are serious about conserving viable populations of farmland-dependent organisms, then we need agricultural methods that allow wildlife to survive within our farmed landscape.

Biotechnology could enable even greater agricultural intensification by making agro-chemicals more efficient and easier to use. The current generation of genetically modified herbicide-tolerant (GMHT) crops are designed to enable higher levels of weed control, and the few com-

Reprinted with permission from *Nature Biotechnology*, by Brian Johnson and Anna Hope, "GM Crops and Equivocal Environmental Benefits," vol. 18, no. 242 (1 March 2000). Copyright 2000 Nature Publishing Group.

parative studies available[1] appear to confirm this. Not only do GMHT crops make weed control cheaper and more efficient, but they are also likely to prove attractive to farmers who, because of climatic or soil problems on their land, find high levels of weed control difficult to achieve economically. These are the very areas that are currently some of the last refuges of U.K. farmland biodiversity.[2] On the basis of our analysis of the available data on the link between herbicide use and declines in farmland insects and birds, we believe that increasing herbicide efficiency by widespread use of GMHT crops on U.K. farmland will further damage natural biodiversity.

Weed control in GMHT crops is achieved by spraying broad-spectrum herbicides, such as glyphosate and glufosinate, during the peak growing season in May and June, when field margin habitats such as headlands, ditches, and hedgerows are in full leaf. Compared with the conventional use of preemergence herbicides and in-crop selective treatments, this increases risks of damage from spray drift, especially for crops at a growth stage that demands nozzles to be set high.[3] This is already an agricultural (and liability) problem in the United States, where drift has caused considerable damage to non-GMHT crops growing alongside GM crops sprayed with broad-spectrum herbicides.[4]

It is claimed that these herbicides are more environmentally benign in terms of direct toxicity to humans and wildlife, and that less herbicide would be used. There is some evidence that broad-spectrum herbicides are directly toxic to small invertebrates,[5] and we contend that the volumes of herbicide used are not related to environmental damage; it is the ecological impact of the type of herbicide and its application methods that should be considered. There is too much confusion in this debate between inputs and impacts.

Until recently, risk assessments for the commercial release of GMHT crops in the United Kingdom did not consider the likely "indirect" effects on biodiversity. These risks to biodiversity from the widespread use of GMHT crops cannot yet be fully assessed because the basic ecological data needed to do so have not been gathered. We can find no comparative studies of ecological effects of GMHT crops before 1998 either in Europe, or in the United States and Canada, where they have been in widespread use since 1997. The U.K. government has now put in place a regulatory system that addresses these issues, and it has initiated a series of studies comparing the ecological effects of growing GMHT oilseed rape (canola), fodder maize, and beet. These three-year field-scale trials using split-field methodology will generate data enabling a full assessment of ecological risk, and may also identify any remedial measures that might be possible if these crops are given consent for commercial release. Other field-scale research is being carried out in France and Germany.

The ecological risks from using GMHT crops do not apply only in the European situation. In many tropical and subtropical countries similar situations exist, wherein important biodiversity is dependent on traditional farming techniques. Farmed wetlands, such as the grazing marshes of northern Europe and rice-growing areas of the Middle East, are vital to large populations of overwintering birds that depend on invertebrates living within the wet fields. The use of GMHT crops on these areas woud risk direct toxicity and indirect effects on the invertebrates and plants that support such bird populations. These are serious risks that need to be properly assessed before consent is given for the use of herbicide-tolerant crops in such areas.

Attempts have been made to manipulate herbicide-tolerant cropping systems to increase biodiversity in the field—for example, work on sugar beet at Brooms Barn in the United Kingdom. Initial results suggest that it may be possible, by carefully timing applications, to allow weeds to grow (and perhaps even flower) before removing them from the crop, although delaying herbicide applications may well result in an unacceptable yield penalty, which would deter farmers. The need for the damaging herbicide atrazine in maize and other crops might also be avoided, by applying a broad-spectrum herbicide over the crop followed by undersowing with a forage crop. But there is no regulatory mechanism yet in place to ensure that any potential benefits could be delivered at the farm level. Experience over the past forty years has shown that farmers will use herbicides to their full potential, reducing weed populations to as low a level as economically possible.

This is an issue not about biotechnology as a technique, but about how we use it in agriculture. A more sustainable approach to weed control in arable crops might be to produce GM crops with greater tolerance to weeds, perhaps by achieving more competitive growth and the production of inhibitors from roots, allowing a proportion of wild plants to survive within the crop and its margins. There is considerable potential in GM insect-resistant crops for minimizing environmental impacts of arable agriculture, although it may be necessary to ensure genetic isolation from sexually compatible native species.

The challenge to biotechnologists is to produce crop varieties that move us away from chemically dependent agriculture, while maintaining yields and sustaining farmland-dependent biodiversity. This is surely the real goal of agricultural sustainability, which should be about sustaining levels of production while simultaneously minimizing the environmental impacts of agriculture. So far, the industry has produced no convincing evidence that GMHT crops will contribute to the latter.

Notes

1. M. A. Read and M. N. Bush, "Control of Weeds in GM Sugar Beet with Glufosinate Ammonium in the UK," *Aspects of Applied Biology* 52 (1998): 410–406; J. D. A. Wevers, "Agronomic and Environmental Aspects of Herbicide-resistant Sugar Beet in the Netherlands," *Aspects of Applied Biology* 52 (1998): 393–99.

2. M. A. Lainsbury, J. G. Hilton, and A. Burn, "The Incidence of Weeds in UK Sugar Beets," *Proceedings of the 1999 Brighton Conference on Weeds* 3 (1999): 817–22.

3. J. B. Sweet and R. Shepperson, "The Impact of Releases of Genetically Modified Herbicide-tolerant Oilseed Rape in UK," *Acta Horticulture* 459 (1998): 225–34.

4. J. A. Springett and R. A. J. Gray, "Effect of Repeated Low Doses of Biocides on the Earthworm *Asporrectodea caliginosa* in Laboratory Culture," *Soil Biology Biochemistry* 24, no. 12 (1992): 1739–44; "Collateral Damage?" *New Scientist* 164 (23 October 1999): 27.

5. Springett and Gray, "Effect of Repeated Low Doses of Biocides on the Earthworm *Asporrectodea caliginosa* in Laboratory Culture."

35

Much Food, Many Problems

Anthony Trewavas

Whatever the reasons for the current furor in the United Kingdom over genetic manipulation of plants, an abundance of food is certainly one of them. We are inundated with new foods, and supermarkets respond to every consumer whim. People in the West eat a much healthier diet now than at the turn of the century, thanks to cheap, plentiful food provided by modern intensive agriculture. But wealth brings its own problems, not least the acceleration in technological change. With an abundance of food and long life has come the demand for a risk-free world. Under these circumstances, the public is little interested in new ways of producing the same food, especially if there is even a minute health risk.

Attempts to introduce genetically modified (GM) foods have stimulated, not a reasoned debate, but a potent negative campaign by people with other agendas who demonize the technology. These opponents ignore common farming practice and well-investigated facts about plants, or inaccurately present general problems as

Reprinted with permission from *Nature*, by Anthony Trewavas, "Much Food, Many Problems," vol. 17, no. 11 (1999): 231–32. Copyright © 1999 Nature Publishing Group.

being unique to GM plants. Almost without exception, opponents of GM foods are not plant biologists.

As a plant biologist myself, I have little time for big, insensitive agribusiness. The "green revolution" of the 1970s, which developed crops of direct value to many people, particularly those in the developing countries, was publicly funded. A decade ago, most plant biologists supposed that genetic manipulation would be used to the same end, for example, to produce crops resistant to yield-destroying diseases such as rust or rice blast, or pests such as locusts. Herbicide-resistant plants would have been near the bottom of my list of priorities, especially using the effective and innocuous herbicide glyphosate.

Superweeds

The British Medical Association[1] and green (environmental) activists have objected to GM rape containing resistance to glyphosate, claiming that it will generate superweeds by gene flow of the one specific transgene into three or four weedy relatives. Yet weeds (and crops) resistant to one or another specific herbicide have been known for fifty years, and ecological studies of the spread of resistance investigated this in detail twenty years ago;[2] such weeds can be eliminated by using another herbicide. (Perhaps Synchrony beans, bred from soybean individuals naturally resistant to sulphonyl urea herbicides, should be relabeled "superbean" to make this point.)

Introduced plants are the real superweeds. Three thousand alien species in the United Kingdom alone, mainly introduced by gardeners, now outnumber the 1,500 or so indigenous species and cause serious environmental damage (for example, *Rhododendron ponticum*) and/or are resistant to virtually all herbicides (for example, Japanese knotweed, *Fallopia japonica*). At least sixty aliens have hybridized with indigenous species, producing additional environmental contamination from the unpredictable consequences of mixing thousands of new genes in a continuing process of illegitimate gene flow. Yet, although environmental activists label gene flow as unacceptable genetic pollution,[3] there is no trampling of flowers or demonstrations at the international flower shows that are potent sources of new foreign pollen, nor demands for barriers miles thick at such shows to prevent cross-pollination by bees. Nor have there been requests for strict laws to prevent people introducing new foreign seeds into their gardens on the grounds that this would cause serious environmental damage by new hybridization and subsequent gene flow.

Another common argument is that, once a transgenic plant is released, it can never be withdrawn. But all domesticated crop plants lack the genetic variability and weedy characteristics of wild plants and quickly disappear from fallow fields.

Just what do anti-GM-food environmentalists really care about? Rachel Carson in her 1962 classic book *Silent Spring* documented the urgent need to reduce pesticide applications to crops. She argued that mankind's best chance for long-term survival depends on minimal impact on planetary ecosystems, and that the biodiversity on which we are ultimately interdependent must be maintained.

On average, conventional farming uses five or more broad-spectrum pesticide applications on crops each year. The technology is ultimately self-limiting because pest resistance rapidly emerges, much as overprescription of antibiotics by the medical profession is hastening the end of useful drugs. The *Bacillus thurigiensis* (Bt) insecticidal proteins (a family of 130 proteins) selectively kill some beetles and caterpillars[4] and target insects that eat crops. Expression of Bt protein into cotton and corn has reduced the application of specific, highly toxic pesticides by more than 80 percent, allowing a substantive return of wildlife to crop fields.[5]

Even though the technique is not perfect, it is surely better than killing virtually all field insects with pesticides. Yet opponents of GM technology not only have not welcomed this giant stride toward Carson's goal, but have completely rejected it. A laboratory study on the effects of GM crops on the monarch butterfly[6] was exaggerated out of all proportion by the media, whereas more realistic assessments[7] are ignored. Negative propaganda against genetic manipulation ensures that pesticide treatments of U.K. farmland will continue. Field-insect numbers remain low and many songbirds die prematurely. Who are the environmentalists now?

Spread by Pollen

The concern that GM plants could contaminate far-off fields with huge numbers of transgenic plants has arisen because, using the polymerase chain reaction test, minute quantities of GM pollen have been detected kilometers away from GM trial fields.[8] But pollen is known to move much further than genes. Pollination distances beyond which growing siblings are not produced were painstakingly measured by ecologists twenty years ago[9] and, as a result, a separation distance of about 50 meters is used internationally to maintain separate lines of the same crop at greater than 99.5 percent purity. Very rarely, individual seeds can be spread by birds over long distances (as pointed out by Darwin in the nineteenth century), but the transgenic crop seeds so far produced have low fitness with poor survival possibilities.

Another way to prevent the distribution of transgenes by pollen is to insert the gene into chloroplast DNA, as the pollen of most crop plants contains no chloroplasts and thus no plastid DNA. Transformation systems for chloroplasts are well established.[10] Very high expression of transgenes can be achieved, and there is just one insertion site. But if chloroplast GM plants were produced commercially, I would expect many opponents to maintain their opposition. Is this simply GM technophobia (see box, overleaf)?

The achievements of the negative GM-food campaign have been to engender unsubstantiated fear among the U.K. public about GM food. One, no-doubt-unintended casualty of such a mood was Axis Genetics, a small company making medical GM products such as the cholera-vaccine-expressing banana. Cholera kills millions of young people in the Third World every year. A banana tree expressing cholera vaccine in each village would have offered the possibility, now denied to many, of a full life.

The international charity Christian Aid, in a report on farming and hunger in the developing world,[11] attacked "green revolution" agriculture and genetic manipulation, stating that politics, and not lack of food, is the main cause of hunger. Yet the report ignored the population explosion, claiming that more people were malnourished after the green revolution than before. According to the United Nations' Food and Agriculture Organization, in 1970 there were 935 million malnourished people in Africa, South America, and the Far and Near East; after twenty years of green revolution agriculture their numbers had been reduced to 730 million.[12] These numbers are still unacceptable, but in the same twenty years the population in these areas almost doubled from two billion to about four billion.[13] The green revolution fed this increase by a doubling of wheat and rice yields per hectare. At least one billion extra people had sufficient food to eat who would have starved using traditional agriculture. Further increased cereal yields via conventional breeding are now unlikely.[14]

The Christian Aid report also claimed that genetic manipulation is not needed because the world can grow enough food without it. But no food supply is guaranteed. Devastating crop diseases occurred in U.S. corn in 1971 and led to mass starvation in Ireland in the 1840s. By 2025, the world population will have increased by a further 2.3 billion, which will require an average annual increase in food production of 1.3 percent.[15] In the past few years, the increase in world food output has dropped below that crucial figure, and many poorer countries are living on the residual excess of the green revolution.

The future is threatened by global warming and unpredictable climate change. The old enemies of locusts, floods, disease, drought, and pests still exist. In the face of these adversaries, diversity in technology becomes a strength and a necessity, not a luxury. We have developed genetic manipulation of food and plants only just in time. Companies and scientists may fumble in its use, but now is the time to experiment, not when a holocaust is upon us. Opposition to the technology is both short-sighted and potentially dangerous.

The Organic Way

As populations rise, inefficient farming will destroy a much greater quantity of wilderness and its associated wildlife than is necessary. Greenpeace has announced its bid for the future, but in reality, this is a lurch into the past. It has opted for pre-1950 agriculture—"organic" farming—in which average crop yields on a variety of soils are about half those of intensive farming.[16]

For very obscure reasons, organic farmers eschew the use of most minerals.

Is there a test for GM technophobia?

- Natural glyphosate resistance exists in rape, and the resistance gene has been isolated. Fields of naturally resistant rape can be grown. Gene flow from such a field would carry the resistance gene into relatives. Pollen containing the resistance gene could be detected on organic farms kilometers away. Reduction in weed number on treatment with glyphosate would curtail insect populations and reduce associated songbird numbers.
- If that isolated gene is now inserted by genetic manipulation into rape and an equivalent field grown and treated, in what way are gene flow, pollen distribution, and ecology different from those in the natural field? Based on previous experience, activists and organic farmers would complain about the second field but not the first.
- Is this simple technophobia? The second GM field can be produced with one-tenth of the expenditure and one-tenth of the time it takes to produce the naturally resistant rape because of the enormous backcrossing required to eliminate undesirable traits.
- Synchrony beans, naturally resistant to sulphonyl urea herbicides, indicate the reality of this process. Fungus- and pest-resistance genes isolated from resistant individuals are also being inserted back into the same crop species by genetic manipulation.

Instead, cow manure is used as the primary fertilizer. Extra land is needed to support the required cow population, so land-use efficiency of organic crops is further reduced. Going organic worldwide, as Greenpeace wants, would destroy even more wilderness, much of it of marginal agricultural quality.[17]

The organic philosophy is negative and restrictive in its rules and regulations. It started as a movement simply to eliminate pesticides from food, and it is indeed beneficial to use pesticides sparingly, as organic farmers do. But the philosophy was founded on a fallacy. Food was thought to be somehow pure and pristine to which dangerous man-made chemicals were added by intensive agriculture. But almost all (99.99 percent) of the carcinogens routinely consumed by people are made by plants to inhibit predation, and are present in variable amounts in all food.[18]

Most organic rules and treatments have never received proper biological investigation as to human safety. In one case in which they have—the use of Bt spores as an insecticide—there are reports of potentially serious consequences for human health.[19] Mycotoxin contamination, and infection from the potentially lethal *Escherichia coli* 0157, are additional problems.[20]

Organic farmers reject GM plants, regarding them as unnatural. In the past fifty years, however, "unnatural" combinations between many different species of crops have been constructed using embryo rescue and cell culture.[21] Well-known examples include crosses between wheat and rye (to produce triticale, grown on one million hectares worldwide), between rice and sorghum, agropyron and wheat. Organic farmers do not reject these "unnatural" plants.

Greenpeace regards organic approaches as working with the grain of nature, a truce in a hypothetical war of humans with nature. This concept is simply wrong. We live on a planet where the ecology is constructed on a competition for resources. Competition is the vital spark that energizes evolution and generates vitality and creativity. We are not at war with nature, but striving toward a new dynamic equilibrium in which our unique biological characteristic, the human spirit, must take its place.

Looking to the Future

The future will demand agriculture to be both flexible and diverse in technology, but efficient in land use. Farmers will have to be highly skilled at using technologies that must sustain farming for thousands of years.[22] Increasingly, farm resources will need to be recycled; green manure and crop rotation will underpin soil fertility. Integrated pest-management systems and zero tillage will be essential to minimize losses due to pests and weeds, and to limit soil erosion. Water will become an increasingly expensive commodity, and a premium will emerge on crops that use water efficiently without loss of yield.[23] In all, this future agriculture, genetic manipulation has a unique and intimate role. Let's have some ideas.

Notes

1. British Medical Association Board of Science and Education, *The Impact of Genetic Modification on Agriculture, Food, and Health* (London: British Medical Association, 1999).

2. R. J. Holliday and P. D. Puwain, "Evolution of Herbicide Resistance in *Senecio vulgaris*: Variation in Susceptibility to Simazine between and within Populations," *Journal of Applied Ecology* 17 (1980): 779–91.

3. Alan F. Raybould, "Transgenes and Agriculture—Going with the Flow," *Trends in Plant Science* 4, no. 7 (1999): 247–48.

4. Ruud A. de Maagd, Dirk Bosch, and William Stiekema, "*Bacillus thuringiensis* Toxin-Mediated Insect Resistance in Plants," *Trends in Plant Science* 4, no. 1 (1999): 9–13.

5. De Maagd, Bosch, and Stiekema, "*Bacillus thuringiensis* Toxin-Mediated Insect Resistance in Plants"; Gary Fitt, Daniel Llewellyn, and Cherly Mares, "Field Evaluation and Potential Ecological Impact of Cottons (*Gossypium hirsitum*) in Australia," *Biocontrol Science and Technology* 4, no.4 (1994): 535–48; www.econ.ag.gov/whatsnew/issues.biotech.

6. John E. Losey, Linda S. Rayor, and Maureen E. Carter, "Transgenic Pollen Harms Monarch Larvae," *Nature* 399, no. 6733 (1999): 214.

7. For example, Tanya H. Schuler et al., "Parasitoid Behavior and Bt Plants," *Nature* 400, no. 6747 (1999): 825–26.

8. Raybould, "Transgenes and Agriculture—Going with the Flow."

9. A. John Richards, *Plant Breeding Systems* 2d ed. (London: Chapman & Hall, 1997).

10. Henry Daniell et al., "Containment of Herbicide Resistance through Genetic Engineering of the Chloroplast Genome," *Nature Biotechnology* 16, no. 4 (1998): 345–48; Madhuri Kota et al., "Overexpression of the *Bacillus thuringiensis* (Bt) Cry2Aa2 Protein in Chloroplasts Confers Resistance to Plants against Susceptible and Bt-resistant Insects," *Proceedings of the National Academy of Science USA* 96 (1999): 1840–45.

11. *Selling Suicide Farming, False Promises, and Genetic Engineering in Developing Countries* (London: Christian Aid, 1999).

12. *World Food Supplies and the Prevalence of Chronic Under-Nutrition in Developing Regions* (Rome: Food and Agriculture Organization, 1992).

13. Maarten J. Chrispeels and David E. Sadava, *Plants, Genes, and Agriculture* (Boston: Jones & Bartlett, 1994).

14. Ibid.

15. Charles C. Mann, "Crop Scientists Seek a New Revolution," *Science* 283, no. 5400 (1999): 310–14.

16. Ibid.; Lloyd T. Evans, *Feeding the Ten Billion: Plants and Population Growth* (Cambridge: Cambridge University Press, 1998); David Tilman, "The Greening of the Green Revolution," *Nature* 396, no. 6708 (1998): 211–12.

17. Denis Avery, "The Fallacy of the Organic Utopia," in *Fearing Food: Risk, Health, and Environment,* eds. Julian Morris and Roger Bate (Oxford: Butterworth-Heinemann, 1999), 3–18.

18. Bruce N. Ames, Margie Profet, and Lois Swirsky Gold, "Dietary Pesticides (99.99 Percent All Natural), *Proceedings of the National Academy of Sciences USA* 87 (1990): 7777–81.

19. Debora MacKenzie, "Red Flag for Green Spray: Even Organic Farming Is Not Immune to Health Scares," *New Scientist* 162, no. 2188 (1999): 4.

20. Avery, "The Fallacy of the Organic Utopia."

21. Michael Baum, Evans S. Lagudah, and Rudi Appels, "Wide Crosses in Cereals," *Annual Review of Plant Physiology and Plant Molecular Biology* 43 (1992): 117–43.

22. Chrispeels and Sadava, *Plants, Genes, and Agriculture.*

23. Evans, *Feeding the Ten Billion: Plants and Population Growth.*

Glossary

Agrobacterium: A genus of naturally occurring soil bacteria capable of inserting **DNA** into plants. Scientists use it as a vehicle to transfer desired genes to plants.

Agroecology: The study of agriculture in its ecological context.

Alleles: Alternative forms of a **gene** that are assigned to a specific location on a **chromosome**.

Allergenicity: The capacity to induce an **allergic reaction**.

Allergic reaction: An aggressive immune system response, usually to foreign proteins.

Antibiotic resistance gene: A gene sequence that confers resistance against microbial infection. It is sometimes coupled with other transgenes because the antibiotic resistance gene is a readily identifiable **gene marker**.

Antibiotics: Compounds that kill or interfere with the growth of microorganisms.

Antigen: A compound, usually a protein, that will elicit an **allergic reaction**.

Apomixis: The reproduction of diploid organisms without sexual reproduction.

Arabidopsis: A small weed plant in the mustard family used as a model organism for studying plant genetics.

Asexual reproduction: Propagation without sexual reproduction, that is, by one parent alone.

Bacillus thuringiensis (Bt): A naturally occurring soil bacterium that produces an insecticidal protein used in organic farming and in transgenic crops.

Backcrosses: Controlled fertilizations of modified offspring by the original varieties.

Biodiversity: The inherent variety that exists in all living things, measured as the number of types of organisms in a given area.

Bioethics: A discipline concerned with the application of ethics to biological problems, especially in the field of medicine.

Biotechnology: Technological processes (agricultural, manufacturing, medicine) that involve the use of biological systems.

Bovine growth hormone: A growth hormone used commercially in cattle to increase dairy yields, also known as Bovine Somatotropin (BST).

Capacity building: Providing frameworks for project identification, formulation, and implementation while making the maximum use of existing skills and resources.

Chromosome: A length of **DNA** on which **genes** are arranged in a linear fashion.

Classical Mendelian plant breeding: Selective breeding of plant varieties based on the expression of characteristics determined by **Mendel's laws** of segregation and independent assortment.

Cloning: The production of a number of genetically identical **DNA** molecules by inserting the chosen **DNA** into a **virus** or **plasmid** vector.

Consequentialism: The view in ethics that right action is determined by evaluating the consequences of that action, and comparing those against other options, or doing nothing at all.

Conventionally bred: Refers to organisms that are bred naturally without recourse to **biotechnology**, particularly any form of **genetic engineering** that alters or augments their genome.

Cross: The sexual breeding of two organisms.

Deontology: The view in ethics that right action is prescribed by duties, not by the evaluation of an action's consequences.

Diploid: An organism or cell whose somatic cells have two chromosomes (two **alleles** for each locus).

DNA: Deoxyribonucleic acid, the macromolecule that serves as the genetic material for all living organisms.

Epistemology: The branch of philosophy that investigates the origin, nature, methods, and limits of human knowledge.

Ethics: The branch of philosophy that investigates the nature of the good life and what constitutes right action.

Farm-scale evaluations (FSE): Assessments of the performance and environmental impact of **GM crops** at farm-scale sizes.

Feral populations: Formerly domesticated organisms now existing in a wild state.

Fitness: The relative reproductive success of an organism attributed to its ability to withstand selective pressures.

Food security: A term that agglomerates the safety, sustainability, and capacity for a long-term, adequate food supply.

Frankenfoods: A pejorative term for genetically modified foods.

Gamete: A germ cell (for example, egg or sperm) that is usually **haploid** and fuses with another gamete during sexual fertilization.

Gene: A unit of heredity and function that consists of a stretch of **DNA**.

Gene markers: Specific gene sequences (such as **antibiotic resistance genes**) that are easily identifiable and, when coupled with another gene, can be used to track the insertion of that gene into another organism.

Gene-splicing: Recombinant **DNA** technique that consists in using restriction enzymes to cut **DNA** at specific gene sequences.

Genetic engineering: The manipulation of the **DNA** content of an organism to alter the characteristics of that organism.

Genetically modified foods (GM foods): Foods that are derived in whole or in part using **recombinant DNA** technology.

Genetically modified organism (GMO): An organism that has been modified by the application of **recombinant DNA** technology.

Genome: The entire genetic constitution of an organism.

Genomics: The study of organisms' **genomes**.

Glyphosate: An ingredient contained in the herbicide Roundup that kills plants by inhibiting a chloroplast enzyme required for the biosynthesis of essential aromatic amino acids.

GM (genetically modified): Refers to modification through **recombinant DNA** technology.

GMHT (genetically modified herbicide-tolerant): A crop that has been modified for herbicide resistance through biotechnology.

GMO: Genetically modified organism; also, **LMO**: Living modified organism.

Growth hormone: A hormone that enhances growth and development.

Haploid: An organism or cell that has one set of chromosomes (one **allele** for each locus).

Herbicide: Any substance that is toxic to plants, usually used to selectively target weeds.

High-yielding varieties: Hybridized crops with above-average growth performance.

Homology: The degree of similarity between two gene sequences, parts of organisms, or between two organisms.

Horizontal gene transfer: The direct transfer of genetic material from one species to another, unrelated species, contrasted with **vertical gene transfer**.

Hybrid: The offspring of two genetically dissimilar parents, often showing superior **fitness**.

Indigenous knowledge: The information proper to the culture and way of life of indigenous groups.

Insecticide: A substance used to control populations of insect pests by either killing them outright or interfering with some aspect of their reproductive cycle.

Interspecific hybridization: A method of **gene** transfer between species for the development of improved cultivars.

Introgression: The introduction of new **genes** from one population to another.

Invasiveness: The ability of an organism to spread beyond its traditional domain into new locations.

IPRs (intellectual property rights): The rights that afford legal ownership and control over intangible inventions.

Keratomalacia: An eye disease caused by a vitamin A deficiency and a leading cause of blindness.

Lutein: A carotenoid molecule that appears to prevent free-radical damage in the macula and retina of the eye.

LMO: Living modified organism; also, **GMO:** Genetically modified organism.

Macula: A lutein-rich and oxidant-sensitive area of the eye susceptible to age-related degeneration and cataracts.

Meiotic abnormalities: Aberrations in the sexual differentiation of gametes.

Mendel's laws: Named after the Moravian monk Gregor Mendel, the laws predict that for diploid organisms, the two alleles comprising a single chromosome locus will segregate into even ratios in gametes. Gametes contain many alleles, sorted independently from one another.

Micronutrient: An essential dietary element (vitamins and minerals) required in very small quantities.

Microprojectile: A **DNA**-coated particle that is fired into living cells to transfer new material.

Molecular biology: The chemistry and physics of the molecules that constitute living things.

Molecular markers: See **gene markers**.

Nontarget organism: Organisms that are unintentionally affected by **GMOs**, or pesticides or herbicides used in conjunction with **GMOs**.

Normativity: With reference to the law, it refers to the law's capacity to obligate.

Novel antigens: **Antigens** that have not yet been encountered by an organism's immune system.

Novel foods: **Conventional** or **GM** foods whose combination of genetic material and characteristics are like no other **conventional** foods.

Organic philosophy: The belief that crops should be **conventionally bred** and cultivated without **herbicides** and **pesticides**.

Participatory technology assessment (PTA): A participatory democracy form of technology assessment involving extensive public consultations.

Patent: A time-limited, government-issued right to ownership and the control of rights over an invention.

Pesticide: Any substance that is toxic to insects, usually used to selectively target pests.

Phenotype: The physical and behavioral characteristics of an organism.

Plant transformation: The modification of a plant by transgene insertion or bacterial vector.

Plasmid: A circular **DNA** molecule that resides outside of some bacteria's chromosome, which can be transferred to other bacteria, and especially to plants. See **Agrobacterium**.

Polishing: A rice-production process in which the fibrous outer layer is removed and only the endosperm (the rice grain) remains.

Precautionary principle: A line of reasoning which dictates that measures should be taken when an activity raises threats of harm to human health or the environment, particularly if the scope or magnitude of those risks are scientifically uncertain.

Prevention principle: A line of reasoning which dictates that it is better to prevent harm from happening than to cure the effects of harm having happened.

Recombinant DNA: Hybrid **DNA** produced by joining pieces of **DNA** from different sources.

Resistance management: The control of the acquisition of pesticide resistance by crops and weeds.

Risk: The chance that one might fall victim to uncontrollable harms.

Risk assessment: An attempt to predict the likelihood of exposure to risk.

Roundup Ready: Characterizes crops that are modified to tolerate Roundup herbicides, which are nonselective among green plants.

Selective breeding: Controlled interbreeding of varieties, usually with the intention of producing offspring that share valued characteristics from each parent.

Somatic: Refers to cells in an organism's body that do not produce **gametes**.

Sovereign jurisdiction: The exclusive right of a national body to determine its own affairs.

Species: The basic evolutionary unit and unit of classification, composed of related individuals often thought to be reproductively isolated.

Species barrier: Species form natural barriers to gene flow because they are generally thought to be reproductively isolated from one another.

Stewardship: The approach that manages natural resources or property for sustainable use and development.

Superweeds: Plant varieties that acquire resistance to one or more herbicides.

Terminator technology: A patented process that engineers crops to kill their own seeds in the second generation, thus making it impossible for farmers to save and replant seeds.

Terms of use agreement (TUA): A contract between a farmer and a seed company that authorizes, and specifies the terms of, the use of the company's seed.

Tissue culture: The controlled and systematic growth of plant tissue.

TPRs (technical property rights): A subcategory of **IPRs** dealing with inventions by technique and process.

Traditional breeding techniques: Cultivation by **conventional cross-breeding** and **selective breeding**.

Traditional foods: Crops not subjected to genetic modification.

Trait: One of many observable properties or characters of an organism; see **phenotype**.

Transgene: A **gene** that is taken from one organism and inserted into the **genome** of another.

Transgene stacking: The accumulation of a number of foreign genes into a plant variety.

Transgenic plant: A plant that has incorporated foreign **DNA** into its **genome**.

Unsustainable management: The administration and regulation of resources in a manner that does not promote their protection or viability.

Utilitarianism: A version of **consequentialism** in which right actions are those that tend to maximize happiness.

Vertical gene transfer: The direct transfer of genetic material to members of the same species, contrasted with **horizontal gene transfer**.

Virus: An intracellular parasite that uses a host organism's molecular machinery to reproduce and proliferate.

Volunteerism: An agricultural problem where uncollected seeds from the last year's crop germinate and grow within the current crop.

Weediness: The character of crops that spread beyond the control of conventional herbicides.

Xerophthalmia: The excessive drying of the conjunctiva and cornea; may be due to local disease or vitamin A deficiency.

Zeaxanthin: A rose-colored carotenoid that colors the retina's macular pigment and functions in vision.

Suggestions for Further Reading

Anderson, Terry L., and Bruce Yandle. *Agriculture and the Environment: Searching for Greener Pastures.* Stanford: Hoover Institution Press, 2001.

Bauer, Peter. *From Subsistence to Exchange.* Princeton: Princeton University Press, 2000.

Center for Environmental Quality and Office of Science and Technology Policy Assessment. *Case Studies of Environmental Regulation for Biotechnology.* Washington, D.C.: GPO, 2000.

Comstock, Gary. *Vexing Nature: On the Ethical Case against Agricultural Biotechnolgy.* Dordrecht, Neth.: Kluwer, 2001.

Conway, Gordon, and Vernon W. Ruttan. *The Doubly Green Revolution: Food for All in the Twenty-first Century.* Cambridge: MIT Press, 1999.

DeGregori, Thomas R. *Agriculture and Modern Technology: A Defense.* Ames: Iowa State University Press, 2001.

Ecological Society of America. "ESA Recommends Cautious Approach to Releasing GMOs into the Environment." Press Release, Washington D.C., 2001.

Food and Agriculture Organization. *Report of the Panel of Eminent Experts in Food and Agriculture.* Rome: FAO, 2000.

Food and Agriculture Organization and World Health Organization. *Safety Aspects of Genetically Modified Foods of Plant Origin.* Geneva: WHO, 2000.

Frontline/NOVA Special Presentation. *Harvest of Fear.* 2001, Video and transcript available at http://www.pbs.org/wgbh/harvest/.

Jesse, Laura Hansen, and John J. Obrycki. "Field Deposition of Bt Transgenic Corn Pollen: Lethal Effects on the Monarch Butterfly." *Oecologia* 125 (2000): 241–48.

Juma, Calestrous. *Science, Technology, and Economic Growth: Africa's Biopolicy in the Twenty-first Century*. Tokyo: United Nations University Press, 2001.

Kneen, Brewster. *Farmageddon: Food and the Culture of Biotechnology*. Gabriola Island: New Society Publishers, 1999.

Leiss, William, and Christina Chociolko. *Risk and Responsibility*. Montreal: McGill University Press, 1994.

Magnus, David, Arthur Caplan, and Glenn McGee. *Who Owns Life?* Amherst, N.Y.: Prometheus Books, 2002.

Manning, Richard. *Food's New Frontier: The Next Green Revolution*. New York: Farrar Straus & Giroux, 2000.

McHughen, Alan. *Pandora's Picnic Basket: The Potential and Hazards of Genetically Modified Foods*. Oxford: Oxford University Press, 2000.

Paarlberg, Robert L. *Governing the GM Crop Revolution: Policy Choices for Developing Countries*. Washington D.C.: International Food Policy Research Institute, 2000.

Pimentel, David S., and Peter H. Raven. "Bt Corn Pollen Impacts on Nontarget *Lepidoptera*: Assessment of Effects in Nature." *Proceedings of the National Academy of Sciences USA* 97, no. 15 (2000): 8198–89.

National Academy of Sciences/National Research Council. *Genetically Modified Pest-Protected Plants: Science and Regulation*. Washington D.C.: National Academy Press, 2000.

Niiler, Eric. "GM Corn Poses Little Threat to Monarch." *Nature Biotechnology* 17, no. 12 (1999): 1154.

Nuffield Council on Bioethics. *Genetically Modified Crops: The Ethical and Social Issues*. London: Nuffield Council on Bioethics, 2000.

Reiss, Michael J., and Roger Straughan. *Improving Nature? The Science and Ethics of Genetic Engineering*. Cambridge: Cambridge University Press, 1996.

Rifkin, Jeremy. *The Biotech Century: Harnessing the Gene and Remaking the World*. New York: Penguin Putnam, 1998.

Rissler, Jane, and Margaret Mellon. *The Ecological Risks of Engineered Crops*. Cambridge: MIT Press, 1996.

Royal Society of Canada. *Elements of Precaution: Recommendations for the Regulation of Food Biotechnology in Canada*. Ottawa: Royal Society of Canada, 2000.

Royal Society of London et al. *Transgenic Plants and World Agriculture*. London: Royal Society of London, 2000.

Sears, Mark, Heather Mattila, and Diane Stanley-Horn. *Preliminary Report on the Ecological Impact of Bt Corn Pollen on the Monarch Butterfly in Ontario*. Nepean, Ontario: Canadian Food Inspection Agency, 2000.

Secretariat of the Convention on Biological Diversity. *Cartegena Protocol on Biodiversity to the Convention on Biological Diversity*. Montreal: Convention on Biological Diversity, 2000.

Shiva, Vandana. *Stolen Harvest: The Hijacking of the Global Food Supply*. Cambridge, Mass.: South End Press, 2000.

Thompson, Paul B. *Food Biotechnology in Ethical Perspective*. New York: Aspen Publishers, 1997.

United Nations. *Convention on Biological Diversity,* United Nations, Rio, 1992.

United Nations Development Program. *Making New Technologies Work for Human Development*. New York: UNDP, 2001.

Wambugu, Florence M. *Modifying Africa: How Biotechnology Can Benefit the Poor and Hungry, A Case Study from Kenya*. 2001. Available through www.modifyingafrica.com.

Contributors

BOB B. BUCHANAN is a researcher with the Department of Plant and Molecular Biology, University of California, Berkeley.

ARTHUR CAPLAN is the director of the Center for Bioethics, University of Pennsylvania, Philadelphia.

CHARLES THE PRINCE OF WALES is a long-standing advocate of organic farming.

GARY COMSTOCK is coordinator of the Bioethics Program and a professor of philosophy and religious studies at Iowa State University.

GREGORY CONKO is director of food safety policy at the Competitive Enterprise Institute, Washington, D.C.

GORDON CONWAY is president of the Rockefeller Foundation, New York.

KEITH CULVER is a professor of philosophy at the University of New Brunswick, Fredericton.

FLORENCE DAGICOUR is an independent research in Montreal.

RICHARD DAWKINS holds the Charles Simonyi Professorship of Public Understanding of Science, Oxford University.

KURT EICHENWALD is staff writer for the *New York Times*.

NORMAN C. ELLSTRAND is a professor of genetics at the University of California, Riverside.

KAREN FANG is with the Center for International Development at Harvard University.

CARL FEIT is Ades Professor of Health Science at Yeshiva University in New York.

INDUR M. GOKLANY is with the U.S. Department of the Interior's Office of Policy Analysis.

E. RICHARD GOLD is a professor of law at McGill University, Montreal.

MARY LOU GUERINOT is in the Department of Biological Sciences at Dartmouth College, Hanover, New Hampshire.

BRIAN JOHNSON and **ANNA HOPE** are biotechnology advisors to the British statutory conversation agencies based at English Nature, Taunton, U.K.

CALESTOUS JUMA is at the Belfer Center for Science and International Affairs at the Kennedy School of Government, Harvard University.

DAVID MAGNUS is senior fellow at the Center for Bioethics, University of Pennsylvania, Philadelphia.

ALAN McHUGHEN is a professor of botany and plant sciences at the University of California, Riverside.

HENRY I. MILLER is a fellow at the Hoover Institution, Stanford University.

JOE N. PERRY is a research scientist at the Plant and Invertebrate Ecology Division, Rothamsted Experimental Station in the United Kingdom.

GABRIELLE J. PERSLEY is with the World Bank in Washington, D.C.

INGO POTRYKUS is a researcher at the Institute for Plant Sciences, Swiss Federal Institute of Technology, Zurich, Switzerland.

WILLIAM SAFIRE is a regular columnist with the *New York Times*.

AMBUJ SAGAR is at the Belfer Center for Science and International Affairs at the Kennedy School of Government, Harvard University.

MARC A. SANER is an environmental policy consultant in Ottawa, Canada.

JAMES N. SIEDOW is a professor in the Department of Botany, Duke University, Durham, North Carolina.

VANDANA SHIVA is director of the Research Foundation for Science, Technology, and Ecology in New Delhi, India.

PETER SPENCER is the former editor of *Consumers' Research* and currently works for the Energy and Commerce Committee, U.S. House of Representatives.

PAUL B. THOMPSON is professor of philosophy at Purdue University, West Lafayette, Indiana.

NICK TOMLINSON is with the Food Standards Agency in the United Kingdom.

ANTHONY TREWAVAS is a research scientist with the Institute of Cell and Molecular Biology at the University of Edinburgh, Scotland.

ROBERT TRIPP is a research fellow at the Overseas Development Institute, London, U.K.

WOLFGANG VAN DEN DAELE is with the Wissenschaftszentrum Berlin für Sozialforschung in Berlin, Germany.

FLORENCE WAMBUGU is director of the International Service for the Acquisition of Agri-Biotech Applications in Nairobi, Kenya.

JACK WILSON is professor of philosophy at Washington and Lee University, Lexington, Virginia.

XUDONG YE is a researcher at the Institute for Plant Sciences, Swiss Federal Institute of Technology, Zurich, Switzerland.